GALILEO AND THE ART OF REASONING

BOSTON STUDIES IN THE PHILOSOPHY OF SCIENCE

EDITED BY ROBERT S. COHEN AND MARX W. WARTOFSKY

VOLUME 61

MAURICE A. FINOCCHIARO

Dept. of Philosophy, University of Nevada, Las Vegas, U.S.A.

GALILEO
AND THE ART OF
REASONING

*Rhetorical Foundations of Logic
and Scientific Method*

D. REIDEL PUBLISHING COMPANY

DORDRECHT : HOLLAND / BOSTON : U.S.A.

LONDON : ENGLAND

Library of Congress Cataloging in Publication Data

Finocchiaro, Maurice A.   1942—
    Galileo and the art of reasoning.

        (Boston studies in the philosophy of science ; v. 61)
        Bibliography: p.
        Includes index.
        1.   Galilei, Galileo, 1564—1642—Logic.   2.   Logic, Modern—
History.   3.   Reasoning.   4.   Science—Methodology—History.
5.   Galilei, Galileo, 1564—1642. Dialogo . . . dove . . . si discorre sopra i
due massimi sistemi del mondo.   6.   Solar system—Early works to
1800.   7.   Astronomy—Early works to 1800.   I.   Title.   II.   Series.
B785.G24F56   160   80—15232
ISBN 90—277—1094—5
ISBN 90—277—1095—3 (pbk.)

Published by D. Reidel Publishing Company,
P.O. Box 17, 3300 AA Dordrecht, Holland.

Sold and distributed in the U.S.A. and Canada
by Kluwer Boston Inc., Lincoln Building,
160 Old Derby Street, Hingham, MA 02043, U.S.A.

In all other countries, sold and distributed
by Kluwer Academic Publishers Group,
P.O. Box 322, 3300 AH Dordrecht, Holland.

D. Reidel Publishing Company is a member of the Kluwer Group.

Printed in The Netherlands

TABLE OF CONTENTS

PART III: THEORY OF REASONING

EDITORIAL PREFACE

The work of Galileo has long been important not only as a foundation of modern physics but also as a model — and perhaps the paradigmatic model — of scientific method, and therefore as a leading example of scientific rationality. However, as we know, the matter is not so simple. The range of Galileo readings is so varied that one may be led to the conclusion that it is a case of *chacun à son Galileo*; that here, as with the Bible, or Plato or Kant or Freud or *Finnegan's Wake*, the texts themselves underdetermine just what moral is to be pointed. But if there is no canonical reading, how can the texts be taken as evidence or example of a canonical view of scientific rationality, as in Galileo? Or is it the case, instead, that we decide *a priori* what the norms of rationality are and then pick through texts to find those which satisfy these norms? Specifically, how and on what grounds are we to accept or reject scientific theories, or scientific reasoning? If we are to do this on the basis of historical analysis of how, in fact, theories came to be accepted or rejected, how shall we distinguish 'is' from 'ought'? What follows (if anything does) from such analysis or reconstruction about how theories *ought* to be accepted or rejected?

Maurice Finocchiaro's study of Galileo brings an important and original approach to the question of scientific rationality by way of a systematic reading and reconstruction of Galileo's project and of the modes of his reasoning. In effect, Finocchiaro suggests that the standard picture of the scientific revolution may be fundamentally mistaken. Where, on older views, the scientific revolution was seen as a radical innovation in method as well as in content, and even as marginal to the dominant cultural themes of the Renaissance — largely literary, political, aesthetic, rhetorical — Finocchiaro proposes that the scientific revolution, in one of its major figures, was characterized fundamentally by the same emphases on dialectic and rhetoric which marked the dominant themes of Renaissance humanism. Thus, the scientific revolution was as much a matter of persuasion as of discovery, its success *as science* crucially a matter of the triumphs of practical reasoning and persuasion in other modes beyond the usually acknowledged one of mathematical rationality in terms of proofs and demonstrations. If this is correct, not mathematical rationality alone but scientific judgment, the instruments of persuasion and of practical argument, the alogical components of discourse, played a major role in the novel mode of scientific reasoning.

Finocchiaro argues that "rhetoric is sometimes crucial in science; and hence rhetoric has an important role to play in scientific rationality and the rhetorical aspects of science should not be neglected". His argument is based on a detailed and careful examination of the text of the *Dialogues* .... What Finocchiaro argues will, we are sure, become a basic proposal for enlargement and revision in the canonical program of logical reconstruction which has so strongly marked and nearly dominated modern philosophy of science. He offers an alternative rational reconstruction of Galileo's scientific thought. Logic is also included, of course, but in a wider sense: "I am advocating", he writes, "a practice oriented logic or theory of reasoning, and thus it is logical principles that are to be tested by and derived from appropriately selected actual reasoning and not the other way around" (291, fn. 8). And so "a *general* logic seems possible, if we concentrate on the critical understanding of actual reasoning, which concentration carries along with it an empirical, historical, concrete or contextual, practical and critical appreciation" (305).

In this book, Finocchiaro not only offers us a sustained analysis of the *Dialogues* in terms of the structure of the work, Galileo's reasoning, his uses of rhetoric. Beyond this, which is the heart of his work, he develops a critical and appreciative discussion in depth of the principal Galileo studies: those of Koyré, Feyerabend, Drake, Clavelin, Kuhn, Shapere. Further, Finocchiaro gives us a series of sustained studies of topics in the 'science of rhetoric' as an initial sketch of a theory or reasoning which enlarges scientific rationality beyond the limits of formal or deductive reason. We may say that Finocchiaro rounds out the traditional approaches to the 'logical syntax' and semantics of scientific discourse with a serious construction of the (long neglected) pragmatics of this discourse as an instrument of argument and persuasion. In a sense, his work is an interesting and promising supplement to the syntactic and semantic reconstruction of the classic Vienna Circle achievement. Nor has Finocchiaro merely appended 'externalist' factors to the 'internalist' logical empiricism; rather, for him the normative is embodied within the actual and practical reasoning of science, internal to what we could call 'scientific culture'. Understanding scientific reasoning must include, then, practical scientific judgment as part of the 'science of rhetoric,' not just the rhetoric of science.

Center for the Philosophy and History of Science   ROBERT S. COHEN
Boston University   MARX W. WARTOFSKY
May 1980

PREFACE AND ACKNOWLEDGMENTS

The investigations in this book are interdisciplinary in the sense that different parts are likely to appeal primarily to individuals in different fields. A layman will find in the Introduction reasons why he ought to be interested in Galileo Galilei's *Dialogue Concerning the Two Chief World Systems*, and in Chapters 1—5 discussions of the aspects of the book most relevant to him. These chapters are also addressed to scientists, who of course will find greater meaning therein. Rhetorical scholars will find Chapters 1—3, 8—9, and 12—17 of some interest; cognitive psychologists may appreciate Chapters 11 and 16, linguists Chapters 13 and 16. Religious historians will find here the most exhaustive study ever published of the book that caused "the greatest scandal in Christendom", Chapters 1—5, 6 and 16 being of greatest relevance to them. These Chapters may also interest cultural historians. Logicians will appreciate Chapters 11—17 the most, while philosophers of science can read Chapters 8—9 first and then Chapters 4—7. For historians of science the critique of Koyré in Chapter 9, and of Strauss, Favaro, Drake, and Clavelin in Chapter 10 can serve as the introduction to Chapters 1—5 and 16.

Accordingly this fact multiplies the number of potential misunderstandings and objections on the part of readers, but this is a risk well worth taking, not only because of the interdisciplinary nature of Galileo's own work, but also because I agree with Karl Popper that a perfect division of labor in scholarship and scientific research would soon stop progress. I was also encouraged to pursue these interdisciplinary investigations by the success achieved by my earlier *History of Science as Explanation* (1973), which also deals with problems at the interface of science, philosophy, and history. If the modesty of that success is due to the specialism of that subject matter (the historiography of science), then one may hope that the universal appeal of the present one (Galileo's *Dialogue* and the art of reasoning) will lead to a success commensurate with it. Be that as it may, it will be a useful guide to the reader to elaborate several themes with respect to which these inquiries may be read.

From a general cultural point of view, this work constitutes a study of a book that may be listed among the classics and of a number of human activities of general interest and relevance, namely the arts of logical reasoning, rhetorical persuasion, philosophical reflection, and scientific investigation.

These four arts are discussed both in their relation to one another, and in their relation to Galileo's *Dialogue*. From this point of view, the Introduction is a discussion of the general cultural relevance of Galileo's book, by reference to which all succeeding chapters concretely study those four activities. Chapter 8 is a general introduction to the four activities, by means of a discussion of the role of reasoning, of rhetoric, and of methodological reflection in science. Chapter 9 distinguishes reasoning from one of the things with which it is often confused, namely a priori rationalism. Chapter 1 is a discussion of the difference between reasoning and relying on religious authority, and of the problems engendered by this difference in the practice of rhetorical persuasion. Chapter 2 discusses the role of the logical structure of reasoning and of the presentation of scientific evidence in the art of rhetorical persuasion. Chapter 3 studies the role of artistic expression and of emotions in reasoning and in persuasion. Chapter 4 relates several different conceptions of science to the *Dialogue*. Chapter 5 discusses the interplay between scientific practice and methodological reflection. Chapter 6 examines some similarities and some differences between methodology and philosophy. Chapter 7 studies the connections between methodology and logic by interpreting reflections on knowledge and method in general as aspects of reflection upon reasoning. Chapter 10 illustrates some problems that the science of reasoning encounters when it studies the historical material in which human reasoning leaves a record. Chapters 11–13 discuss various ways of studying reasoning, namely the science of formal logic, psychology, linguistics, inductive logic, and rhetoric, as well as the informal involvement in reasoning. Chapter 14 elaborates a number of principles useful for the understanding of reasoning, the central idea being that reasoning involves the inter-relating of ideas in accordance with a definite structure. Chapter 15 discusses the problem of determining how good or bad reasoning is, and it suggests that practical involvement in and development of the reasoning under consideration is needed. Chapter 16 is a detailed analysis of many instructive arguments found in the *Dialogue*. Chapter 17 discusses the importance in reasoning of criticism, evaluation, interlocking steps, and several varieties of invalidities.

From the point of view of rhetorical scholarship, these investigations constitute an attempt to develop concretely a rhetoric of science and a science of rhetoric. Our example of science consists of Galileo's *Dialogue*, and the concreteness derives from the fact that all discussions refer to it. The rhetoric of science refers to the rhetorical dimension of science, and the science of rhetoric refers to the systematic study of rhetorical activities. Chapter 1 then deals both with the rhetorical dimension of the *Dialogue*, in the sense of rhetoric that refers to external appearance and pretended effects of verbal discourse, and also with another important distinction for rhetorical science, namely the difference between relying on authority and engaging in rational argument. Chapter 2 deals with the book's rhetoric, in the sense of the term that refers to argumentation, and also with the importance of the concept of argument structure in a science of rhetoric. Chapter 3 studies directly the book's rhetoric, in the sense of the term that refers to emotional appeals and beautiful-sounding expressions, and indirectly the role of these concepts in rhetorical science. Chapters 4–7 deal directly with various aspects of the scientific methodology of Galileo's book and indirectly lay the foundations for determining what it would be for rhetorical phenomena to be studied scientifically. Chapter 8 discusses in a general way the role of rhetoric in scientific rationality. Chapter 9 is both a concrete analysis of one rhetorical aspect of Galileo's book, in the general sense of rhetoric that refers to argumentative techniques, and also a concrete analysis of an important feature of one of the most fundamental concepts in the science of rhetoric, namely reasoning and how it differs from rationalism. Chapters 10–17 develop a number of fundamental topics in the science of rhetoric. Chapter 10 is a critique of a number of scholars, suggesting that the historical record of reasoning cannot be left entirely in the hands of mere historians. Chapters 11–13 examine the nature of and inter-relationships among the rhetorical and the psychological, logical, linguistic, and informal study of reasoning. Chapter 14 is a formulation and illustration of a number of basic principles for the understanding of reasoning. Chapter 15 is a critique of the fallacy approach to the evaluation of reasoning and suggests an alternative practical-oriented approach. Chapter 16 is a detailed analysis of the logical and the rhetorical features of about half of all the arguments in Galileo's book. Chapter 17 draws a number of theoretical conclusions from the data of the previous chapter.

From the point of view of philosophy, these investigations constitute concrete studies in the logic of science and the science of logic. Here the logic of science may be taken to refer to a distinctive approach to the philosophy of science, which emphasizes reasoning as opposed to other concepts such as knowledge, explanation, theory, and progress; while the science of logic refers to the scientific study of reasoning. The mutual inter-dependence of the two is obvious: in order to study the nature of science from the point of view of reasoning, the logic of science must sooner or later give an adequate articulation of the concept and activity of reasoning; and in order to study reasoning scientifically, the science of logic must use or develop an adequate view of what science is and of what it is to be scientific. Chapter 8 argues that a change of emphasis from epistemology (e.g., theory and observation) to logic (reasoning) is the most fruitful move for contemporary philosophy of science; the argument is grounded on an elaboration and correction of Paul Feyerabend's main doctrines. Chapter 9 shows that the study of actual scientific reasoning has not been neglected by the best historians of science (e.g., Koyré), but that important corrections need to be made if their work is to become useful and acceptable. Chapters 1–7 develop a concrete logic of science, around the case study of Galileo's *Dialogue*. Chapter 1 argues that this book is a complex multi-dimensional argument in support of Copernicanism. Chapter 2 elaborates one aspect of this argument, namely the one relating to the presentation of the available evidence and the organization of one's claims by means of an inferential structure. Chapter 3 discusses the emotive and aesthetic aspects of Galileo's argument. Chapters 4–7 discuss its methodological aspect: first its content for present-day scientists (Chapter 4), then the methodology inherent in this scientific content (Chapter 5), then the philosophical significance of this methodology (Chapter 6), and finally the logical significance of this methodology, that is, its significance for the clarification of the concept of reasoning (Chapter 7). Chapters 10–17 are a series of studies in the science of logic. Chapter 10 argues that the historical approach needed in the science of reasoning requires better scholarship than that found even among the best historians. Chapter 11 explores the value and the limitations of formal logic and of experimentation in the study of reasoning; this is done by a detailed discussion of recent work in the psychology of reasoning. Chapter 12 discusses the importance of logical analysis in relation to rhetorical analysis, and it clarifies the concept of question-begging; this is done by a critical elaboration of Perelman's 'new rhetoric' and by using Galileo's tower argument as an example. Chapter 13 brings into focus the problem of a theory of scientific reasoning and of a scientific theory of reasoning by discussing certain problems at the foundations of inductive logic; by discussing some methodological questions in the linguistics of reasoning and elaborating certain suggestions made by Y. Bar-Hillel; by exploring the possibility of a 'practical' approach to the study of reasoning and elaborating certain suggestions by Michael Scriven; and by a critical analysis of Stephen Toulmin's views on logic. Chapter 14 formulates and illustrates a number of basic principles for understanding the structure of reasoning, in accordance with our approach to logic. Chapter 15 makes a specific suggestion for the study of the other main dimension of reasoning, namely its worth or goodness, in accordance with the same approach. Chapter 16 constructs a data basis by giving detailed reconstructions, analyses, and evaluations of sixteen complex arguments (consisting of more than 200 simple arguments) in Galileo's book. Chapter 17 draws a number of theoretical conclusions from this data basis, suggesting that the two main types of arguments are critical and constructive; that several different types of invalidities can be defined; and that reasoning normally consists of complex rather than of simple arguments.

From the point of view of historical scholarship and erudition, these investigations constitute a commentary to a crucially important work. Each chapter may then be viewed as studying a particular aspect of Galileo's book: the Introduction may be viewed as a discussion of the general cultural significance and impact of the *Dialogue*; Chapter 1 as an analysis of the book's religious and theological rhetoric; Chapter 2 as an elaboration of the book's logical structure; Chapter 3 as an analysis of its emotive, aesthetic, and literary content; Chapter 4 as an examination of its content relevant to scientists; Chapter 5 as a study of its methodological, epistemological, and philosophical content; Chapter 6 as a discussion of its significance for philosophers; Chapter 7 as an examination of its logical content and significance, in the sense of the theory and practice of reasoning in general; Chapter 8 as a discussion of how the book is a fruitful example of scientific rationality; Chapters 9—10 as a critique of certain aspects of the book's historiography; Chapters 11—13 as an argument for its relevance to formal logicians and other students of reasoning, such as linguists, psychologists, rhetorical scholars, inductive logicians, and scholars who stress informal logic; Chapter 14 as a discussion of the book's value for the illustration of a number of basic principles for the understanding of reasoning; Chapter 15 as a suggestion of its value for the illustration and formulation of principles for the evaluation of reasoning; Chapter 16 as a detailed analysis of about half of all the arguments in the book; Chapter 17 as a number of theoretical conclusions that can be drawn from the previous chapter's study of the book's logical content. The Appendix is an analysis of the pagination in several important editions of the book.

Finally, from the editorial-literary point of view, these investigations group themselves into three somewhat autonomous parts, each of which may perhaps be appreciated independently of the others. The first part consists of Chapters 1—5 and deals with various aspects of Galileo's *Dialogue*, the second consists of Chapters 6—13 and deals with logical and methodological critiques of various scholars, the third consists of Chapters 14—17 and is a sketch of various aspects of the theory of reasoning.

My investigations received the support, encouragement, comments, or constructive criticism of a number of people and institutions, to whom I should like to express my gratitude: the National Science Foundation for a research grant (SOC76–10220), and the University of Nevada at Las Vegas for a sabbatical leave, in the academic year 1976–77, during which most of this book was written; and Joseph Agassi (Boston, Bielefeld, and Tel-Aviv), Harold I. Brown (Northern Illinois), Stillman Drake (Toronto), Paul Feyerabend (Berkeley), Mary Hesse (Cambridge), P. N. Johnson-Laird (Sussex), Henry W. Johnstone, Jr. (Pennsylvania State), Larry Laudan (Pittsburgh), Marvin Loflin (Alaska), Arthur Millman (Colorado), Michael Scriven (San Francisco), William A. Wallace (Catholic University), Craig Walton (Nevada, Las Vegas), and Robert S. Westman (UCLA).

As for claiming responsibility for my own errors, I shall refrain from being redundant and insulting the reader's intelligence by paying lip service to the ritualism that authors usually engage in on such occasions; instead I shall refer him to what Michael Dummett says on the matter in *Frege: Philosophy of Language*.

The book was conceived and written as a single sustained argument, but publication delays made it inevitable for various parts to find their way separately into journals and conference proceedings. Thus acknowledgments are hereby made to the editors and publishers of the journals and books where those parts first appeared, as follows: the nucleus of Chapter 2, in *Logique et Analyse* **22** (1979), 159–80; parts of Chapter 4, in *Journal of College Science Teaching*, Sept. 1980; the bulk of Chapter 5, in *Scientia* **112** (1977), 95–118, and 371–85; the third section of Chapter 6, in *PSA 1976*, Vol. 1, edited by F. Suppe and P. D. Asquith (East Lansing: Philosophy of Science Association, 1976), pp. 30–39; substantial parts of Chapter 8, in *PSA 1978*, Vol. 1, edited by P. D. Asquith and I. Hacking (East Lansing: Philosophy of Science Association, 1978), pp. 235–46; the core of Chapter 9, in *Physis* **19** (1977), 5–27; the first three sections of Chapter 10, in *Synthese* **40** (1979); the nucleus of Chapter 11, in *Philosophy of the Social Sciences* **9** (1979), 277–91; the last three parts of Chapter 12, in *Physis* **16** (1974), 129–48; the first section of Chapter 15, in *American Philosophical Quarterly*, Library of Philosophy volume on *Studies in Epistemology*.

Acknowledgments are also due to the publishers for permission to quote from the following books: G. Galilei, *Dialogue Concerning the Two Chief World Systems*, Berkeley: University of California Press, 1967; R. P. Feynman *et al.*, *The Feynman Lectures on Physics*, Vol. 1, Reading, Mass.: Addison-Wesley Publishing Company, © 1963, California Institute of Technology; B. Croce, *History: Its Theory and Practice*, 1920; New York: Russell and Russell, 1960; and A. Koyré, *Études Galiléennes*, Paris: Hermann, 1966.

GALILEO'S *DIALOGUE* AND WESTERN CULTURE

Certain individuals and certain books perform a unique role in human civilization. In world history in general, Socrates, Buddha, Confucius, and Jesus may be regarded as four paradigmatic individuals of incomparable significance, as Karl Jaspers has argued.[1] Restricting ourselves to Western history and to literary culture, the works of such men as Plato, Augustine, Galileo, and Marx emerge as having unique significance and value. This is especially true for certain particular books of theirs, which have become classics. Thus one does not have to be a professional philosopher to be able to enjoy, and profit from Plato's Socratic dialogues; one does not have to be a Christian in order to appreciate Augustine's *Confessions*; one does not have to be a scientist in order to understand Galileo's *Dialogue Concerning the Two Chief World Systems*; nor does one need to be a communist in order to value Marx's *Capital*. We may say that the universal relevance of such books derives partly from their self-contained readability, and partly from historical circumstances; paradoxically, it also derives partly from their individuality, that is, from the fact that each one of these books plays a unique role in a specific, but fundamental and primary human endeavor: morality and education, religion, science, and social theory and practice, respectively. Moreover, it is interesting to reflect on the pattern of their influence. From the chronological point of view Socrates' influence is of course the most firmly established since it has been operative the longest — twenty four centuries; Augustine's influence makes up in intensity what it lacks in age, insofar as his effect on Christianity has been deeper than Socrates' effect on any comparable group, culture, or tradition; Galileo's makes up in universality what it lacks in age vis-à-vis Socrates and what it lacks in intensity vis-à-vis Augustine, insofar as Galileo may be regarded as the father of modern science and insofar as modern science has become the most widespread cultural force that the world has ever known and the only one to have penetrated all other subcultures of the earth; finally, though Marx's influence is barely a century old, though Marxism has not (yet) prevailed the world over, and though in countries where Marxism has prevailed its influence has not *generally* reached the depths of the human soul, nevertheless the success and diffusion of Marxism has been the most spectacular, so that Marx's influence has made up in rapidity what it lacks in other respects.

Next, restricting our attention to Galileo, I want to argue that, though his specialty is science, from a more general point of view his work can be taken as exemplary for the fundamental human desire of the search for truth. Generalizing in the same manner, Socrates would be a model in the search for the good and the right; Augustine in the search of the holy and the supernatural; Marx in the search for the practical.[2]

The paradigmatic value of Galileo's work among scientists is, of course, well known and will be elaborated later. What needs discussion now is the attitude of other groups, for example, laymen and professional thinkers. Regarding Galileo's influence on laymen, here the effect probably works by means of the image of Galileo being persecuted and tried by the Catholic Church. The image was portrayed with classic incisiveness by John Milton, when in his *Areopagitica* he recalls his visit to Florence by saying, "There it was that I found and visited the famous Galileo, grown old, a prisoner to the Inquisition for thinking in Astronomy otherwise than the Franciscan and Dominican licensers of thought".[3]

To be sure, in 1968 Pope Paul VI announced a revision of Galileo's 1633 trial by the Inquisition. This is a very interesting cultural-historical fact not only because it reflects a liberalizing trend within the contemporary Catholic Church, but also because of the fact that such an announcement strikes us generally as telling, or promising to tell, more about the Church than about Galileo. This reaction − typical of layman, scholar, and scientist alike − in turn is an indication that Western culture has come a long way in the last three and one-half centuries.

We may agree, as Ortega y Gasset aptly expressed it, that the trial originated "more in the small intrigues of private groups than in any dogmatic reservations of the Church".[4] Nevertheless, the historical significance of an event depends on the magnitude of its *effects*, rather than of its causes, and this is why the whole Galileo affair came to be perceived as "the greatest scandal in Christendom".

Since the occasion for the trial was Galileo's publication of his *Dialogue* in 1632, this book would seem to deserve both inspirational reading for the truth-seekers, and careful scholarly study. The former it certainly has received, but it is one of the greatest scandals of modern scholarship that no exhaustive, systematic study of the book in all of its aspects and implications has ever been made. This is not meant to deny the existence of countless partial studies, which are very informative, insightful, and helpful: the Galileo literature is so immense, and the *Dialogue* and the trial so crucial in Galileo's life, that it could hardly be otherwise. Nevertheless, it is equally undeniable that there is no study of Galileo's book in all of its aspects, scientific, methodological, philosophical, logical, rhetorical, theological, religious, literary. Such a study is undertaken below.

Before doing so, let us look at some general cultural influences on people other than scientists and laymen.

David Hume's *Dialogues Concerning Natural Religion* is a classic discussion of arguments about the existence and nature of God. The first arguments it discusses are arguments based on experience: for example, that the universe must have an intelligent Maker because the universe is like a machine and we know that machines are made by human beings who have intelligence; or the universe is an ordered arrangement, and all ordered arrangements can be observed to derive from human intelligence, and therefore the universe must have an intelligent cause. Here then the problem arises of whether it is at all possible to prove the existence or attributes of God with arguments based on experience, since in order to check such reasoning we would have to have experience of the origin of universes. To help solve this problem Hume asks whether to prove by experience the origin of the universe is any different from proving the earth's motion by experience. He answers that there is a difference because we do have observational knowledge of the motion of other 'earths', namely the moon and planets, and these bodies are similar to the earth. In making this last point Hume adds a discussion about the structure of Galileo's argument in the *Dialogue*, namely that he first tries to establish an analogy between the earth and the heavenly bodies in his attempt to prove the earth's motion.[5]

I am not saying either that Hume's reference to Galileo is absolutely accurate, nor that the mere reference to Galileo completely proves his own critique of the possibility of proving the existence of God by experience. It suffices that Hume's reference to Galileo is there, that it is correct to *some* extent, and that the reference *strengthens* Hume's case. This in turn shows how Hume used Galileo as a model on one important occasion.

A second example is Immanuel Kant. In the Preface to the Second Edition of the *Critique of Pure Reason* he gives a sketch of the main achievement in his book. He claims to have brought about a revolution in the field of metaphysics "in accordance with the example set by geometers and physicists".[6] No geometers are named, but concerning the latter he expresses himself with words that are so explicit and eloquent as to deserve extended quotation:

When Galileo caused balls, the weights of which he had himself previously determined, to roll down an inclined plane; when Torricelli . . . ; when Stahl . . . a light broke upon all students of nature. They learned that reason has insight only into that which it produces after a plan of its own, and that it must not allow itself to be kept, as it were, in nature's leading-strings, but must itself show the way with principles of judgment based upon fixed laws, constraining nature to give answers to questions of reason's own determining. . . . Their success should incline us, at least by way of experiment, to imitate their procedure, so far as the analogy . . . . . . to metaphysics may permit. Hitherto it has been assumed that all knowledge must conform to objects. But all attempts to extend our knowledge of objects by establishing something in regard to them *a priori*, by means of concepts, have on this assumption ended in failure. We must therefore make trial whether we may not have more success in the tasks of metaphysics, if we suppose that objects must conform to our knowledge.[7]

As in the case of Hume, Kant's reference to Galileo need be neither completely accurate nor completely decisive in order to have its intended effect.[8]

Third, in *Man and Crisis* José Ortega y Gasset pays to Galileo a double homage, so to speak. First, since Ortega's subject matter is cultural history, he takes Galileo as the paradigm example of a thinker of the modern age, thus giving his own book a Galilean dimension in a substantive sense. Second, Ortega feels that his approach in his historical investigation must be patterned after Galileo's, thus making his own book Galilean in a methodological sense. Ortega's statement of his first point can hardly be excelled:

The greatest crisis through which the European destiny has ever passed ends with Galileo and Descartes – a crisis which began at the end of the fourteenth century and did not taper off until the early years of the seventeenth century. The figure of Galileo appears at the end of this crisis like a peak between two ages, like a divide that parts the water. With him modern man enters into the modern age.

It is therefore a matter of supreme interest for us to take this crisis and this entrance into a new period under very close consideration. Every act of entering into any place, every coming out from any corner has about it a bit of the dramatic; at times it has a great deal – hence the rites of the doorway and the lintel. The Romans believed in special gods who presided at that condensation of enigmatic destiny which is the act of going out or of coming in. The god of going out they called *Abeona*; the god of coming in, *Adeona*. If, in place of a pagan god, we speak in Christian terms of a patron saint, nothing would seem more justifiable than to make of Galileo both the *Abeona*, patron saint of our departure from modernity, and the *Adeona*, patron saint of our emergence into a future still palpitant with mystery.[9]

After an analysis of Galileo's principle of inertia,[10] Ortega applies his conception of Galileo's method to his own inquiry, which he calls history, the science of human lives:

Every science which is concerned with reality, whether of the body or of the spirit, must be not merely a mirror of the facts, but a genuine construction. Because the science of physics in the time of Galileo was thus developed, it has endured as a model science and a norm of knowledge for the whole of the modern age.

History must adopt a similar decision and prepare itself to construct. It goes without saying that this comparison between physics as it is and history as it is and as it ought to be is only valid, for the moment, at this single point – the element of construction as such. The other characteristics of physics are not such as to be desirable for history. Take exactitude, for example . . . [11]

Then by using Galileo's method in the subject matter of human lives Ortega arrives at two principles which he takes to be the historical analogues of the concept of inertia: "first that every man's life starts with certain basic convictions about what the world is and what man's place in it is – it starts from there and moves within them; second, every life finds itself in surroundings which include more or less technical skill or control over the material environment".[12] For Ortega, this means that life both has a structure and is a drama: "human life always has a structure – that is to say, it consists in man's having to cope with a predetermined world of which we can sketch the profile. The world presents certain problems which, relatively speaking, are solved, and raises others, thus giving a definite aspect to man's struggle for his own fate".[13] Moreover, "life is not solely man, that is to say, the subject which lives. It is also the drama which arises when that subject finds himself obliged to fling his arms about, to swim shipwrecked in that sea which is the world".[14] Then Ortega goes on to apply these ideas to the details of Western cultural history. As before, one need not agree with Ortega's reading of Galileo, though the interpretation is not implausible; still less need one accept his view of "man and crisis". Nevertheless, *Man and Crisis* is perhaps the best example of the kind of influence and relevance that I am here trying to articulate.

NOTES

[1] *The Great Philosophers*, Vol. I.
[2] Benedetto Croce's elaboration of the universal category of the 'useful' may be taken as such a generalization of Marx's work. *Cf.* Croce's *Historical Materialism and the Economics of Karl Marx* and his *Philosophy of the Practical*, Antoni's *Commento a Croce*, and my review of Gentile's *Filosofia di Marx*.
[3] Quoted in S. Drake, 'Galileo in English Literature of the Seventeenth Century', p. 423.
[4] J. Ortega y Gasset, *Man and Crisis*, p. 9.
[5] D. Hume, *Dialogues Concerning Natural Religion*, pp. 141–152, especially pp. 150–152.
[6] I. Kant, *Critique of Pure Reason*, p. 25 (Bxxii).
[7] *Ibid.*, pp. 20–22 (Bxii–Bxvi).
[8] Kant complicates his discussion by referring to Copernicus and the Copernican revolution of changing the point of view from attributing motion to the heavenly bodies to attributing it to the spectator. This presents no problem for my argument not only

because I need not claim exclusive reference to Galileo in this context, but also for the following reason. Kant uses the reference to Copernicus in two ways, first in order to elucidate the *content* of his own metaphysical revolution, insofar as the change from knowledge conforming to objects, to objects conforming to our knowledge, is analogous to the change from motion subsisting in heavenly bodies to motion subsisting in the observer; hence this Copernican reference does not play the role of a methodological justification in Kant's discussion. Kant recognizes that Copernicus' achievement was to have "dared, in a manner contradictory of the senses, but yet true, to seek the observed movements not in the heavenly bodies, but in the spectator". (*Ibid.*, p. 25, footnote a; Bxxii); hence Kant goes on to explain that the proper *methodological* similarity between his own situation and Copernicus' is insofar as, *in the Preface*, Kant is conjecturing a revolutionary metaphysical thesis, which he will then establish in the course of the book (*Ibid.*). In short, I would say that Kant refers to Copernicus for an elucidation of his thesis, and to Galileo for a justification of his approach and consequently for a partial justification of his thesis.

9 Ortega, *op. cit.*, p. 11.
10 *Ibid.*, pp. 11–19.
11 *Ibid.*, p. 20.
12 *Ibid.*, p. 27.
13 *Ibid.*, pp. 27–28.
14 *Ibid.*, p. 28.

PART I

GALILEO'S *DIALOGUE*

CHAPTER 1

FAITH VERSUS REASON:
THE RHETORICAL FORM AND CONTENT OF
GALILEO'S *DIALOGUE*

The *Dialogue* is written in the form of a dialogue among three characters and contains numerous indirect references to a fourth person, a Lyncean Academician, common friend of the interlocutors, who may be identified with Galileo himself. Moreover, the book is full of passages containing disclaimers by the supposedly Copernican spokesman, Salviati, to the effect that his Copernican beliefs are not necessarily as strong as they may appear when he is involved in the discussion, so that he really ought to be regarded as being a good actor who is wearing a Copernican mask. Other passages continuously remind the reader that the aim of the discussion is not to decide the Copernican issue one way or the other, but rather to present all the arguments for and against, so that people in a position of authority, who are entitled to make such decisions, can do so intelligently. The *Dialogue* is also full of emotionally charged passages, where various persons, books, and ideas are criticized not dispassionately but rather with ridicule, scorn, and contempt. In other passages the interlocutors play tricks on each other, trying to trap one another into saying something so that they can win the argument; such trickery is complicated by the fact that these maneuvers are usually exposed in the course of the discussion and by the fact that they are (at least apparently) perpetrated almost always at the expense of the Peripatetic spokesman, Simplicio. These and other similar features, all of which are rather obvious even to a casual reader of the *Dialogue*, may be labeled 'rhetorical'. The specialist in *Geisteswissenschaften* will be interested in such rhetorical matters merely because they exist, and he will try to understand their nature, how they function, how they relate to other verbal and mental processes, and to other human activities, etc.

This is not, however, the only reason for studying the rhetoric inherent in the *Dialogue*. For even a superficial acquaintance with the book's origin and fate makes it obvious that the book had, and was meant to have, more than a purely intellectual content: in writing it Galileo was also being a man of action, interested and concerned in bringing about certain practical effects, namely to persuade the appropriate Church officials of the truth of Copernicanism and thereby to cause the repeal of the condemnation of 1616. Moreover, as everybody knows, what the book caused instead was Galileo's trial and conviction by the Inquisition for heresy and for disobedience. In other words, the book originated in part as a practical act, and was so judged by the Church. Hence the study of those parts of the book that have or seem to have practical import can shed light on what some came to call "the greatest scandal in Christendom". This practical dimension of the *Dialogue* will also be labeled 'rhetorical' since it too, like the dramatic, emotive, and moral aspects, refers to nonintellectual matters.

A third reason for studying the rhetorical content and structure of the *Dialogue* is that nearly all scholars recognize a polemical and propagandistic dimension in the *Dialogue,* however greatly they may differ about the importance of this aspect of the book and about its relationships with other aspects; hardly anyone, however, has seriously studied this topic. It is not clear why this should be so. The first thought that comes to mind is that rhetorical analysis (as we may call the scholarly study of verbal propaganda) requires skills and sensibilities that science-oriented scholars seldom possess, and, conversely, rhetorically competent scholars lack proficiency and interest in scientific topics. This, however, cannot be the whole story because there is an analogous disjointness between science and poetry or literature, and yet the literary-practical aspects of the *Dialogue* have received more serious study than its rhetoric. Part of the rest of the story is that rhetorical studies have been generally neglected in modern times, whereas aesthetics has gained increasing popularity and depth. But this raises a new problem, for after all the last several centuries are also the era when science has gained unparalleled attention and dominance; the problem is, why is it that in the last three or four centuries science has flourished but rhetorical studies have decayed? Is it because there is an intrinsic conflict between the scientific and the rhetorical points of view? What is clear is that most scholars *believe* in this conflict; in fact, when they mention the so-called propagandistic aspect of the *Dialogue,* they typically do so by contrast with its scientific value.[1]

In the last ten or fifteen years the situation has changed somewhat, and some, e.g. Feyerabend, have not been afraid of concluding that propaganda and rhetoric play an essential role in science.[2] To be sure, Feyerabend's language (his 'rhetoric', as it were!) is still that of the opposition between science (or at least reason) and rhetoric; however, the logic of his position is different, and I would describe it as follows. The context of the discussion has been the problem of scientific rationality and the question of whether the transition from one scientific theory to another can be made in a rational manner.[3] The problem arose from the realization that, in the case of very fundamental scientific developments such as the Copernican revolution and the transition from classical to modern physics, purely logical considerations, rational argumentation, and appeals to the rules of scientific method are not enough to make a scientist change his mind. This realization was then generalized to conclude that the same limitation applies to all, or at least most, or at least the important and interesting, scientific developments. The generalization is I believe illegitimate,[4] but it has generated a lot of discussion and confusion: the more conservative philosophers of science, feeling that the rationality of science was being threatened, have tended to counterargue that transitions from one theory to another *can* be and have historically been made in a rational manner, using as evidence the admittedly less problematic minor transitions or at least the less problematic aspects of the major transitions; on the other hand, the more revolutionary philosophers have tended to reiterate their arguments. Moreover, these latter philosophers have drawn further conclusions from their generalization about the limits of reason in scientific changes. Some have concluded that the only sense in which science is rational is that the occurrence of scientific developments is explicable, that is that we can see how and why it comes about that one theory is abandoned and another one emerges in the scientific community. Other philosophers have concluded that the only sense in which science is rational is that scientific changes can be justified after the fact, in terms of an abstract ahistorical scheme which shows how and why the later theory is an improvement of the older one. Some other philosophers have simply concluded that science is an irrational enterprise, or at least no better than myth, magic or witchcraft.[5] It is possible, however, in fact preferable, to draw the following conclusion:[6] in addition to argument, rhetoric is sometimes crucial in science; and hence, rhetoric has an important role to play in scientific rationality and the rhetorical aspects of science should not be neglected. To what extent this conclusion can be grounded on an analysis of the rhetoric in the *Dialogue* is one of the things we will try to determine here.

To conclude, a rhetorical analysis of the *Dialogue* is needed because the rhetoric is *there,* because its writing was partly a practical act, because this aspect of the book has never been seriously examined though it has been universally recognized, and because it can provide a case study in the role of rhetoric in science; by rhetorical analysis is meant examination of the non-intellectual (*and* non-literary-aesthetic, to be exact) content, structure, and aspects of the book.

RELIGIOUS RHETORIC

Immediately after the title page, the *Dialogue* has a two-page letter of dedication to the Grand Duke of Tuscany, and immediately after this we find a three-page preface entitled 'To the Discerning Reader'. Reading the latter by itself could make one think that the book is an apology of Catholic piety. One of the most obvious features of this preface is that it was printed in a type different from that of the book's text.[7] This fact was cited as one of several accusations in the indictment compiled against Galileo soon after the book was published,[8] the alleged crime being that the different printing had "rendered it useless inasmuch as it was alienated from the body of the book".[9] Though the fact was not mentioned in the final sentence pronounced against Galileo,[10] its being mentioned in the original indictment is indicative of the rhetorical effect (or lack of effect) that the preface may have. I am inclined to agree that the preface is indeed alienated from the text by the printing, however one cannot conclude that it is thereby rendered rhetorically 'useless', but only that its rhetorical effect is thereby diminished. The reasons for this are that the print of the preface is of the same type as the letter of dedication, the postils printed in the margin of almost every page of the text, the two-page errata section immediately following the last page of the text, and the 30-page appendix which is analogous to the index in a modern book and which (on the very last page) is referred to as "Table of the most notable topics which are contained in this book". Another feature that must be counted as lessening the rhetorical alienation of the preface is the fact that the preface was left unsigned, unlike the dedication. It is true, of course, that Galileo may have left it unsigned for the (otherwise) very good reason that the preface had been imposed upon him by the church censors;[11] on the other hand, the preface probably incorporated suggestions made by Galileo himself in the course of his discussions with them. At any rate the lack of a signature to this section of the book, which after all is not titled 'preface' but merely 'To the Discerning Reader', does make it more of an integral part of the body of the work.[12]

The preface consists of seven paragraphs[13] and contains three elements: a statement of the aim of the book, a description of its content, and an explanation of its dialogue form. The last need not concern us here. The stated aim is a refutation of the rumor that the anti-Copernican decree of 1616 was based on ignorance, and a proof that Catholic acceptance of the earth's motionlessness derives from piety, religiousness, knowledge of divine omnipotence, and awareness of the weakness of the human mind. The rumor will be refuted by showing that Catholics are at least as knowledgeable about the forbidden topic as Protestants, for Galileo claims that in 1616 the Church had consulted him before reading its decision. The motive of piety will be established by showing that, from the point of view of evidence and arguments, the Copernican position can be made to appear superior; here the text adds the qualification that this does not mean that the thesis of the earth's motion is superior in an absolute sense, but only that the Copernican supporting *arguments* are better than the Aristotelian geostatic ones. Then three of the book's main results are stated: first, that no terrestrial experiments can prove its motion; second, that celestial phenomena favor the Copernican hypothesis; and third, that new light can be shed on the problem of the cause of tides, if we assume the earth's motion.

The following points should be noted. The preface gives Galileo partial responsibility for the decree of 1616; independently of the historical correctness of this claim, its rhetorical effect is simultaneously to increase the intellectual respectability of the decree and the religious credentials of the present book. The claim has the nature of a compromise, with each side both gaining and losing a little.

Second, the justification of the motive of piety presupposes a commitment to a limited rationalism, which it is very easy to confuse with irrationalism. The presupposition seems to be that the rational acceptability of an idea is only one consideration for deciding upon beliefs, and that one may decide to overrule reason and believe what religious authorities dictate. However, to be more exact, one would have to mention another operative presupposition, namely that religious authorities are usually right; that is, in speaking of rational acceptability one would have to distinguish between individual rational acceptability, and rational acceptability to the relevant authorities. Hence, the words in the text are not really an expression of irrationalism but merely of piety, in the sense of general willingness to accept the conclusions that religious authorities deem rationally acceptable, even when this goes against one's individual judgment. In this sense the preface is preaching religious piety.

Third, the summary of the book's results is expressed with the proper emphasis. Regarding the terrestrial arguments, the stress is on the fact that they cannot prove the earth's motion, though this is only one side of the coin, the other being that they cannot prove that the earth stands still either. Regarding celestial evidence, the superiority of Copernicanism is qualified by the explicit use of the word 'hypothesis' in speaking of the 'Copernican hypothesis', and by the qualification that this hypothesis seems to prevail "for facility in astronomy, not for necessity of nature" (F30). Regarding the argument from the tides, the qualifications are even more obvious: the text calls it an 'ingenious fantasy', which only 'sheds light' on the problem of tides, and which exhibits "the probability which would render it [the phenomenon of tides] comprehensible, given that the earth were in motion" (F30).

It is well known, however, that this 'rhetoric' fooled no one, or at least no one besides the two or three censors who gave permission for the book to be printed. What is not clear is why this rhetoric did not work. Concerning the book's content, the preface makes four points: (1) that the Copernican *arguments* are shown to be better than the Aristotelian ones; (2) that terrestrial arguments are shown to be incapable of proving the earth's motion; (3) that astronomical arguments favor the Copernican hypothesis; and (4) that a novel, original argument in favor of Copernicanism can be formulated based on the cause of tides. Each of these four points is indeed accurate as a description of the book's content. Therefore, what all readers must have refused to do is to draw the conclusions, from these four points, that (A) therefore geostatic beliefs must be taken on (religious) faith and cannot be accepted on the basis of reason, and that (B) therefore to the extent that the decree of 1616 was made in the light of such knowledge it was not arbitrary. Instead readers must have concluded that (A') therefore geostatic beliefs are irrational and should be abandoned, and that (B') therefore to the extent that the decree of 1616 was made in the light of such knowledge it was irrational. In other words, the religious rhetoric of the preface consists of a logically invalid justification, and no one accepted this rhetoric because no one was willing to infer the required apologetic conclusions on the basis of the (more or less) agreed-upon premises descriptive of the book's content.

In short, what the preface claims was universally rejected because of its internal incoherence, and it provides us with evidence that people will not always believe what they are told to believe, or what they hear, or what they read; in particular, they will not accept it if it does not make logical sense to them.

At the other end of the rhetorical spectrum, and it so happens at the other end of Galileo's book, we find an example of implicit rhetoric, where it is not so much the words that are important, but rather the actions, facts, and deeds. I am referring to the fact that the skeptical argument favored by the pope is put in the mouth of Simplicio, who is supposedly a simpleton. The argument is the following:

I know that both of you, asked if God with his infinite power and wisdom could give to the element of water the reciprocal [tidal] motion which we observe in it by some other means than by moving the containing vessel, I say that I know that you would answer that He would have the power and wisdom to do this in many ways, even if unthinkable by our intellect. Therefore I immediately conclude that, because of this, it would be excessively bold if someone wanted to restrict and force divine power and wisdom to a particular fancy of his [F488, my translation].

Galileo had been required by the censors, as a pre-condition for publication, that at the end of the book he should include such an argument which the pope found very impressive.[14] Galileo did include the argument, which appears on the very last page of the text in the 1632 edition.[15] Though the argument does constitute Simplicio's last word, it is followed by a few short remarks by Salviati, after which Sagredo in turn utters a few sentences. After the book was published, some people found fault with the rhetorical force of such an ending. In fact, part 2 of the sixth accusation of the original indictment charges Galileo with "having placed the medicine of the end in the mouth of a fool and in a section which can be located only with difficulty, in such a way that it is later approved by the other interlocutor in a cold manner and only by casually mentioning and not stressing its worth, which is to say unwillingly".[16] The pope himself seems to have at first accepted this interpretation and thus felt personally ridiculed, a feeling which helps to explain his intransigence about the trial, conviction, and sentence.

In this passage the rhetoric is allegedly contained not in what is explicitly said, but in how it is said; this manner of saying includes: (1) the failure to state the argument in a physically separate section of the book, under some such label as 'Ending' or 'Conclusion'; (2) the fact that the argument is given by the least intelligent of the three interlocutors; and (3) the fact that the most intelligent of them, Salviati, accepts the argument somewhat incidentally and casually. These three points are accurate when stated in this more objective manner, which changes the characterization of Simplicio from 'fool' to 'least intelligent', and of Salviati's attitude from 'unwilling' to merely 'casual' acceptance. There is no question that it is often rhetorically effective to insinuate claims, without expressing them explicitly. Also there is no question that Galileo did not like the kind of ending he was forced to include. However, the rhetorical problem is to determine whether the text is indeed such as to convey the impression that the skeptical argument is foolish.

In deciding this problem, it seems necessary to say a few words about the logical merits of the argument. Is the argument *per se* a foolish argument? Obviously, if Simplicio is giving a good argument, then to that extent *his* intelligence appreciates in the eyes of the reader. In other words, if the argugument is correct, then the ending provides evidence that Simplicio is not the fool that some would like to believe. I think that the argument is in fact unanswerable. Though expressed in what might be called the mythological language of God and divine omnipotence and omniscience, the demythologized version of the argument has become the stock-in-trade of modern epistemologists and methodologists. In modern terminology we might say: no scientific theory is ever conclusively proved since a scientific theory can be proved only relative to a finite amount of data, and for any finite set of data there is always more than one scientific theory about them, whether we view the theory as being inductively inferred from the data or as being an hypothetico-deductive explanation of the data. This is true basically because of the so-called problem of induction, or because of the formal invalidity of affirming the consequent.[17] Moreover, the force of the argument was felt in Galileo's time as well. In fact, a very clear statement of the argument can be found in a book published in 1629 by Augustinus Oregius, a theologian who was later (in 1633) one of the three consultants asked by the Inquisition to report on the content of the *Dialogue* as it related to the decree of 1616.[18] In his book Oregius reports on a discussion between Galileo and the pope and states that the pope

having agreed with all the arguments presented by that most learned man [Galileo], asked if God would have had the power and wisdom to arrange differently the orbs and stars in such a way as to save the phenomena that appear in heaven or that refer to the motion, order, location, distance, and arrangement of the stars. If you deny this, said Sanctissimus, then you must prove that for things to happen otherwise than you have presented implies a contradiction. In fact, God in his infinite power can do anything which does not imply a contradiction; and since God's knowledge is not inferior to his power, if we admit that he could have done so, then we have to affirm that he would have known how. And if God had the power and knowledge to arrange these things otherwise than has been presented, while saving all that has been said, then we must not bind divine power and wisdom in this manner. Having heard these arguments, that most learned man was quieted, thus deserving praise for his virtue no less than for his intellect.[19]

Finally, there is evidence that Galileo himself felt the force of the argument. In fact, in the Third Day discussion of the motion of sunspots (F372–83, D345–56) after Salviati has given a Copernican explanation, Galileo not only has Simplicio state the point in logically clear terminology (F379–80) but takes it seriously enough to have Salviati present an alternative (geostatic) explanation, though the latter is rejected because of its *ad hoc* character.[20] Galileo shows a similar awareness of the logic of theoretical explanation in his discussion of the stellar parallax objection (F409–14, D383–7), where it is Salviati who uses to his rhetorical advantage these logical facts,[21] and in the Second Day discussion of the arguments in favor of the earth's diurnal motion, where Salviati admits that the Copernican side is weakened by the merely probable force of arguments based on hypothetical explanations (F139–50, D114–24).[22] A fourth relevant discussion occurs in the Fourth Day, after Salviati has finished his explanation of the tidal diurnal period, when Simplicio once again gives a relatively clear statement of the logic of theoretical explanation and applies it to the basic tidal argument: Salviati feels obliged to give a detailed answer. (F462–70, D436–44).[23]

Thus according to Galileo, his contemporaries, and modern philosophy of science, 'the argument of Urban VIII' is a very good one. Galileo would not care for the theological way of expressing the point, any more than we moderns would. But if someone wants to speak the logical language, then neither Galileo nor we can object to the substance of the point being made. It follows that the rhetorical effect of the ending is not that claimed in Galileo's original indictment. Actually, the rhetorical import is favorable to religion insofar as it shows that to think in theological terms leads one to the same conclusion reached by nontheological, methodological arguments. In fact, no one believed this charge, except the pope who was personally involved; instead the fact that no one accepted the conclusiveness of Galileo's tidal argument is further evidence the 'medicine of the end' worked, though it must have worked because of its logical content and not because of its authoritative source.

If there is any rhetorical problem with this final passage in the book, it is its pointlessness. In other words, the point has already been made at least four times in the book (F139–50, F372–82, F409–14, F462–70), including once in the Fourth Day, so it ought to be logically superfluous. The logically sensitive reader, like Galileo, feels bored by the passage. Nevertheless, it may be given the rhetorical function of reemphasizing one last time an important point, and of doing so in language more easily understood to the man in the street with whom Galileo wanted to communicate. This may help explain why Galileo made the final argument part of the text of the Fourth Day, instead of having a short special section devoted to it. This separate section which some of the censors favored would then have been rhetorically less effective from their own point of view, or at least less effective if their aim had been to impart some genuine understanding, however rudimentary, to the masses. Unfortunately, their aim was more likely to be one of belief-control by mere obedience and submission of will. From this point of view, of course, a separate section, containing a more explicit theological qualification, would have been better since then people could have just looked at the 'Ending' and would have been informed that the motion of the earth had *not* been proved in the preceding book; they would have thereby saved themselves the trouble of reading it, and if their understanding of the problem would have been none the better for it, their actual beliefs would have been the 'right' ones.

Thus, from the rhetorical point of view, the controversy over the book's preface and ending is very enlightening. It illustrates two different conceptions of rhetoric, a Galilean kind according to which one tries to induce assent by arguments which may have to be simplified and expressed in appropriate language so that it can have an effect on the mind and understanding of the audience, and an authoritarian kind according to which one's utmost concern is to specify clearly what is or is not to be believed so that one's will can easily act upon it. Nothing illustrates this better than the contrast between the logically incoherent and rhetorically ineffective preface drafted by the pope's secretary, and the logically superfluous but rhetorically appropriate ending that Galileo came up with. It seems that, in trying to end the book in accordance with what the Church wanted, Galileo could not help doing so in a competent manner; whereas in explicitly compiling the preface, the pope's secretary could not avoid doing so in an incompetent manner. The fact that neither the crude rhetoric of the preface, nor the more elegant one of the ending, helped Galileo in the short run is perhaps an indication that he was doomed all along, a measure of the magnitude of the tragedy.

THE TITLE, OR THE RHETORIC OF INDECISION

So far I have considered what may be called the religious rhetoric contained in the *Dialogue*. Another equally obvious level of rhetoric relates to the problem of deciding on the relative merits of the two sides, the Copernican and the Peripatetic. The emphasis here is on decision, rather than on discussion; that is, there is throughout the book a recurrent theme that its author is merely discussing the relative merits of the two positions, rather than deciding on them, which decision is something to be reached by the proper authorities, who however may use the book's discussion as a basis. This is the rhetoric inherent in the book's full title, which it is useful to translate literally: Dialogue by Galileo Galilei, Lyncean Academician, Extraordinary Mathematician at the University of Pisa, and Philosopher and Chief Mathematician to the Most Serene Grand Duke of Tuscany; where in meetings over the course of four days one discusses the two Chief World

Systems, Proposing indeterminately the Philosophical and Natural reasons for the one as well as for the other side. The same theme is explicitly stressed on at least nine occasions, which it is useful to discuss one by one.

At the beginning of the Second Day, Sagredo's first speech summarizes the First Day by saying that the day before they had examined which of two opinions was "more probable and reasonable" (F132): the Aristotelian division of the universe into a celestial and a terrestrial part, or a second opinion according to which the earth is a globe very much like the other planets. Sagredo's summary ends by saying that the day before they had "concluded that this second opinion has more verisimilitude than the other" (F132). Upon hearing this, Salviati is quick to add the disclaimer that

I did not conclude this, just as I am not deciding upon any other controversial proposition. My intention was only to adduce those arguments and replies, as much on one side as on the other — those questions and solutions which others have thought of up to the present time (together with a few which have occurred to me after long thought) — and then leave the decision to the judgement of others. [D157]

A second disclaimer is found in the course of the preliminary statement of the objections to the earth's diurnal motion. Here, at one point Sagredo stresses that Salviati seems to understand very well those geostatic arguments; this fact has impressed the Aristotelian Simplicio himself, who has let those arguments be expounded by Salviati even though the latter does not accept them and will afterwards criticize them. After Sagredo's speech, before going on to give a statement of some other geostatic arguments, Salviati notes:

Before going further I must tell Sagredo that I act the part of Copernicus in our arguments and wear his mask. As to the internal effect upon me of the arguments which I produce in his favor, I want you to be guided not by what I say when we are in the heat of acting out our play, but after I have put off the costume, for perhaps then you shall find me different from what you saw of me on the stage. [D131; cf., F157–8.]

Since Salviati is indeed the Copernican spokesman, and since in this discussion he is acting as the expounder of Aristotelian arguments, it seems at first that his justification is misplaced, and that instead he should call attention to the fact that he is now wearing an *Aristotelian* mask. However, upon reflection, since in the present discussion his Aristotelian mask is obvious, what he is probably doing is to take this opportunity to call attention to the less obvious fact that his whole Copernican position is an act in a play. Indeed this is so much *less obvious,* that it may not be true at all; it certainly is not convincing, and there is no evidence other than his words that it is so. At any rate, such is Galileo's rhetoric.

The third relevant passage is one at the end of the critique of the objection that if the earth rotated that would mean that our senses would be deceived. Salviati is trying to show that there would be no deception of the senses, since we are all agreed upon what the sensory appearances are; the problem is really with the conclusions we draw therefrom. It is wrong for some Peripatetics to make it sound as if on a moving earth we would experience terrestrial phenomena differently, or as if the believers in the earth's motion are seeing things differently; in giving a reason for this, Salviati inserts the disclaimer that "just as I (who am impartial between these two opinions, and masquerade as Copernicus only as an actor in these plays of ours) have never seen nor ever expect to see the rock fall any way but perpendicularly, just so do I believe that it appears to the eyes of everyone else" (D256; cf., F281).

At the end of the Second Day, its main conclusion is summarized. Simplicio asserts that all that has been shown is that the geostatic arguments are not compelling, and that so far no compelling argument has been given to convince one of, or to prove, the earth's motion. Salviati replies:

I have never taken it upon myself, Simplicio, to alter your opinion; much less should I desire to pass a definite judgment on such important litigation. My only intention has been, and will still be in our next debate, to make it evident to you that those who have believed . . . [in the earth's motion] were not blindly persuaded of the possibility and necessity of this. Rather they had very well observed, heard, and examined the reasons for the contrary opinion, and did not airily wave them aside. [D274; cf. F298]

Next, in the Third Day, after Salviati has finished giving both Copernican and geostatic explanations of the apparent motion of sunspots, each of the three interlocutors expresses his feelings on their relative merits. Simplicio confesses that the decision between the two is too important for him to make (F382–3). Sagredo thinks that he has never heard anything more convincing, other than the demonstrations of pure mathematics. Salviati states:

I do not give these arguments the status of either conclusiveness or of inconclusiveness, since (as I have said before) my intention has not been to solve anything about this momentous question, but merely to set forth those physical and astronomical reasons which the two sides can give me to set forth. I leave to others the decision. [D356]

On the next page, they begin examining the arguments contained in a book by one of Galileo's contemporaries, where they find, among others, the religious and Biblical objections that Copernicans are allegedly forced to say that Christ *ascended* into hell and *descended* into heaven, and that when Joshua ordered the sun to stand still, it was instead the earth that stood still. Salviati takes exception to this writer's "having mixed passages from the ever venerable and mighty Holy Scriptures among these apish puerilities, and his having tried to utilize sacred things for wounding anybody who might, without either affirming or denying anything, philosophize jokingly and in sport, having made certain assumptions and desiring to argue about them among friends" (D357).

Another disclaimer is found at the end of the discussion of the stellar dimensions required in a Copernican universe by the lack of apparent annual parallax. Sagredo has just finished a disquisition against the arrogance, indeed the 'insanity' of those who seem to argue that, because we don't know how Jupiter or Saturn is of service to us, therefore they are superfluous, or even they do not exist (F438). Salviati tries to calm him down, saying that "there is no need, Sagredo, to probe any further into their fruitless exaggerations. Let us continue our plan, which is to examine the validity of the arguments brought forward by each side without deciding anything, leaving the decision to those who know more about it than we" (D369).

At the end of the Third Day, in concluding the discussion of magnetism, Sagredo refutes the objection that a simple body like the earth cannot have several natural motions by arguing that the earth is probably not a simple body, and that anyway it probably is a loadstone, and loadstones have several natural motions. In his criticism at the very end of that discussion, Sagredo comes down pretty hard on the Aristotelians. Salviati then makes the following plea for moderation:

Sagredo, please, let us weary ourselves no longer with these particulars, especially since you know that our goal is not to judge rashly or accept as true either one opinion or the other, but merely to set forth for our own pleasure those arguments and counterarguments which can be addressed for one side and for the other ... therefore we shall suspend judgment, and leave this in the hands of whoever knows more about it than we do. [D413]

Finally, at the end of the Fourth Day, just before the skeptical "medicine of the end" is given by Simplicio, Salviati somewhat apologetically asserts that "I do not claim and have not claimed from others that assent which I myself do not give to this invention, which may very easily turn out to be a most foolish hallucination and a majestic paradox" (D463).

This rhetoric of indecision has never impressed anyone. The problem with it is that it is merely cosmetic. One might say that if Galileo's arguments were really equally balanced, there probably would be no need for these constant explicit disclaimers. I do not think that the task described in the title is intrinsically impossible; rather it is valid to distinguish between an impartial *presentation* of arguments for two opposite sides, and a decision as to which side is better. The problem with the (full) title of the book is that in the *Dialogue* the presentation of the arguments is such that there can be no question as to which side is more plausible or probable, as Galileo's enemies and the inquisitors were quick to point out. It is not so much that the presentation is biased, but rather that it *determines* a decision.

It is well known that such rhetoric was imposed on Galileo by the Church censors. In fact, Galileo would have liked at least to mention the ebb and flow of tides in the book's title, as is shown by his correspondence between the years 1624—1631.[24] However, such a title was vetoed by the interpreters of the pope's desires.[25] What seems to have happened is that Galileo's better rhetorical instincts were channeled in the wrong direction by Church officials.

To begin unraveling this problem let us examine the view, widely prevalent among scholars, that Galileo's motivation for the original title was his belief that the tidal argument of the Fourth Day provided a conclusive proof of the earth's motion, and that the Church's motivation for the change was its unwillingness to put a stamp of approval on the argument.[26] Such a view is (a) based on no direct evidence, (b) rhetorically naive and implausible, and (c) a misinterpretation of the crucial idea of hypothetical reasoning.

(a) The only scholar who gives any evidence is Shea. He refers to Galileo's letter to Elia Diodati of August 16, 1631 and quotes the statement that "I have not been able to obtain permission to mention the tides in the title of the book although it is the principal argument that I consider in it".[27] However, this quotation is taken out of context, for after a few sentences the letter ends with the assertion, "I think that, if the book had been entitled on the ebb and flow, that would have been to the greater profit of the printer. But after a while the news will spread by means of the first ones who will have read it."[28] Thus if this letter proves anything (and it may not), it is that he is concerned with the book's circulation and popularity. He seems to feel that more people would be likely to buy or read a book whose title claimed to be dealing primarily with tides. Is Galileo's judgment right here? A point in its favor is that whereas the Copernican theory was the subject of the prohibition of 1616, the tides were not the subject of any such decree.

(b) Second, in the light of the fact that the anti-Copernican decree was still standing, it seems implausible that Galileo would want to call attention, in the very title of his book, to an alleged proof of its philosophical incorrectness. Whatever private or verbal assurances may have been given by Church officials or even the pope himself, Galileo would not have been so rash as to advertise his violation of the decree, for in 1616 he had wanted to

be safe enough as to obtain a letter from Cardinal Bellarmine certifying that
Galileo had *not* been the object of any disciplinary proceedings in connection
with the decree.[29] Moreover, this same letter which Galileo had kept stated
explicitly that the Copernican theory could not be 'held or defended'. In
other words, it would have been out of character for Galileo, and intrinsically
imprudent, to want the tides mentioned in the title out of a desire to stress
the argument from tides.

But if we agree that Galileo was far too prudent to have such a wish, isn't
it a fact that it was dangerous for him to mention the tides, in the light of
the tidal argument? It all depends on the nature of this argument. There is no
question that Galileo thought that this was a very good argument, indeed his
strongest one in favor of the earth's motion. However, there is also no
question of Galileo's awareness of the hypothetical-explanatory character of
this argument; as mentioned above, he refers to this *explicitly* twice in the
Fourth Day, once when the logic of reasoning *ex suppositione* is mentioned
(F462), and once with theological language at the very end; and, as emphasized above, this only reflects an attitude that Galileo expresses on at least
three other occasions in the book. Moreover, the *actual structure* of the
Fourth Day is hypothetical in the sense that a problem is presented, various
solutions criticized, and an original one proposed. I conclude that his desire
to have the tides mentioned in the title stems from his desire to emphasize
the hypothetical character of his book, and thus conform to the decree.

This conclusion is reinforced by the various descriptions of the title found
in Galileo's correspondence between 1624 and 1630.[30]

(c) A third point requires discussion now, relating to the distinction among
three different ways of treating a topic, which we may call absolute, hypothetical, and indeterminate. This distinction is mentioned in the original
indictment against Galileo,[31] where it is charged that the book's treatment of
the earth's motion is either absolute or at least indeterminate, but not hypothetical. To treat the earth's motion absolutely means to discuss it in such a
way that its reality is asserted unconditionally; this requires[32] arguments
which are alleged to be deductively valid *and* based on premises which are
alleged to be indisputably factual. An indeterminate treatment of the earth's
motion is one where all available arguments for *and* against it are presented
without a final decision being reached concerning the truth or falsity of the
conclusion. An hypothetical treatment is one where the supporting arguments
are such that the conclusion (earth's motion) is presented not as a logical
consequence of known truths but as a conjectural explanation of them, in
the sense that *if* the conclusion is assumed to be true *then* the known truths

can be seen to follow. The book's actual title and the above-mentioned corresponding rhetoric throughout its pages claim that an indeterminate treatment of the earth's motion is being given. This is pretty obviously false because it is immediately clear even to a casual reader that Galileo does show that the favorable arguments are better than the unfavorable ones; hence he is *in fact* deciding the question; hence, the explicit disclaimers to the contrary are 'mere rhetoric', that is, they have no real rhetorical force. Unfortunately, even if the book had really been an indeterminate treatment, it still could have been damned, and hence the title is still rhetorically infelicitous; for it was pointed out in the original indictment[33] and in the final sentence[34] that it is wrong to regard as undecided the question of whether or not the earth moves, which is decided in the Bible and anyway had been decided by the decree of 1616.

The same is not true of an hypothetical treatment, which may not have violated the spirit of the decree, and which therefore might very well have been tolerated, though of course it certainly is a way of teaching the subject, and it also is a weak kind of defending an idea. Therefore, a title mentioning the tides, which would have stressed the hypothetical character of the discussion of the earth's motion, would have been rhetorically superior from Galileo's point of view. This is the sense in which his better rhetorical instincts were misdirected by the censors.

THE RHETORIC OF STRICT DEMONSTRATION

In view of the presence in the *Dialogue* of such obvious rhetoric of religious piety and of indeterminacy, it would seem incredible that anyone would seriously claim to find in it a rhetoric of rigorous proof, from which point of view the book would be a conscious attempt to give a strict demonstration of Copernicanism.[35] Once again, the emphasis here is on the 'rhetoric', for the same scholars who stress this aspect of the book are the first ones to point out that it did not succeed in the goal, partly because it is an unrealizable goal, partly because the book's arguments are not in fact strictly demonstrative, and partly because several of the arguments are erroneous in one way or another. The basic problem here is not so much that the book completely lacks explicit expressions of the desirability of strict demonstrations and implicit claims of the conclusiveness of its arguments, but rather that the former are very rare and the latter never occur in the same context as the former, so much so that this kind of rhetoric becomes visible only with the tunnel vision which it is possible to acquire through certain kinds of scholarly discipline.

By explicit expressions of the desirability of strict demonstrations, I mean passages like the following:

[1] I take it to be definite and certain that for the proof of a true and necessary conclusion there are in nature not merely one but very many powerful demonstrations . . . . I believe on the other hand that to make a false proposition appear true and convincing, nothing can be adduced but fallacies, sophisms, paralogisms, quibbles, and silly inconsistent arguments full of pitfalls and contradictions. [D130; cf. F156.]

[2] It is not possible within the bounds of human learning that the reasons adopted by the right side should be anything but clearly conclusive, and those opposed to them, vain and ineffective. [D356; cf. F383.]

It is not easy to find any other such passages.

By implicit claim of the conclusiveness of arguments I mean claims, *occurring in the course of the presentation* of an argument, emphasizing the strength of the inferential links, and the certainty of the conclusion. For example:

[1] Having established, then, that it is impossible to explain the movements perceived in the waters and at the same time maintain the immobility of the vessel which contains them, let us pass on to considering whether the mobility of the container could produce the required effect in the way in which it is observed to take place. [D424]

[2] You see, gentlemen, with what ease and simplicity the annual motion – if made by the earth – lends itself to supplying reasons for the apparent anomalies which are observed in the movements of the five planets, Saturn, Jupiter, Mars, Venus, and Mercury. It removes them all and reduces these movements to equable and regular motions; and it was Nicholas Copernicus who first clarified for us the reasons for this marvelous effect.

But another effect, no less wonderful than this, and containing a knot perhaps even more difficult to untie, forces the human intellect to admit this annual rotation and to grant it to our terrestrial globe. This is a new and unprecedented theory touching the sun itself. For the sun has shown itself unwilling to stand alone in evading the confirmation of so important a conclusion, and instead wants to be the greatest witness of all to this, beyond exception. So hear this new and mighty marvel. [D344-5]

I am stressing here the fact that such remarks occur during the statement of arguments; it is this that gives these remarks their implicit character. The contrast to keep in mind is to those remarks made after the argument has already been stated where, as we saw in part in our discussion of the rhetoric of indeterminateness, contrasting reflecting judgments are passed upon the argument: Simplicio remains usually unimpressed and states various reservations, Sagredo is usually overimpressed and finds the arguments very strong and compelling, whereas Salviati pays his lip service to the indeterminacy-claim inscribed in the book's title.

The important thing to notice about these two kinds of passages is that they never occur together. Hence, the implicit claims must be regarded as an integral part of the normal rhetoric of argumentative prose, rather than as elements of a 'strict demonstration' theory of science instantiated by the actual arguments in the statement of which these claims are made. On the other hand, the few explicit claims about strict demonstration occur in other contexts. The first one of these explicit claims is actually made by Simplicio and is part of a quasi-Socratic, somewhat 'ad hominem'[36] metalogical discussion trying to show to Simplicio that the consideration of arguments both for and against a view, far from leading to confusion of mind, is the only way of arriving at the truth. The 'strict demonstration' aspect of Simplicio's remarks is completely incidental; it is either an exaggerated way of expressing the contrast between good and bad arguments, or a reflection of the Aristotelian theory of science of *Posterior Analytics.* The second explicit passage, on the other hand, is part of Salviati's disclaimer at the end of the discussion of sunspots in the Third Day. The remark is actually a disclaimer within a disclaimer, for Salviati first says that, in accordance with his usual policy, he is not going to label the argument from sunspots as either conclusive or inconclusive, but then adds that ultimately the authoritative determination cannot be itself inconclusive since one of the two sides is true and the other false, and the arguments for the truth must be as conclusive as those for the falsehood are inconclusive.

And this brings us to an examination of the argument which leads certain scholars to find a rhetoric of strict demonstration in the *Dialogue;* it turns out (as if Saviati's disclaimer within a disclaimer were true) that this argument is as unsound as its conclusion is false, which I have just shown. Let us consider the account given by Ernan McMullin, who is perhaps its leading proponent.[37] He begins with the following quotation from the Third Day:

However well the astronomer might be satisfied merely as a calculator, there would be no satisfaction and peace for the astronomer as a scientist . . . . (Copernicus) very well understood that although the celestial appearances might be saved by means of assumptions essentially false in nature, it would be very much better if he could derive them from true suppositions . . . . One of the arrangements must be true and the other false. Hence it is not possible within the limits of human learning that the reasons adopted by the right side should be anything less than clearly conclusive, and those opposed to them vain and ineffective.[38]

This beginning is an inauspicious one, for the first two sentences of this quotation are separated from the last two by fifteen pages of text, as McMullin's own footnote indicates.[39] This questionable practice has the effect of connecting by fiat two things which do not have the requisite connection, namely Galileo's epistemological realism and his alleged rhetoric of strict demonstration. The first is indeed a basic feature of Galileo's work, but the second is an insignificant one (at least as far as the *Dialogue* is concerned). We may agree that if someone is a 'strict demonstrationist' then he would have to be an epistemological realist; however the converse is not true since one can be a realist and a probabilist. Hence the quotation as given supports a rhetoric of demonstration only if its last two sentences do. Unfortunately these two sentences are injudiciously taken out of context, which is that of Salviati's disclaimer of a disclaimer. Moreover, though McMullin's last two sentences are an accurate quotation from the translation from which it is being quoted, it so happens that the translation of this particular passage is clearly incorrect precisely at the place needed as evidence for 'strict demonstration'. In Santillana's revision of Salusbury the last sentence reads, "It is impossible that (always confining ourselves within the limits of human doctrine) the reasons alleged for the true hypothesis should not manifest themselves as conclusive as those for the contrary vain and ineffectual".[40] The original Italian (F383) makes it even clearer that what is meant is a comparison of the degree of conclusiveness of one side with the degree of inconclusiveness of the other.

A second difficulty with McMullin's account in general is the unclarity of the notion of a demonstration of truth; that is to say, it is difficult to see what he is talking about in putting forth his claim. It will not do to use the clear terminology of formal logic and say that it is a formally valid argument with true premises. In fact, if the argument's conclusion is an empirical proposition (as in the present case), then at least one of the premises has to be empirical, and hence its truth is always question*able* even if it is unquestion*ed*. Since the success of a demonstration so conceived depends both on the truth of the premises and on the validity of the inference steps, it follows that it could always be questioned, which is to say that such demonstrations are impossible for empirical claims. Now McMullin cannot be excused by saying that all he is doing is to report Galileo's beliefs, intentions, and aims; for if he wants to do this in the sense of 'merely reporting', then he is not conforming to his own aim since it is obvious he is also interpreting and trying to make sense out of that subject matter; but in doing the latter it is methodologically undesirable to use an intrinsically incoherent concept.

A third difficulty is that the view is largely based on events and documents relating to the years 1615–1616, which climaxed in the anti-Copernican decree, the main document being Galileo's *Letter to the Grand Duchess Christina*. However, this letter is three steps removed from the *Dialogue*, first chronologically; second, insofar as it belongs to the context of theory rather than practice, that is, it contains explicit remarks on science, but no actual scientific arguments; third, insofar as even within the letter there are two conflicting theories of science being put forth, and only one of these includes the 'strict demonstration' ideal.[41]

Fourth, McMullin argues that, since there is evidence that "Galileo was a thoroughgoing Aristotelian",[42] therefore, as an Aristotelian trying to convince other Aristotelians, Galileo was required by the theory of science of *Posterior Analytics* to present a strict demonstration of his thesis in order to get a hearing. As a historical interpretation of the *Dialogue*, McMullin's argument is at best circular (which is not to say that his conclusion is false or his argument logically incorrect), since what is presently in question is whether this book is Aristotelian in the required manner. At any rate, the direct evidence from this work is that Galileo develops, by both words and deeds, a theory of science different from the Aristotelian one and such as to obviate the need for a strict demonstration (at least for the moment). Galileo does this by examining the 'strict demonstration' given by the Aristotelians in support of their geostatic view, finding that their 'strict demonstration' is faulty in several ways, concluding that the Aristotelians have *not* proved the geostatic view, presenting and formulating the best arguments he can in support of the contrary view, and throughout this argumentation never avoiding but always discussing relevant epistemological and methodological issues with an eye toward justifying his procedure.[43]

THE JUSTIFICATION OF COPERNICANISM

If the effective rhetorical content of the *Dialogue* is not that of an apology for the decree of 1616, nor that of an indeterminate pre-decision study of the evidence, nor that of a strict demonstration of Copernicanism, in spite of the 'rhetoric' to the contrary, what else is it? It is pretty obvious that it is a *justification* of Copernicanism. Agreement concerning this can be found in the action and words of the theologians and Church officials involved in the prosecution of Galileo, in the mostly favorable reaction by seventeenth century scientists and philosophers, and in the judgment of modern scholars. By a justification of Copernicanism I mean an attempt by verbal means and techniques to induce or increase adherence to Copernicanism.

The justification is carried out at three conceptually distinct (though not completely separable) levels, logical, philosophical, and rhetorical. The logical aspect of the justification is the one that involves presentation, analysis, and evaluation of the evidence and arguments for and against Copernicanism; the philosophical aspect is the one that involves the discussion and clarification of the epistemological and methodological concepts inherent in this attempt to show the evidentiary superiority of Copernicanism over geostaticism; the rhetorical aspect, in a more restricted sense of 'rhetorical' than the general one used so far in this chapter, involves the attempt to view in a favorable light Copernican theses, concepts, and procedures, and in an unfavorable light the Aristotelian counterparts, and to do so by means other than logical, that is primarily by arousing the emotions of the reader in the appropriate way.

What I am calling here the logical dimension of the justification of Copernicanism corresponds to what is often called the 'scientific' aspect of Galileo's problem. In subsuming the analysis of evidence, under the concept of 'logical', I am using the label in a somewhat wider meaning than is customary, but I do so in order not to pre-judge questions like whether there are two kinds of logic and of reasoning (namely, deductive and inductive), and if so which is more important and prevalent. I think that 'logical' is preferable to 'scientific' because of the more time-dependent connotations of the latter, that is, because 'scientific' is often plausibly taken to mean "pertaining to present-day scientific knowledge". Though it is important and useful to examine Galileo's *Dialogue* from the scientific point of view in this sense (as I do in Chapter 4), it is equally important to examine it from the point of view of an objective analysis and presentation of evidence and arguments (which I do in the next chapter); this is what I am labeling the 'logical' dimension.

There is universal recognition of the need for a philosophical discussion, besides a logical one, in a situation like that faced by Galileo, which involved very fundamental changes in our cognitive and intellectual structure. For everybody recognizes that such changes involve not only changes in specific beliefs and conclusions, but also changes in general, philosophical attitudes and perspectives. Hence the philosophical dimension of the *Dialogue* has been widely studied, though no one has done so systematically (as I do in Chapter 5) with the result that even the best scholars have misinterpreted its philosophical content (Chapter 10)[44] or failed to appreciate its true philosophical significance (as I argue in Chapter 6).

As regards the restrictedly rhetorical level of Galileo's justification (in an alogical, but not illogical sense of 'rhetoric'), this fact is generally admitted but seldom appreciated. As suggested above, Paul Feyerabend is the only recent scholar who has perceived the necessity of such (alogical) rhetoric, though his point is obscured by his own 'rhetoric', which is one of irrationalism.[45] His point is simply that for certain very fundamental intellectual changes (like the scientific revolution of the 17th century), logic and even philosophy are not enough, and one needs rhetoric in the sense under discussion. However, as I argue later,[46] one cannot even begin to understand this rhetoric unless one has a clear understanding of the book's logical structure. Hence we shall proceed to examine the latter before going on to the analysis of the means, above and beyond the logical-scientific and the philosophical-methodological, whereby Galileo justifies Copernicanism.

NOTES

[1] A. Koyré, *Études Galiléennes*, pp. 212–215; L. Geymonat, *Galileo Galilei*, pp. 132–135; and S. Drake, *Galileo Studies*, pp. 253–255.

[2] P.K. Feyerabend, *Against Method*, esp. Chapters 1, 12, and 18. Cf. also my discussion of Feyerabend's views in Chapter 8 below.

[3] Besides Feyerabend, the problem has been discussed in T. S. Kuhn, *The Structure of Scientific Revolutions*, especially pp. 144–160; and in I. Lakatos and A. Musgrave (eds.), *Criticism and the Growth of Knowledge*, especially pp. 1–24, 91–196, 197–278.

[4] As I have argued in my *History of Science as Explanation*, pp. 180–198.

[5] These three approaches are discussed in more detail in my Essay-review of Lakatos's *Criticism and the Growth of Knowledge*. The third conclusion, about the irrationality of science, corresponds, of course, to Feyerabend's *rhetoric*.

[6] This is the conclusion that Feyerabend ought to be drawing, and the one that we may take him to be *really* drawing. Another very interesting conclusion to draw would be that important segments of actual, historical science exhibit features of what is labeled 'philosophy' in H. W. Johnstone, Jr., *Philosophy and Argument*. The coincidences are, in fact, remarkable: for example, Johnstone speaks of conflicting philosophical statements lacking the property of being 'logically commensurate', whereas Kuhn and Feyerabend speak of scientific theories being incommensurable; Johnstone holds that the arguments for and against a philosophical statement are part of its meaning (p. 32), whereas Kuhn and Feyerabend speak of meaning variance and hold that the meaning of scientific terms is theory-dependent. This opens the way for vindicating the rationality of scientific revolutions by making use of Johnstone's theory of philosophical argumentation.

[7] A fascimile of the 1632 edition is now published by Culture et Civilisation, 115 Avenue Gabriel Lebon, Brussels.

[8] G. Galilei, *Opere*, ed. Favaro, **19**, 326; cf. G. de Santillana, *The Crime of Galileo*, p. 211.

[9] *Opere*, **19**, 326; my translation.

[10] *Ibid.*, pp. 402–426.

[11] *Ibid.,* pp. 327–330.

[12] Here one cannot help but commenting on the *apparent* inconsistency of the charge numbered (6.2) in the original indictment (*Opere* 19, 326) for this charge, besides accusing Galileo of having alienated the preface from the body of the text, also charges him with having made the prescribed ending (the so-called 'medicine of the end', or 'the argument of Urban VIII') a completely integral part of the text (in the last few pages). It must be admitted, however, that the inconsistency is more apparent than real, for the charge is accusing Galileo of excesses in *both* directions, that is, in the case of the preface, setting it too much apart from the text, and in the case of the ending, integrating it so much in the text as to be difficult to find. One might add, however, 'Difficult to find for whom?' The answer would seem to be, "For people and readers interested only in beginnings and endings". But this means for people primarily interested in form or appearance or in 'mere' rhetoric.

[13] *Opere* 7, 29–31. References to this book will be made in the text by preceding the page number(s) by 'F', short for 'Favaro'.

[14] *Opere* 19, 327, 330; cf. G. Galilei, *Dialogo sopra i due massimi sistemi,* edited by L. Sosio, pp. 548–549, n. 1.

[15] G. Galilei, *Dialogo* (Florence, 1632), p. 458; cf. F488–489, and G. Galilei, *Dialogue Concerning the Two Chief World Systems,* tr. Drake, pp. 464–465. Subsequent references to the last book will be given in parenthesis within the text, by prefixing the page number(s) with a 'D'.

[16] *Opere* 19, 326; my translation.

[17] See, for example, C. G. Hempel, *Philosophy of Natural Science,* pp. 3–18; and A. Pap, *An Introduction to the Philosophy of Science,* pp. 139–173.

[18] *Opere* 19, 348; cf. Santillana, *The Crime of Galileo,* p. 245.

[19] Quoted in Galilei, *Dialogo,* edited by L. Sosio, pp. 548–595, n. 1 from Augustinus Oregius, *De Deo uno* (Rome, 1629), pp. 194–195.

[20] Cf. Chapter 5, passage F372–383.

[21] F413–414; cf. my reconstruction in Chapter 5.

[22] Cf. my reconstruction in Chapter 5.

[23] Cf. my reconstruction in Chapter 5.

[24] *Opere* 13 and 14.

[25] *Opere* 19, 327; cf. 14, 289.

[26] E. McMullin, 'Introduction: Galileo, Man of Science', p. 34; A. Koestler, *The Sleepwalkers,* p. 480; W. R. Shea, *Galileo's Intellectual Revolution,* pp. 173–174; and G. de Santillana, *The Crime of Galileo,* p. 183.

[27] Quoted by Shea, *Galileo's Intellectual Revolution,* p. 187, n. 12. Cf. *Opere* 14, 289.

[28] *Opere* 14, 289, my translation.

[29] Santillana, *The Crime of Galileo,* pp. 125–144; cf. *Opere* 19, 348.

[30] *Opere* 13, 236, and 282; 14, 49, 54, 66, and 130.

[31] *Opere* 19, 325, lines 38–48; and 19, 326, lines 95–100.

[32] This *conception* can be found in the report of one of the consultants to the Inquisition: *Opere* 19, 356, lines 24–28.

[33] *Opere* 19, 326, lines 99–100.

[34] *Ibid.,* p. 404, lines 62–68.

[35] See, e.g., E. McMullin, 'Introduction: Galileo, Man of Science', pp. 31–35; and idem, 'The Conception of Science in Galileo's Work', pp. 247–51.

[36] In the old, seventeenth century meaning of this term.
[37] McMullin, 'Introduction: Galileo, Man of Science'.
[38] Quoted in McMullin, pp. 31–32 from Galilei, *Dialogue,* tr. Drake, pp. 341, 356.
[39] McMullin, *op. cit.,* p. 40, n. 70.
[40] Galilei, *Dialogue on the Great World Systems,* tr. Salusbury-Santillana, p. 366.
[41] Cf. McMullin, pp. 33–34.
[42] *Ibid.,* p. 32.
[43] For more details, see Chapters 2 and 5.
[44] In the part dealing with Clavelin.
[45] Cf. *Against Method,* and Chapter 8.
[46] See Chapter 8.

CHAPTER 2

FACT AND REASONING:
THE LOGICAL STRUCTURE OF GALILEO'S ARGUMENT

Here I shall provide a solution to a problem which very much deserves discussion, which I shall call the problem of the structure of the *Dialogue*. The first aspect of this problem derives from the fact that there is a scholarly need for a descriptive outline of the book. Such an outline would facilitate communication among scholars insofar as it would provide a standard way of referring to a particular passage or discussion. Moreover, it would provide an easy way of locating a particular passage that one may be looking for; at present this is rather difficult because, though the book is very long, the only signposts are the division into four 'Days', the few diagrams which are interspersed in the text, and the numerous marginal notes by the author himself. The usefulness of the latter is seriously impaired by the fact that they are *not* formulated from a single point of view. In fact, what one reads in these marginal notes is not always an idea propounded by Salviati and opposed by Simplicio, but sometimes one propounded by the latter and opposed by the former, sometimes an idea accepted by both, sometimes one rejected by both, and sometimes a speculative idea being considered only hypothetically. Now, in order to provide an adequate descriptive outline of the whole book, one needs to study and ascertain its structure, since one needs to consider the relationships among the various sections.

The second aspect of the problem is the book's *apparent* lack of structure, or lack of explicit structure. The *Dialogue* is such that it is very natural and easy for the reader to feel that, in de Santillana's eloquent words, it "meanders at ease across the whole cultural landscape of the time, carrying in its broad sweep strange material of various origin. As a composition, it looks unfinished, unpolished, at times inconsistent. This is partly nature, partly art. It has no unity except that of life itself".[1] This leads one to ask whether the apparent lack of structure, which is undeniable, is real; whether the book really lacks unity; or, if its unity is 'that of life itself', what this amounts to. It will be shown below both that the proposition 'the earth moves' provides the book with unity and exactly how this is so.

The third aspect of the problem is the book's wealth of content; the content is so rich that it creates confusion. By wealth of content I mean the fact that the book deals explicitly with and has obvious implications for philos-

27

ophy, physics, astronomy, literature, practical polemics, and pedagogy. Koyré is making this point when he says at the beginning of his discussion of Galileo's *Dialogue,* in the third part of *Études Galiléennes:*

The *Dialogue on the Two Chief World Systems* pretends to be an exposition of two rival astronomical systems. But, in fact, it is not an astronomy book, nor even one of physics. It is above all a book of criticism, a work of polemic and struggle; it is at the same time a pedagogical work, and a philosophical work; it is finally a history book, 'The History of Galileo's Mind'.[2]

This remark needs qualification as regards the astronomical and physical content of the book; nevertheless it conveys the right idea. This wealth of content is, of course, one of the things that makes the book uniquely important and importantly unique in world history, but it can be confusing. The wealth of content is pretty well displayed by Shea in his book *Galileo's Intellectual Revolution.*[3] Shea's account is adequate, competent, and helpful and it probably represents the best that can be accomplished by following the procedure of retaining in one's account the mixture of topics and aspects found in the original work. Yet a mere mixture has no structure. To find structure along these lines one must do the following; one must begin by distinguishing several points of view: physical, philosophical, astronomical, literary, rhetorical, dramatic, pedagogic; then one must choose one point of view, and then reconstruct (reconstruct, not merely summarize) the whole book from that point of view. A very good example of a reconstruction from a single point of view is Clavelin's account in Chapters 4 and 5 of his book *The Natural Philosophy of Galileo;*[4] and in spite of the title of Clavelin's book, the point of view in these chapters is that of physics. There are two problems with this approach to the study of structure. The first is that one is likely to carry out the analysis from only one or a few of the possible points of view, the most common being physics, philosophy, and astronomy. Of course, this is not an insurmountable problem; there is no reason why the literary, polemical, and pedagogic aspects should be neglected; yet it is a historical fact that no one account of the *Dialogue* takes all the points of view in turn. The second problem with this multifaceted approach is that the book's structure from the various points of view is different. The question then arises whether these different structures are really in the book or whether they are being imposed on it. I do not think that this problem is insurmountable either, but I see no way of solving it other than dismissing it as a problem.

This may motivate one to follow another approach. The only other way of finding structure, and thus avoiding confusion, and hence gaining under-standing, is to examine the interrelationships of the various parts and various statements found in the book, and to keep as close as possible to the original wording in one's reformulation. In a sense this, too, is a point of view, namely the internal point of view; that is, the point of view of the book itself. This is not the point of view of the title of the book, or of the author's preface or introduction, or of the author's stated intention as found in other documents; it is not even the point of view of the author's table of contents, if there is one, though it can be very helpful. It is the point of view of the whole book in its entirety, and of the interrelationships of *all* its parts. The following analysis of Galileo's *Dialogue* is offered in part as an attempt to delineate its internal structure.

The logical structure of the *Dialogue* is that of an argument designed to show that the earth moves. The critiques of the various geostatic arguments can be easily integrated into such an argument as follows. Those critiques are designed to show that there are no sound reasons for thinking it false that the earth moves, or ontologically expressed, that there is no real evidence against the earth's motion. The reconstructed argument would then start as follows: We may conclude that the earth moves since there is no real evidence or sound reasons against its motion while there is considerable evidence and various reasons in its favor; the former is true because all arguments against the earth's motion are incorrect; the latter is true because there are sound arguments for the earth's motion. Then would come all the details of the arguments for and against the earth's motion, in such a way that virtually every main topic men-tioned in the book would be integrated into this main argument.

However it will be almost impossible to appreciate this integration, or to understand the details of the argument, or to give a statement of it, or to check its accuracy without some fairly precise way of referring to the various portions of the book. What is needed is an outline, something that historians and scholars also need, for their own reasons, as mentioned above. So, I have constructed the outline given below with page references to the standard scholarly edition by Antonio Favaro,[5] to Stillman Drake's English transla-tion,[6] and to Pietro Pagnini's edition,[7] which is the most valuable and easily available one in Italian. The various outline subdivisions are partly grounded on the book's literary structure and so will remain unaffected by possible changes in the outline headings; good examples of this are IIC and IIIA. The subdivisions are also grounded on the subject matter or topical unity of a given section, and the various headings are usually descriptions of this subject matter; a good example of this is IC1, 2, 3, and 4. Moreover, both the headings and the subdivisions are partly grounded on the logical unity, clarity, or beauty of the various paragraphs making up what I call below the Analytical Summary. However, it is primarily the analytical summaries that originate from the outline subdivisions and descriptions. It should be noted that the outline headings are in every instance the grammatical subjects of the first (or only) sentence in the corresponding paragraph of the Summary. The Analytical Summary constitutes my reconstruction of the main steps of the argument in the *Dialogue*. It thus defines what may be called the logical structure of this work. Of course, many paragraphs of the summary could themselves be expanded in such a way as to integrate (and thus number) almost literally every proposition in the book.

My Analytical Summary is thus open-ended in a downward direction, so to speak. That is, there are many subarguments (long, complex, important, and interesting arguments) of which I have stated only the conclusion in order to use it as a premise of the subarguments that I did state. It is clear, however, that this open-endedness does not make the summary incomplete but is a necessity because a summary should be a summary and not a complete reconstruction. An Analytical Reconstruction of the *Dialogue* would include the whole iceberg of which my Analytical Summary is the tip. Since it is obvious that the various paragraphs of my Analytical Summary vary in logical complexity (some even being mere propositions, not arguments) I should mention that I have followed rather rigorously a criterion of comprehensibility in deciding how much to include. That is, I have left out the reasoning and stated merely the conclusion in those sections where the reasoning is either too complicated or too controversial or relatively easy. Examples of excessive complexity are IA3, IB3, IIB6; examples of controversy are IA3, IIB2c; examples of relative ease are IB2, IC1, IVB3. In some cases (e.g., IIB2a, IIB3e), I have given reasoning which is relatively complex because it is particularly beautiful and it is in part this logical beauty that justifies making that portion of the text into a distinct outline subdivision.

Besides being thus open-ended in a downward direction my Analytical Summary is open-ended in an upwards direction, namely insofar as it leaves undetermined other steps in the over-all argument which are not too explicit in the text and which would serve to integrate into it those sections which are not as obviously relevant as the examination of the objections and the statement of the favorable arguments. The sections I am referring to primarily are IIIA and I. In summarizing IIIA1, IIIA2, and IIIA3 as I have in IIIA, I have already taken a step toward integrating into the main argument this section of the *Dialogue* which at first looks like a digression. However, it would have been inaccurate as a summary to go too far in that direction; but as part of what might be called the latent structure we could add the following intended consequence: Hence it is not improper to consider astronomical evidence nonquantitatively (as it is done in IIIB1, IIIB2, and IIIB3). The relevance of the First Day (I) is more or less obvious. It is an examination of the main conceptual objections to the earth's motion, and this can be made clear by the following argument: The earth's motion is not a conceptual impossibility because neither the objection from natural motion nor the objection from the earth-heaven distinction is sound. The objection from natural motion is that the natural motion of the earth parts is straight toward the center of the universe (and hence the whole earth stands still therein); this objection is groundless because the empirical argument that the earth parts move toward the center of the universe is circular (IA4) and because any conceptual justification of the same premise would involve all the problems besetting the concept of natural straight motion (IA2, IA3). The objection from the earth-heaven distinction is that the earth's annual motion would involve the conceptual absurdity of placing the earth in (the third) heaven; this objection is groundless because the earth-heaven distinction is untenable and false (IB, IC).

Finally, in my Analytical Summary I have some semi-technical terminology which it will be good to explain briefly. *The objection from A* (where *A* is a noun phrase) refers to the argument from *A* against *B* (where it is contextually obvious what *B* is). *The argument from A* refers to the argument whose main premise is some proposition $p(A)$ constructed from *A* (in a contextually obvious manner) and whose conclusion is some proposition $p(B)$ whose identity is contextually obvious and unproblematic. For example, in the argument from the motion of sunspots: *A* is the motion of sunspots; *B* is the earth's annual motion; $p(A)$ is the proposition that sunspots move as described in (a), (b), and (c) of IIIB3 in the Summary; and $p(B)$ is the proposition that the earth has the annual motion. *The argument from A is false* means that the proposition $p(A)$, constructed from *A* in a contextually obvious and unproblematic way, is false. *The argument from A is groundless* means that the proposition $p(A)$ is not in the context supported by any sound argument. *The argument from A is invalid* means that, whether or not $p(A)$ is true, it does not support the conclusion of the argument. *The argument from A is unsound* means that the argument from *A* is false, groundless, invalid, and/or subject to some other problem.

OUTLINE

ANALYTICAL SUMMARY

I. *First Day: The Unity of the World*

A. *Natural motions* are the same for all bodies.

1. *The perfection of the world* is better grounded on its being the work of God rather than on its three-dimensionality.

2. *Aristotle's classification of motions* into straight and circular, simple and mixed, and natural and violent is untenable *because* (a) his equation of natural circular motion with motion around the center, and of natural straight motion with motion toward or away from the center, is conceptually unjustified (and *hence*

prejudicial); (b) his idea of mixed motion is incoherent; and (c) his distinction between simple and mixed motion is viciously circular.

3. *Straight and circular motions* are not two distinct instances of (simple) natural motion but rather two different stages of natural motion: straight motion can be acquired naturally but cannot naturally continue forever, whereas circular motion can naturally continue forever but cannot be acquired naturally without straight motion.

4. *The evidence from the senses,* namely the up and down motion of terrestrial bodies, constitutes a methodologically significant but ultimately unsound objection to the idea of natural circular motion; it is methodologically significant *because* the idea seems to conflict with that evidence (and *hence* to violate one of the most fundamental principles of philosophizing); and it is ultimately unsound *because* that up and down motion is likely to be either not straight or not natural or not peculiar to the earth. It may not be straight *because* the fact that it appears to be straight does *not imply* that it really is straight (as it will be *shown* later). It may not be natural *because* it has not been shown to be toward and away from the center of the universe, and it is more likely to be primarily toward and away from the center of the earth. And it may not be peculiar to the earth *because* that motion is likely to exist on each of the celestial bodies.

B. *The terrestrial-celestial dichotomy* is untenable and false.

1. *The argument from contrariety* is that celestial and terrestrial bodies are very different because change derives only from contrariety, contrariety exists only among terrestrial bodies, and *hence* change exists only among terrestrial bodies. It uses an assumption (namely, the connection between change and contrariety) which is more questionable than its main intended consequence (namely, the motionlessness of the earth). It may be self-contradictory *because* it implies that celestial bodies are changeable as well. It is groundless *to the extent that* the contrariety of rarity and density exists among celestial bodies and *insofar as* the contrariety of straight-up and straight-down does not exist among terrestrial bodies only. And it is ambiguous *because* the various mentioned 'bodies' sometimes refer to *whole* bodies, sometimes to *parts* of bodies.

2. *The a posteriori justification* of the unchangeability of the sky is that no celestial changes have ever been observed, and it is both invalid and factually false.

3. *The teleological argument* for the unchangeability of the sky is that celestial changes would be superfluous and useless, and it is unsound.

C. *The moon and the earth* do not differ in their nature.

1. *Similarities* between the moon and the earth include: shape, darkness and opacity, solidity, unevenness of apparent brightness, phases, reciprocal illumination, and reciprocal eclipsing.

2. *The roughness of the lunar surface* may be *justified by* the fact that it is visible at all and by its mountainous appearance through the telescope.

3. *The reflecting power of the earth* may be *justified by* the fact that during daylight both terrestrial objects and the moon appear equally bright and that during the night the moon has a secondary light (whose cause can only be the light reflected by certain parts of the earth).

4. *Differences* between the moon and the earth do exist *because* the moon has (a) no water, (b) a night-and-day period of one month, (c) no significant seasons, and (d) no rain, and *hence* no life similar to ours. Other unimaginable differences are bound to exist *because* the human mind cannot pretend to be a measure of what can occur in nature.

I. *Second Day: The Earth's Diurnal Motion*

A. *The problem* with the idea of the earth's motion is that it goes counter to Aristotle's authority, that the arguments favoring it though plausible are indirect and only probable, and that there are apparently insuperable objections to it.

1. *Aristotle's authority* deserves respect but is abused and harmed by his followers *because* they accept blindly and construe prejudicially his words; what is needed in the philosophical search for truth is not an authority but rather independent-mindedness.

2. *The arguments in favor* of the earth's diurnal motion are plausible but merely probable *because* they derive from the problems besetting the contrary view, namely that a celestial diurnal motion would (a) violate the principle of simplicity, (b) imply that each planet has two contrary circular motions, (c) violate

the law of periods of revolution, (d) imply that the fixed stars have incongrously unequal orbits and velocities, (e) imply that each fixed star keeps on changing its orbit and velocity, (f) make it inexplicable why the circular motion transferred from one celestial sphere to the one below it is not transferred to the earth, and (g) force the Aristotelians to postulate the existence of a fictitious *primum mobile.*

3. *The objections* to the earth's diurnal motion are numerous and apparently conclusive, and *hence* great open-mindedness and great rational-mindedness are required for their refutation.

B. *Examination of the classical objections* to the earth's diurnal motion shows that all the phenomena alleged as counterevidence would happen the same way whether the earth is rotating or standing still.

1. *Aristotle's first two arguments* (from violent motion and from the doubleness of circular motions) are equivocations, the first *because* the clause "the parts of the earth would also move circularly" can mean either that those parts would move around their own centers or else that they would move around the earth's center, the second *because* its conclusion could be a denial either of the diurnal motion or of the annual motion.

2. *Falling bodies* provide no evidence against the motion of the earth but rather provide the basis for a new concept of motion, according to which motion has the properties of conservation and composition.

   a. *The objection from vertical fall* is either circular or invalid *because* if it is stated in terms of *actual* vertical fall then it depends on the assumption that apparent vertical fall implies actual vertical fall (which is true if and only if the earth stands still), and if it is stated in terms of *apparent* vertical fall then it depends on the invalid argument that the nonoccurrence of mixed motion on a moving ship implies its physical impossibility.

   b. *The ship analogy argument* is false *because* the body falls at the foot of the mast even when the ship is moving.

   c. *Conservation and composition of motion* are two properties suggested by the criticism of the objection from falling bodies; they mean that motion is conserved if undisturbed and that it may be mixed without its components interfering with each other; and such a concept of motion can be further *justified* as follows: (1) it is the one required for adequately solving the problem of the cause of projectile motion, (2) it has interesting and novel consequences concerning the motion of projectiles, (3) it can explain certain puzzling facts about projectiles, and (4) it fits well with the idea of natural circular motion.

3. *Projectiles* provide no evidence against the motion of the earth but rather illustrations of the new concept of motion with the properties of conservation, composition, and relativity.

   a. *East-west gunshots* provide no evidence against the earth's motion *because* on a rotating earth the range of gunshots in one such direction would still be equal to the range in the opposite direction.

   b. *The relativity of motion* is the concept which *shows* that there would be no denial of the senses if bodies were regarded to move transversally in reality while they were seen to fall vertically.

   c. *Vertical gunshots* provide no evidence against the earth's motion *because* their horizontal motion on a rotating earth would be conserved; rather they provide a clear illustration of the composition of motion.

   d. *North-south gunshots* provide no evidence against the earth's motion, but *not because* shooting on a rotating earth is analogous to hunters' shooting at birds, rather *because* on a rotating earth the cannon ball would have the same eastward speed as the target.

   e. *Point-blank gunshots* provide no evidence against the earth's motion primarily *because* computation shows that the alleged deviation would be imperceptibly small (and *hence* there is no way of knowing that such gunshots are *not* in fact high toward the east and low toward the west), but also *because* on a rotating earth the gun as well as the target is rising (or falling) at the same rate (and *hence* the cannon ball would have the same motion, up or down with respect to the fixed tangent, as the target), and *because* one could argue equally plausibly that if the earth stood still such gunshots would *then* be high toward the west and low toward the east.

4. *The flight of birds* is the basis of a distinct argument against the motion of the earth *because* birds unlike projectiles have the power of self-movement (and *hence* they could not follow the

earth's rotation naturally but would have to do so through their own efforts). This argument is groundless *because* the air within which birds fly would be following the earth's motion (and *hence* they would not have to do so by their own efforts).

5. *A crucial experiment* (to nullify all the evidence alleged against the earth's motion from falling bodies, projectiles, and the flight of birds) could be made below decks on a ship by observing the flight of flies and butterflies, the swimming of fish in an aquarium, the dripping of water, the motion of smoke from incense, and the effort required to jump or throw objects in different directions. You would notice that everything would happen the same way when the ship is moving uniformly as when it is standing still.

6. *The extruding power of whirling* provides no evidence against the earth's motion *because* (a) the argument as ordinarily stated would prove at best only that the earth did not at one time start rotating after having been at rest; (b) the extruding motion would be tangential, and the downward tendency due to the weight is always large enough to overcome the tangential tendency; and (c) the extruding tendency depends not on the linear speed at tne circumference, which is large, but rather on the angular speed, which is small.

C. *Examination of contemporary authors* opposing the earth's motion shows that none of their objections has any force.

1. *The time of fall from the moon* to a rotating earth constitutes no objection to the earth's motion *because* the objection conflicts with (a) the mathematical fact that the radius of a circle is only a fraction of its circumference, (b) the law of squares which yields a time of less than a day rather than six days, and (c) the double-distance rule which yields a much greater terminal speed.

2. *The objection from the inexplicability of the earth's motion* is invalid *because* our relative ignorance of a cause does not imply the non-existence of the effect, and groundless *because* it cannot be shown that the cause of the earth's rotation is neither external nor internal.

3. *The objection from the deception of the senses* is false (a) *because* shared motion is imperceptible (and *hence* there is nothing for our eyes to see about falling bodies besides their downward motion); (b) *because* wind is air moving relative to us (and *hence* there is no perpetual wind due to the earth's rotation

for us to feel); and (c) *because* experience with navigation shows that we can only feel changes in motion (and *hence* the earth's rotation is not something susceptible of being felt). Moreover, the objection is invalid *because* the fact that the senses are to some extent deceived in this and other cases *implies* that one has to be careful about what the senses tell us rather than that they are useless.

4. *The objection from the impossibility of multiple natural motions* in simple bodies is (a) groundless *because* the function of joints in animals is not to allow multiple motions but rather to allow some parts to be moved while others are not (and *hence* it is unjustified to say that bodies without joints cannot have multiple natural motions); (b) irrelevant *because* there is no way for the earth to have joints which would enable it to have its three multiple motions (and *hence* even if it did have joints, it could not have the types of motion it has, and *hence* there is no point in saying that bodies without joints cannot have multiple natural motions); and (c) false *because* Jupiter's satellites and the sun have multiple motions but no joints.

5. *The similarity of motions of similar substances* does not constitute a valid objection to the earth's rotation *because* (a) dissimilar substances like water and air need not have dissimilar motions completely but only to the extent that their dissimilar natures can be inferred from differences in motion or other behavior (and *hence* if the earth rotates the common diurnal motion of water and air would not conflict with their dissimilar natures); and (b) similar are the earth and the planets on the one hand and the sun and the fixed stars on the other, due to their darkness and luminosity (and *hence* the rotation of the earth would not imply that similar bodies — the fixed stars, planets, and the sun — were having dissimilar motions).

6. *The objection that motion causes tiring* is false and invalid: false *because* the cause of animal tiring is the use of parts to move the whole, and *because* much animal motion is violent rather than natural; invalid *because* even if motion caused tiring the earth would not tire any more than the *primum mobile* or stellar sphere do in the Peripatetic system.

III. *Third Day: The Earth's Annual Motion*
  A. *The 1572 nova* provides a good example of how unreliable quantita-

tive astronomical data can be and of how careful and critical one
must be in drawing conclusions from them.

1. *Preliminary discussions* point out that though certain arguments
   are fallacious they deserve discussion for reasons other than their
   logical merit.

2. *The evidence from parallax difference* examined by Chiaramonti
   does not imply that the 1572 nova was sublunary *because* his 12
   computations yield distances differing by as much as a factor of
   1500 and *because* those computations are a biased fraction of all
   those possible from his own evidence. This evidence, if a conclusion must be drawn, supports rather the superlunary location
   of the nova *because* when all his data are taken into account some
   corrections are necessary, and fewer are needed to harmonize the
   data yielding a superlunary location than those yielding a
   sublunary location.

3. *The polar and stellar distances* of the nova can be used to argue
   that it is superlunary *because* the needed observations are very
   simple and are not invalidated by, respectively, the effects of
   refraction and the instrumental difficulties of using the sextant.

B. *The favorable arguments* for the earth's annual motion are very
   cogent.

   1. *The heliocentrism of planetary motions* supports the earth's
      annual motion (*because* the earth is located between bodies that
      go around the sun) and is supported by the planets' pattern of
      changes in their apparent size and shape *because*: (a) each planet's
      distance from the earth varies greatly (and *hence* the earth is not
      the center of their motion); (b) the outer planets are close to the
      earth when they are in opposition, and they are distant and look
      round when they are in conjunction; (c) Venus always stays close
      to the sun and appears horned in shape when large and round
      when small; and (d) Mercury stays even closer to the sun and is
      brighter than Venus.

   2. *Retrograde planetary motion* supports the earth's annual motion
      because it is best explained as resulting from the earth's annual
      motion.

   3. *The motion of sunspots* is best explained in terms of the earth's
      annual motion *because* they appear to move across the solar disc
      along paths which exhibit the following features: (a) they curve
      upwards in the solar disc for half a year and downwards for the

other half; (b) they also slant upwards in the solar disc for half a year and downwards for the other half; and finally (c) both the curvature and the slant are continuously changing in such a way that the paths are straight twice a year when the slant is greatest, and the slant is absent twice a year when the curvature is greatest.

C. *The objections* to the earth's annual motion are inconclusive though they are instructive and cannot be dismissed.

  1. *Biblical passages* cannot properly be used in hypothetical reasoning about natural phenomena.

  2. *The stellar dimensions* implied by the earth's motion and by the lack of annual parallax would not be absurdly great *because* apparent stellar diameters have been universally overestimated by a factor of about 30 (and *hence* stellar sizes have been universally overestimated), and *because* the required stellar sizes and distances, though very great, are not absurd; this is so *because* (a) some Ptolemaic estimates of distances are of the same order of magnitude, (b) size and distance are concepts such that all large ones after a certain amount are unimaginable, (c) there is no justification for saying that the space between Saturn and the fixed stars is useless, and (d) size and distance are relative concepts.

  3. *Tycho's objection* from stellar dimensions is groundless *because* he assumes without investigation that stellar positions show no annual change and *because* he is unclear about the exact changes implied by the earth's annual motion.

  4. *The celestial polar elevation* would not show any annual change if the earth had the annual motion *because* in that case the celestial pole would be defined by the terrestrial one and the elevation of the latter can change only by moving on the earth's surface, not by moving the whole earth as in the annual motion; *hence* the objection that the celestial pole shows no annual change in elevation is invalid.

  5. *The changes in stellar elevations* implied by the earth's annual motion would not be at all comparable to the changes resulting from moving along the earth's surface *because* the former motion occurs on a plane surface while the latter occurs on a relatively highly curved one; *hence* the objection that stellar elevations show no great changes is invalid.

6. *The annual constancy in stellar appearances* constitutes an invalid objection *because* the changes implied by the earth's motion are complicated and no one has systematically tried to observe them, and they are very small and no instruments are available for detecting them (and *hence* the apparent constancy does not imply that the changes do not exist).

7. *The sun's apparent motion* provides no evidence against the earth's annual motion *because* (a) the sun's apparent motion in the order of the signs of the zodiac would be a simple consequence of the earth's own motion in the same direction; (b) the significant seasonal changes in solar elevation and length of nights and days would be consequences of the inclination of the earth's axis and of its unchanging direction; and (c) there is no conflict between the large differences in solar elevation and the small ones in stellar elevations implied by the earth's annual motion.

8. *The properties of loadstones* invalidate the objection that a simple body like the earth cannot have three or four natural motions *because* (a) loadstones have several natural motions; and (b) they have the property of always pointing toward the same place, and there is evidence that the earth is a loadstone (and *hence* its axis is always parallel to itself, and the third motion attributed to the earth by Copernicus does not exist).

IV. *Fourth Day: The Cause of the Tides*
   A. *Previous theories* about the cause of the tides must all be rejected *because* (1) differences in sea depth cannot produce and sustain the motion of the water, (2) lunar attraction could not produce tides in only certain parts of a given sea and not in others, (3) the water in a tide has the same temperature and density as ordinary water (and *hence* lunar heat is inoperative), (4) the miracle explanation is not to be invoked unless one can find no other cause, and (5) the periodic attraction and expulsion of water by the earth through undersea caves could not produce tides in only certain parts of a given sea and not in others.
   B. *The geokinetic theory* explains the main features of the tides as resulting from the interaction of two causes: the primary cause is the combination of the earth's diurnal and annual motions, the secondary cause consists of the fluid properties of water.
      1. *The primary cause* of the tides is the daily accelerations and

retardations produced in every part of the earth as the diurnal component is added or subtracted from the annual component of the earth's motion; this is so *because* water in a container can be made to move like the tides by accelerating or retarding the container.

2. *The fluid properties of water* which act as a secondary cause of the tides are that (a) it tends to oscillate before reaching equilibrium; (b) its oscillations take less, the smaller the length of the basin; (c) the deeper the water, the shorter the period of oscillation; (d) it moves vertically at the extremities and horizontally at the middle of the basin; and (e) different parts of the same body of water can move at different speeds simultaneously.

3. *The basic tidal effects* that can be explained as resulting from the interaction of the primary and secondary causes are (a) the absence of tides in lakes and small seas, (b) the six-hour tidal interval in the Mediterranean and presumably different periods in other seas, (c) the absence of tides in seas that are narrow in an east-west direction, (d) that tides are greatest at the extremities and least at the middle of a gulf, (e) the great currents through certain straits, (f) the violent agitations and vortices in certain straits, and (g) the unidirectional flow of currents through certain straits.

C. *The behavior of winds* provides no evidence against the geokinetic explanation of the tides *because* air unlike water does not have the property of retaining acquired motion (and *hence* the earth's motion could cause the tides without causing a perpetual wind); and *because* the turning of the lunar orb could not produce the prevailing westward winds that do exist, and these could not produce the back and forth motion of the tides (and *hence* these winds are not the link between the diurnal motion of the lunar orb and the tides).

D. *The monthly and annual periods* of the tides must be caused by variations in the velocity changes that cause the diurnal period (namely, variations in the speed of the earth's annual or diurnal motion or both); *hence,* the monthly period is caused by the variations in the speed of the earth's annual motion that occur monthly as the earth-moon system goes around the sun in a circle whose effective radius undergoes monthly variations as a result of the changing relative positions of the earth, sun, and moon; *hence* also, the annual period is caused by the variations in the effective speed of

the earth's unchanging diurnal motion resulting from the inclination
of the earth's axis to the plane of its orbit.

The fact that it is possible to reconstruct the *Dialogue* in the way just
described suggests the following historical and philosophical theses. First, it
is a mistake to regard the book as primarily a defense of the whole Copernican system and then to blame Galileo for neglecting the details of technical
astronomy; instead, the detailed examination of the *whole* book shows that
it is primarily a defense of Copernicanism only to the extent that the proposition 'the earth moves' (with the diurnal and annual motions) is part of
Copernicanism. The book is rich enough in its content, as it is; there is no
good reason why Galileo *should* have discussed the details of technical
astronomy.

Second, it is unfair to fault Galileo for not explicitly discussing the
Tychonic system since from the point of view of the proposition 'the earth
moves' there is no difference between it and the Ptolemaic system. Tycho's
arguments are discussed in IIIC insofar as they are relevant to Galileo's
purpose.

Third, Galileo's alleged commitment to 'natural circular motion', if not
taken out of context, is part of a critique of Aristotle's concept of natural
motion (in IA) and of an elucidation of a new concept of motion (IIB2c).
When seen in this light, there is nothing very obviously wrong with it.

Fourth, it is a mistake to regard Galileo as inimical to or unappreciative of
logic.[8] The possibility and accuracy of the above reconstruction makes
Galileo a logician-in-action or applied logician, that is a skillful practitioner
of logical analysis and explicitly formulated argumentation.

Fifth, if it should turn out, from another type of analysis of the *Dialogue,*
that the book has no unity or prevalent characteristic in terms of some other
epistemological-methodological idea or practice, then we would have to
conclude that Galileo is *first and foremost* a logician. At present I state this
thesis as a conditional, though my conjecture is that the book lacks any other
prevailing methodological characteristic; and this conjecture is quite consistent
with the claim that the book is full of philosophy and methodology. In fact,
I believe that the full extent of the wealth of its philosophical content has not
so far been appreciated. What we seem to have in the *Dialogue* is a philosophical goldmine, but one lacking a unifying theme, other than that of applied
logic exhibited in this chapter.

Finally, there is no conflict between this reconstruction of the *Dialogue*
and the rhetorical interpretation, popularized by Feyerabend,[9] that the book

is a piece of propaganda aimed at winning the argument and at persuading the opponents at all costs, and exploiting their weaknesses. On the contrary, a reconstruction of the internal structure of the *Dialogue* is logically prior to the rhetorical interpretation since the formulation of the latter amounts to claiming that certain arguments actually given in the book are fallacious or deceptive in various ways. The rhetorical interpretation loses its relevance if and to the extent that its reconstructions are inaccurate, or taken out of context.

NOTES

[1] G. de Santillana, *The Crime of Galileo*, p. 174.
[2] A. Koyré, *Études Galiéennes*, p. 212.
[3] W. Shea, *Galileo's Intellectual Revolution*, pp. 109–186.
[4] M. Clavelin, *The Natural Philosophy of Galileo*, pp. 183–267.
[5] G. Galilei, *Opere*, edited by Favaro, Volume 7.
[6] G. Galilei, *Dialogue Concerning the Two Chief World Systems*, tr. S. Drake.
[7] G. Galilei, *Opere*, edited by P. Pagnini, Vols. 2 and 3.
[8] See, for example, Shea, *Galileo's Intellectual Revolution*, p. 88.
[9] P. K. Feyerabend, 'Problem of Empiricism II', pp. 275–353; idem, *Against Method*.

CHAPTER 3

EMOTION, AESTHETICS, AND PERSUASION:
THE RHETORICAL FORCE OF GALILEO'S ARGUMENT

It was argued in a previous chapter that a book like Galileo's *Dialogue* must have a type of content that can best be appreciated after its scientific-logical content has been analyzed. This is what may be called its rhetorical content, in a restricted sense of 'rhetoric', according to which the term refers to actual substance present in the book, and not to merely cosmetic verbal expressions of desires and intentions. The substance actually present in the book from this rhetorical point of view is a type of intellectual content, but one that plays upon feelings and emotions, either directly and explicitly by verbal expressions that have the desired emotive effect, or else indirectly and implicitly by emphatic identification with what is explicitly said or done.

Some of this type of rhetorical content derives from the book's dramatic structure. However, it is an oversimplification to follow the typical scholarly opinion[1] that Salviati is a Copernican spokesman; that Simplicio is an Aristotelian spokesman; and that Sagredo is an intelligent layman, attentive listener to both sides, curious and open-minded, and unwilling to use anything but his own judgment in making up his mind. It is supposed to be more or less obvious that Simplicio is something of a simpleton, as his name suggests; that Salviati is Galileo's mouthpiece; and that Sagredo is a neutral, objective, uncommitted judge.

Such an interpretation would not do justice to the complexity of the book's dramatic structure. The thesis that Salviati is Galileo's only spokesman is at least complicated and perhaps invalidated by the following: (1) there is in the *Dialogue* a fourth character, referred to as 'the (Lyncean) Academician' or 'our common friend',[2] who is obviously Galileo, so that if he is Galileo then Salviati cannot be; (2) frequently it is Salviati who expounds Aristotelian views (e.g., F38–9, F356–63), though the text sometimes indicates that these are not his own beliefs, and that he is doing a job that belongs to Simplicio (e.g., F151–3).

The thesis about Sagredo's objectivity is faced by the difficulties that he always sides against the Aristotelian position; that he tends to exaggerate the strength of the Copernican arguments (e.g., F132–3, F383, F485–6); that he sometimes expresses his judgments somewhat prematurely (e.g., F98); and that occasionally it is he who obviously expresses Galilean views, rather

46

than Salviati, whose statements are less correct than his (e.g., F202–3).

The thesis about Simplicio's simple-mindedness is made problematic by the fact that he often expresses flashes of great logical acuity (e.g., F379–80). In view of these facts, it is best to regard all three interlocutors as being partial spokesmen for Galileo. The important question is not which speaker speaks for Galileo, but rather how the specific details of the dramatic structure of the dialogue contribute to his main purpose, which is that of justifying Copernicanism. Let us then examine all the passages that seem relevant to the concept of rhetoric under investigation.

In the course of the critique of the earth/heaven distinction, Simplicio makes an appeal to fear when he says that a manner of philosophizing that does away with this distinction, tends to the subversion of natural philosophy and of nature herself (F62, D37–8). It is interesting to point out that Simplicio expresses this fear when he is unable to make an intellectual response to Salviati's argument that since it does not make sense to distinguish between two kinds of natural motions (straight and circular), and since all natural motions must be circular, it follows either that there is no difference between heaven and earth from the point of view of perfection, or else that circular natural motion has nothing to do with perfection. It is certainly quite legitimate to express one's emotional feelings about a certain state of affairs in nature; if the resulting emotions are negative, then the responsible philosopher-scientist will not dismiss the matter. In general a fear can be allayed partly by a reiteration that there is no ('rational') reason for the fear, in an attempt to let reason conquer one's emotions, and partly by pointing out an aspect of the situation which would elicit a positive emotional reaction, in an attempt to fight one emotion with another. Galileo does both of these things: with respect to natural philosophy, Salviati says that there is no reason for the fear because natural philosophy can only benefit by such an attempt to subvert, either by making new discoveries, or by strengthening and reinforcing the old doctrines; with respect to nature herself, Salviati points out that in abolishing the earth/heaven distinction the Copernicans are, so to speak, placing the earth in the heavens and hence are giving it more nobility and perfection. This last point obviously has the intent of eliciting some kind of rejoicing, which might make one forget about the previously expressed fear.

In the discussion of the existence and significance of sunspots in the First Day (F77, D56), Galileo does not miss the opportunity of describing the situation with emotive language. Salviati calls sunspots 'importunate' and says that they have come to render the heavens and especially Peripatetic philosophy 'turbid'. Does such name-calling have any function? I think what is

happening here is that Galileo is appealing to our aesthetic-linguistic sensibility
to score some points against the Aristotelians. Galileo is calling the Peripatetic
heavens dirty and the Peripatetic philosophy muddled. Given that from the
Peripatetic point of view sunspots are impurities, the temptation to say that,
from that point of view, the heavens are dirty, is irresistible. Since, as Galileo
argues, there is no way of removing the dirt as long as one remains Aristotelian, the need for a change becomes more intense. Moreover, the linguistic
possibility of using the epithet ('turbid') with the figurative meaning of
'muddled' makes irresistible the inclination of saying that the Peripatetic
philosophy is muddled. Thus, though few would be likely to be swayed
merely by such name-calling, few could resist the temptation of so using
language at the expense of Aristotelianism.

In the First Day, at the end of a discussion on the role of authority in
natural philosophy, Salviati argues very cogently that it would be not only
more correct, but even more in the spirit of Aristotle to argue that heaven is
alterable because the senses so indicate, rather than to argue that heaven is
inalterable because so reasoned Aristotle (F80–81). Galileo is here advocating
a significant change in the practice of natural philosophy. The discussion ends
with a very long passage, full of images, in which Sagredo describes what it
would be like to do without the security of an authority. The fact that the
description is put in the mouth of Sagredo (who asserts that he is thereby
portraying Simplicio's attitude), together with the length of the passage and
the wealth of its images, can only be interpreted as an attempt by Galileo to
show that he can sympathize with the predicament of a committed Aristotelian. This is a very effective and quite legitimate rhetorical technique,
for the pain of unpleasant experiences tends to be softened by the proof
that other people understand us; this is especially true when the expression
of sympathy comes from the people who have something to do with causing the pain, which they regard as a necessary evil. Though there are traces
of irony in Sagredo's speech (F81, lines 21–22), the over-all tone is one of
sympathy.

A few pages afterwards, in the discussion of the teleological argument,
in opposition to the Peripatetic idea that change makes things imperfect,
Sagredo says that he thinks that those who value unchangeability so much,
do so out of a fear of death and a desire to live a long time, and that they
deserve to encounter a Medusa's head which would turn them into statues,
so as to become more perfect than they are (F84, D59). This is a good
example of what some would nowadays call an *ad hominem* argument,
namely an argument which tries to refute an idea by discussing the (usually

unfavorable) motives of the person holding the idea, rather than by discussing the issue on its own merits. Though many logic textbooks regard such arguments as fallacies,[3] such a view is at best superficial.[4] It would indeed be wrong to regard such motivational analysis as a conclusive argument against the truth of the idea, but this is not to say that the argument lacks any force whatever. In fact, such arguments are especially effective when values are involved, as in the present case. The issue here seems to be whether or not change is a good thing. How does one discuss this 'on its own merits'? If people approve unchangeability because they want to remain alive, that is logically relevant to the issue because one can then question their approval by questioning their motivating desire. Would it really be good to live forever? Galileo seems to be casting doubts on this by suggesting that this would cause overpopulation problems, the solution of which might require an end to births, with the consequence that it would be doubtful whether most of us who are now alive would have been born in the first place. Moreover, in his reference to Medusa, Galileo seems to suggest that an immortal life might be a kind of death, and thus a conceptual contradiction. When seen in this light, Sagredo's remarks, though not decisive, acquire some logical value, in the broad sense of 'logical' that includes persuasive force. If there is a problem with Sagredo's remarks, it is their irrelevance, at least contextually speaking; for it emerges from the discussion that the issue is not one of valuing or not valuing unchangeability, but rather whether changes in the heavens would be superfluous and therefore are to be ruled out. However, Sagredo's argument against unchangeability is irrelevant only when viewed from the point of view of the fully articulated teleological argument (i.e. that there cannot be any changes in the heavens because change would make celestial bodies imperfect; and this is so because changes in celestial bodies would be superfluous; and, in turn, this is true because changes in celestial bodies would be of no use or benefit to man);[5] for, from the point of view of coming to discover that unchangeability may not be an intrinsic Peripatetic value, Sagredo's argument has the methodological function of facilitating this discovery.

According to Peripatetic philosophy, some heavenly bodies, especially the sun, caused changes on the earth, while they remained themselves unchanged and unchangeable. This offered Galileo the opportunity for a memorable comparison: he says that this is like placing a marble statue next to a bride and expecting children from such a union (F84–5, D60). This critique could be interpreted as an argument from analogy, possessing a certain degree of plausibility, depending on the strength of the analogy between the two unions. However, such an interpretation would hardly do justice to it, for I believe the argument has a persuasive force much greater than that deriving from the inherent analogical reasoning. The reason is that one of the things being compared is not merely abstractly and objectively false, but emotionally absurd as well; we might say that to expect children from the union is doubly unnatural, that is, against the nature of (physical) things, and against human nature in the sense of human psychology, human wishes, desires, and dreams. In short, not only no children *will in fact* come from such a union, but no children *should* come from it, in the sense that it would be much better if they came in the usual way. To be able to argue on the basis of comparisons that not only have some factual basis but which also strike a resonant chord in human feelings involves more than being merely a logician, or merely a poet; what is involved is the combination of the two which is what is done by the good rhetorician (in the sense presently under discussion).

Another Peripatetic doctrine stated that the heavens consisted of a fifth element, different from the four terrestrial ones, namely earth, water, air, and fire. This fifth element was thought to pervade the whole celestial region, where the heavenly bodies were merely its denser parts; it was also thought to be impenetrably hard but to lack all other ordinary qualities such as visibility and weight. This doctrine had become highly discredited in Galileo's time, primarily because the motion of comets seemed to make impossible the existence of any hard, impenetrable, crystalline substance. Galileo ridicules the doctrine by means of the following exchange:

SAGR. What excellent stuff, the sky, for anyone who could get hold of it for building a palace! So hard, and yet so transparent!
SALV. Rather, what terrible stuff, being completely invisible because of its extreme transparency. One could not move about the rooms, without grave danger of running into the doorposts and breaking one's head.
SAGR. There would be no such danger if, as some of the Peripatetics say, it is intangible; it cannot even be touched, let along bumped into.
SALV. That would be no comfort, inasmuch as celestial material, though indeed it cannot be touched (on account of lacking the tangible quality), may very well touch elemental bodies; and by striking upon us it would injure us as much, and more, as it would if we had run against it.
    But let us forsake these palaces, or more appropriately these castles in the air, and not hinder Simplicio. [D69; cf. F94]

Though this wit may be recognized as brilliant, one may question its logical force as an argument against the impenetrable hardness of the fifth element. I believe that the argument lacks any *intrinsic* logical force, but that it acquires some from the inherent wit; what happens is that one acquires a negative emotional attitude toward this substance because it lends itself to such ridicule, and such an attitude then translates cognitively into the belief that one ought not to have the concept of such a substance in one's system, and that it probably is an unhappy invention, and that it probably does not exist.

At the beginning of the Second Day, Sagredo asserts that he finds the daily rotation of the universe around a motionless earth as absurd as if someone went on top of the cupola of a cathedral to have a view of a city and its surroundings, and then he expected that the city be made to turn around him so that he would not have to make the effort of moving his head (F414, D115). Here we have another passage that could be interpreted as an argument from analogy, with some degree of plausibility, but whose persuasive force is increased beyond the purely logical strength of the analogical reasoning. The increase derives from the emotive repulsiveness of the image of the lazy, arrogant observer.

In the preliminary exposition of the geostatic arguments Salviati is the one who gives a statement of several ones based on gunshots. After Salviati is finished Simplicio expresses his joy at the fact that the truth (that the earth stands still) should be supportable by such 'unconquerable' arguments and by such new evidence (as compared to that available to Aristotle). At this point Sagredo interjects the exclamation, "What a shame that there were no cannons in Aristotle's time! With them he would indeed have battered down ignorance . . . " (D127; cf. F153). This is certainly a witty remark, and it gives the reader great aesthetic pleasure. In this case, however, the wit does not perform any logical-rhetorical function; the gunshot arguments are neither strengthened nor weakened. A rhetorical service is performed by the fact that the anti-Aristotelian Salviati shows such understanding of the Aristotelian position as to be able eloquently to present new evidence in its favor, for this inclines the reader to think more highly of Salviati and his judgment, and indirectly and ultimately of Copernicanism. However, this is a service performed by the dramatic features of the present dialogue, rather than by Sagredo's witticism, which must therefore be attributed a purely literary-aesthetic value.

In the critique of the ship analogy argument (F169–75), Salviati argues that the rock falls at the foot of the mast regardless of whether or not the ship is in motion. He feels that it is unnecessary to appeal to direct sense-experience because he is sure the fact can be proved from other more easily accessible observations; he thinks he can even convince Simplicio, who is sure that the rock falls away from the foot of the mast when the ship is in motion.

Salviati's display of confidence reaches a climax when he boasts, "I am so handy at picking people's brains that I shall make you confess this in spite of yourself" (D145; cf. F171). This arouses even Sagredo, who was about to make a comment, but who now remarks, "the interest aroused in me by hearing you threaten Simplicio with this sort of violence . . . has deprived me of any other desire; I beg you to make good your boast" (D145; cf. F171). It is obvious that the reader's interest and curiosity are aroused as much as Sagredo's, hence the passage has literary-dramatic value. But does its worth extend beyond that? It is true, of course, that displays of confidence are standard rhetorical devices designed to insinuate logical force and thus achieve persuasion, or at least increase assent. It is also true that such displays sometimes have the desired effect, the operative assumption on the part of the audience being that, because of the inherent risk of failure to make good one's boast, people do not normally claim more than they can prove. All the same, it is also common knowledge that for some people, their boasting is mostly an indication of their confident or over-confident nature. Since the general question of the logical worth of displays of confidence is thus somewhat debatable, let us see about the present case. It should be noted that this question of the rhetorical worth of the present boast is independent of the question to what extent Galileo's ensuing argument justifies the boast, for what we are looking for here is some alogical worth, above and beyond the aesthetic, and pertaining to persuasion. I believe that by looking at the dialogue just preceding Salviati's present speech, we can see why there is a rhetorical need for some such boast. In fact, to a naive, uncritical empiricist like Simplicio, the question of what happens on the moving ship is a paradigm example of one to be decided by the senses, and so he does not see any need for further discussion of the matter and is about to give up the dialogue at this point (F170, D145). The only way to prevent this, the only way to prevent his Aristotelian readers from closing the book at this point, is for Galileo to make a boast that would otherwise seem excessive. Whether or not he proves his point to their satisfaction, he has at least held their attention and convinced them that the matter is not so simple, that it does require discussion.

In the discussion of the cause of projectile motion (F175–80, D149–54), Galileo criticizes the Peripatetic idea that when air is in its own region it has the disposition to receive and conserve an impulse (F178, D152). Among the evidence Galileo presents against this idea, there is the following very quaint experiment: "Let us go into that room and agitate the air as much as possible with a towel, then stopping the cloth, have a little candle flame brought immediately into the room, and you will see from [its] quite wandering . . . that the air has been instantly restored to tranquillity" (D152; cf. F178). Here it is the familiarity and accessibility of the experience that gives the experiment a special rhetorical force. Though we may agree that, logically, the force of this counter-evidence is no greater than that of the 'thousand other' experiments that Salviati claims could be made, from a psychological point of view there is a great difference. The ability to choose appropriately familiar examples is a valuable skill in the art of persuasion, of which purely logical reasoning is a part, an important part to be sure, but only a part.

The discussion of conservation and composition of motion (F175–93, D149–67), after the critique of the ship analogy argument, is an elaboration of the concept of motion presupposed by Galileo's critique of this and of the other arguments from falling bodies. It is a relatively long passage, and lacks the relatively more explicit structure provided by the pattern of argument statement, argument analysis, and argument evaluation that characterizes so much of the book. Consequently, this passage might strike the layman or the Aristotelian not interested in the topic for its own sake as being relatively dull, and he might very well omit it. However, in the middle of the discussion we find a very interesting exchange between Simplicio and Sagredo, which has the rhetorical function of preventing an onset of boredom. Sagredo has just finished saying that the doctrine of conservation of motion allows him to understand the puzzling fact that the hoops used by players are sometimes seen to move faster over a later portion of their path than over a previous one. He ends his speech with the question, "Now what would Simplicio say to that?" (D157; cf. F183). The latter replies, "I should say that in the first place I have not observed such things; second, that I do not believe them; and then, in the third place, if you should assure me of them and show me proofs of them, that you would be a veritable demon" (*Ibid.*). To this Sagredo adds, "One like Socrates's, though; not one from hell" (*Ibid*). This is indeed a very effective way of arousing our curiosity and anticipation, besides being, of course, very clever from a literary point of view.

In the discussion of semicircular fall (F188–93) Galileo refrains from considering quantitatively the acceleration of falling bodies, giving as a reason that it would not be relevant for present purposes, i.e., would be a "digression within a digression" (F190, D164). He wants to make sure that the reader does not get the impression that his reason is his acceptance of the Peripatetic view of the relationship between science and philosophy, and so he gives a very brief critique of this idea. This critique consists of a very eloquent but arrogant, and consequently self-defeating statement by Simplicio, together with a remark by Sagredo that this is like putting a philosopher on a royal throne, which is a rather repulsive image, and which therefore makes the view implicitly undesirable. The passage deserves literal translation since the two available ones miss some rhetorically significant phrases:

SIMP. Philosophers occupy themselves principally with universals; they discover definitions and the most common symptoms, then leaving to mathematicians certain subtleties and trivia, which are rather curiosities. In fact, Aristotle was content with defining what is motion in general, and with showing the principal attributes of local motion, namely the differences among natural, violent, simple, compound, uniform, and accelerated; and for accelerated motion he was content with explaining the cause of acceleration, leaving then the investigation of the proportion of such acceleration and of other more specific accidents to the mechanic or to some other inferior artist.
SAGR. All right, my dear Simplicio. But you, Salviati, descending occasionally from the throne of Peripatetic majesty, have you ever played with the investigation of this proportion of the acceleration of the motion of falling bodies? [F189–90; cf. D189–90][6]

I believe that, here, the persuasive force of the criticism derives from the kind of *visual image* we get of the Peripatetic philosopher, namely of someone removed from ordinary reality. I do not think the description is biased; the emphasis is even on qualities which would otherwise be favorable. The problem is with the inappropriateness of these qualities.

   In the critique of the objections from gunshots there is a passage which absorbs and captivates the reader like few others. It is one where Sagredo imagines an artist painting a scene on a ship going from Venice to Alexandretta and notes that the tip of his pen would describe a line which over its total length would vary almost insignificantly from a smooth or straight line (F197–200, D171–4). Yet from such an (almost) smooth line we get a painting; the only part of the actual line that is effective is what from another point of view are insignificant variations. Simplicio describes himself as 'bewitched' by the image (F199, D173), which indeed is bewitching. The example is given as illustration and justification of the idea of the relativity of motion, and the aesthetic magic works in the sense that it enhances the possibility of the doctrine and of its application to the situation on a rotating earth. To emphasize the rhetorical effect of the example, and its logical import, Galileo has the speakers also mention a number of what he calls "very insipid subtleties" (F199, cf. D173), allegedly elicited by Sagredo's description: these are the facts that when a ship is in motion the top of its mast moves more than its bottom, and the fact that when a person walks his head moves more than his feet, in both cases because of the earth's spherical shape.

The next rhetorically significant passage offers a contrast between the Peripatetic approach to natural philosophy and the one Galileo is advocating. In order to appreciate it properly, a literal translation is needed:

SALV. . . . I am beginning to realize that you [Simplicio] have so far been of the herd of those who, to understand how such things occur and to learn the details of natural phenomena, do not go onto boats or around crossbows and cannons, but rather retire into their study to shuffle indices and outlines to find out whether Aristotle has said anything about them, and who, having ascertained the true meaning of the text, neither desire anything else nor believe that anything else could be known.
SAGR. Great happiness, and much to be envied; for if knowledge is naturally desired by all and if it is the same to be as to claim to be, they enjoy a very great benefit and can persuade themselves to know and understand everything, by contrast with those who, knowing not to know what they don't know, and consequently saying that they know only a very small part of what there is to know, kill themselves with lack of sleep and with studies, and waste away around experiments and observations. [F211, my translation; cf. D185]

To be sure, the contrast is exaggerated; however, since it is not a fabrication and does have a grain of truth, it does have the desired effect of arousing our sympathy for the Galilean type of natural philosopher. This emotion is indeed sympathy, for, as Sagredo says, as far as envy goes, it is the Peripatetic philosopher who is to be envied. However, such envy can be felt only at a superficial level, for at a deeper level one feels repugnance for the 'herd' of people who shuffle papers in order to learn about the real world. Here this criticism consists in the mere description, and the strength of the criticism in the intensity of the contrast. How valid is the criticism? Obviously, it applies to some Peripatetics but not to others; Galileo thought that it certainly did not apply to Aristotle himself (F75, F136–7). The caricature does not lose its rhetorical force, however, for once created it acquires a life of its own.

In the critique of the centrifugal force objection (F214–44, D188–218), there is a Socratic discussion between Salviati and Simplicio attempting to show that, though terrestrial rotation would give objects a centrifugal tangential impetus, their weight however small would be sufficient to keep them on the surface. Just before this Socratic discussion is concluded, Simplicio gives an interpretation of the situation according to which a feather would have to fly away from the surface, and "since this does not happen, I say that the earth does not move" (F223, D196). Salviati replies, "Oh, Simplicio, you yourself rise up so fast that I begin to fear more for you than for the feather. Relax a little, and listen" (D197; cf. F223). This is a very pleasing and clever aesthetic image, but I do not think that it has any rhetorical force. It does have, to be sure, what might be called a rhetorical import, insofar as it wants to convey the idea that the Aristotelians often draw hasty generalizations. However, the incident is too parenthetical, and the image of Simplicio flying away faster than a feather on a rotating earth according to the Aristotelian conception, is short-lived. What this means is that the remark has only aesthetic-literary value, but no logical-rhetorical worth; but to say the latter is not to say that it has logical-rhetorical demerit. What I am saying is that in this case the value exists only in the aesthetic domain; this would entail rhetorical demerit only if it could be proved that everything in the text indicates that at precisely this place, Galileo is attempting to score a rhetorical point. Since there is no such evidence, the error would be committed by the critic, if he were to see rhetorical failure. In short, nonrhetoric is not bad rhetoric.

One difficulty that some people found with attributing the daily rotation to the earth was the problem of how the diurnal motion could inhere in both living and inanimate beings. Besides giving a more or less technical answer to the problem through Salviati (F264–5, D238–9), Galileo has Sagredo say that the people who feel this difficulty "must believe that if a dead cat falls out of a window, a live one cannot possibly fall too, since it is not a proper thing for a corpse to share in qualities that are suitable for the living" (D239; cf. F265). This may be interpreted as a *reductio ad absurdum* argument; the rhetorical component here is in the absurdity being referred to, which is not, or not merely, an abstract impossibility, but primarily an emotively ridiculous situation. For those who cannot follow the technical refutation, or for those who want to add ridicule to refutation, the present image is at their service.

Next, we should not neglect a passage at the beginning of the discussion of Chiaramonti's book on the new stars; the wit and sarcasm are memorable, while the fact that the passage seems to lack a rhetorical function may serve as a reminder that we should not confuse lack of rhetoric with rhetorical failure. Galileo's remark is a comment to Chiaramonti's attempt to show that the so-called new stars were sublunary. Chiaramonti's argument was based on an analysis of parallax data, the very same data that had convinced most astronomers that the new stars were superlunary. Galileo says that "since this author brings about such an undertaking, namely to remove the new stars from the heavens into inside the elemental sphere, he deserves to be greatly exalted and himself transferred among the stars, or at least that his name be made eternally famous among them" (F272, my translation; cf. D248).

A few pages afterwards we do have what must be regarded as a rhetorical failure, a small one to be sure, but an error nonetheless. The passage begins with Sagredo remarking that the previous discussion had somewhat settled his stomach which had been upset by the fish and snails he had eaten (F275, D250). This is a humorous remark from an aesthetic point of view; and from a literary point of view it is a clever and successful device to introduce a new topic into the previous discussion which was a critique of the geostatic objection from the deception of the senses, as found in Chiaramonti's book; in fact, Sagredo says that besides settling his stomach the previous discussion has made him think of correcting the erroneous view that the telescope cannot be used as effectively at the top of the mast of a ship as on the deck. The correction proceeds by Sagredo arguing that the only thing that can affect the view is angular displacements of the telescope, which are the same on the mast as on the deck, not parallel displacements which are too small even on the mast. Throughout the discussion Salviati plays the role of someone who finds the erroneous view initially very plausible, and who is receiving enlightenment by Sagredo in a semi-Socratic discussion; Salviati ends the discussion by saying that Sagredo's correction illustrates how very careful one must be before affirming or denying a proposition, and that just as this erroneous belief is excusable because of its *prima facie* plausibility, so is the belief of those who cannot understand how the apparently vertical path of a freely falling body might not really be straight (F278, D253). Thus, because of Salviati's role and his conclusion, the rhetorical intention and pretension of this passage is to soften Galileo's criticism of the objection from the deception of the senses and to make us sympathize with the author presenting the objection. However, most of the evidence from the rest of the passage criticizing this objection (F272–81, D247–56) points in the opposite direction, namely that Chiaramonti is not only wrong, but foolish in putting forth this objection. The most that can be said in favor of Galileo's rhetorical attempt here is that it indicates ambivalence, since there are traces of a respectful attitude toward Chiaramonti; for example, at the beginning of the passage Simplicio is made to introduce him as "a consummate philosopher and also a great mathematician" (F272, my translation; cf. D247), and at the end Sagredo is made to describe him as a "philosopher, who seems to me a cut above most of the followers of these doctrines" (D256; cf. F281). However, the ambivalence is not sustained either, and in the rest of the discussion of Chiaramonti (F281–346, D256–318) the criticism becomes more intense, and the contempt more apparent. Therefore, in any case, Sagredo's correction must be regarded as a rhetorically ineffective digression.

The next rhetorically striking passage occurs toward the end of the Second Day, at the end of the critique of the objection that similar substances must have similar motions. The geocentric universe, together with the idea that the heavens are noble, perfect, and incorruptible, whereas the earth is base and full of imperfections and impurities, lends itself to the criticism that this is like having a lazaretto in the center of town rather than away from it in the country (F292—3, D268). The effect of this comparison is very strong; it creates in us a very powerful feeling of the inappropriateness of a geocentric arrangement, probably powerful enough to overcome the positive feeling of its being cozy. Realizing that the geocentric universe was not only support-able by evidence and arguments but also emotionally satisfying, Galileo is here calling attention to a feature that ought to make it emotionally *dissatis-fying*.

At the end of the Second Day, the objection is stated and criticized that a moving earth could not move forever without tiring, since animals become tired even though they move by an internal principle (F293—9, D269—76). Besides answering this objection with serious criticism, Galileo feels that it also deserves rhetorical ridicule, for it seems to Sagredo "that there are also animals which refresh themselves from weariness by rolling on the ground, and that hence there is no need to fear that the earth will tire; it may even be reasonably said that it enjoys a perpetual and tranquil repose by keeping itself in an eternal rolling about" (D271; cf. F295). This is *ad hominem* reasoning (in the 17th century sense of the term),[7] where the Aristotelian principles used for the formulation of the objection are used to answer it and show that it is not impossible for the earth to move. However, the total rhetorical force of the reply is much greater than its logical strength; it is the possibility of creating such wit that adds the extra strength.

At the beginning of the Third Day we find an explicit statement that, besides trying to enlighten the reader, Galileo wants to arouse his feelings. The occasion is a discussion of the location of the 1572 nova and a critique of Chiaramonti's attempt to show that it was sublunary. Salviati tells Simplicio that, once he has understood Chiaramonti's error, he can feel indignation if he notices that Chiaramonti, "covering his cunning with the veil of your naïveté and that of other mere philosophers, tries to insinuate himself into your good graces by gratifying your ear and puffing up your ambition, pre-tending to have convicted and silenced these trifling astronomers who wanted to assail the ineradicable inalterability of the Peripatetic heavens, and what is more, to have struck them dumb and overpowered them with their own weapons" (D285; cf. F309). Galileo is charging Chiaramonti not only with a logical-scientific error but also with a morally reprehensible deception; he is not saying that this deception is intentional, but only that it is real. I believe that the rhetorical effect of this charge is to arouse the curiosity of Simplicio and of the Peripatetic readers of the book, and to motivate them to under-take the effort to understand the time-consuming and elaborate calculations and analyses of quantitative data required to see the error in Chiaramonti's argument. What we have here, then, is an instance of the technique of making a strong claim in order to arouse interest and curiosity.

A few pages later (F313, D289), Salviati calls Simplicio an Archimedes after Simplicio has shown his understanding of a geometrical point in the theory of parallax; Salviati says that he had almost despaired of being able to explain it to a pure Peripatetic philosopher. These remarks obviously have the function of encouraging ordinary Aristotelians that the computations in Galileo's critique are not beyond their intellectual powers, and that they will be able to follow them if they only put their minds to it.

After Galileo has criticized Chiaramonti's examination of the evidence from parallax difference, he goes on to criticize how Chiaramonti explains away the evidence from polar and stellar distances, and the occasion comes for the creation of a poetic image:

When a person finds no defense to be of any avail against his mistake and produces a frivolous excuse, people say that he is reaching for ropes from the sky. This author grasps not at ropes, but at spiderwebs from the sky, as you will plainly see upon examining these two points just mentioned. [D311; cf. F339]

Though the value of this passage is primarily aesthetic, from the rhetorical point of view it functions to hold the reader's attention.

The critique ends with a series of three images:

SAGR. This is as if I were watching some unfortunate farmer who, after having all his expected harvest beaten down and destroyed by a tempest, goes about with pallid and downcast face, gathering up such poor gleanings as would not serve to feed a chicken for one day.
SALV. Truly, it was with too scant a store of ammunition that this author rose up against the assailers of the sky's inalterability, and it is with chains too fragile that he has attempted to pull the new star down from Cassiopeia in the highest heavens to these base and elemental regions. [D318; cf. F346]

From a rhetorical point of view Sagredo's description might arouse sympathy and therefore might be taken as an indication of Galileo's ambivalent attitude toward Chiaramonti that we have previously mentioned. However, Salviati's description attributes a somewhat bellicose attitude to Chiaramonti, and thus destroys that feeling of sympathy. So, rhetorically speaking, the two descriptions undo each other. This time, the aesthetic value is also dubious; for, though each description has value by itself, and though Salviati's two images go well together, it is difficult to picture Chiaramonti, or anyone, both as the 'unfortunate farmer' and as the soldier of heavenly inalterability. The aesthetic worth might be rescued by emphasizing the dramatic situation, that is, by taking Sagredo's image as a description of *his own* attitude toward Chiaramonti, and the other as being Salviati's own state of mind. This would be confirmed by the fact that Salviati has earlier had a similar image (D311, F339), namely of Chiaramonti grasping for spiderwebs from the sky, and the fact that Sagredo had earlier expressed respect for Chiaramonti, in describing him as a cut above most other Peripatetic philosophers (D256, F281).

In the middle of the discussion of the heliocentrism of planetary motions, after the basic Copernican arrangement has been sketched, we find Salviati dismissing certain arguments as being for people whose definition ('rational animals') contains only the genus ('animals') but lacks the species ('rational') [F355].[8] This is great wit, though incredibly subtle, and an unprosaic way of saying what Galileo himself says in a marginal note, namely that these are "utterly childish reasons suffic[ing] to keep imbeciles believing in the fixity of the earth" (D327; cf. F355). The reasons being referred to are: (1) that it does not happen that one is in Constantinople at lunch time and in Japan at supper time; and (2) that the earth is so heavy that it cannot climb up over the sun in order later to roll down again (*Ibid.*). To mention and then summarily dismiss an argument is of course a rhetorical technique to express a negative judgment; however, here Galileo seems to be overdoing it. This is especially unfortunate because Salviati's main point in this context (F354–6, D327–8) is to emphasize the great initial cogency of the objections to the earth's annual motion. It seems that Galileo here was carried away by the literary brilliance of his witticism and became temporarily forgetful of rhetorical effects. In fact, he later, after the book had been published, decided to insert at this point a passage which discusses some of these previously dismissed arguments (F356–62, D328–33), I believe in order to soften the literary excess and rhetorical (counter) effect of his witticism.

Galileo ends his discussion of the heliocentrism of planetary motions with the remark, "Now you see how admirably these three chords, which at first seemed so dissonant, accord to the Copernican system" (F368, my translation; cf. D340). The three 'chords' are the apparent sizes of Mars and Venus, the apparent shapes of Venus, and the geocentrism of the moon's orbit. At first these seemed dissonant with Copernicanism in the sense that with the naked eye no significant variation is observable of planetary sizes or of the shape of Venus, as it is required by the Copernican system; moreover, in being around the earth, the moon's proper motion seemed to go against the general trend alleged for the other planets. The accord is the result of the telescopic observations that Venus and Mars show the required periodic change in apparent size, that Venus shows phases, and that Jupiter's satellites have a proper motion analogous to the moon. Though Galileo's eloquent musical metaphor has great aesthetic value, it is doubtful that it has any rhetorical force; it may make us feel somewhat reluctant to disturb the harmony it creates, but since such reluctance has a purely aesthetic origin, it should be possible to overcome it.

At the beginning of the critique of the objection from the annual constancy in the stellar appearances, Sagredo says to Salviati, "This is a knot which has never passed through my comb, and if you untie it for me I shall consider you greater than Alexander" (D378; cf. F405). This reference to the Gordian knot is a rhetorically effective way of emphasizing the importance of the objection and of its resolution. It is an unprosaic way of making the same point that Galileo makes in a marginal note when he describes it as the "chief objection against Copernicus" (D378; cf. F405). We also have here a suggestion that the Galilean answer to the objection may be more analogous to a cutting than to an untying of the knot, that is, that it constitutes not an actual solution but rather an effective response. In fact, Galileo's answer consists of an admission that no change can be observed and a rather detailed outline of a program of theoretical and experimental research that would be likely to lead to the observation of some stellar variations.

Near the beginning of the discussion of the properties of loadstones, Galileo expresses his criticism of the Peripatetic dogmatic attitude toward new ideas, such as those contained in Gilbert's *De Magnete*. He also ridicules that attitude by asserting that "perhaps Gilbert's book would never have come into my hands if a famous Peripatetic philosopher had not made me a present of it, I think in order to protect his library from its contagion" (D400; cf. F426). The absurdity of the imputed motive, together with the cleverness of the caricature, contributes to instilling an anti-Peripatetic attitude.

On the next page, in order to arouse the curosity of his Peripatetic readers, and to let this curiosity prevail over their fears, Galileo resorts to one of his characteristic exaggerations. Salviati proposes to Simplicio:

If you like, I can make it evident to you that you are creating the darkness for yourself, and feeling a horror of things which are not in themselves dreadful – like a little boy who is afraid of bugaboos without knowing anything about them except their name, since nothing else exists beyond the name. [D401; cf. F427]

Later on, in the discussion of magnetism, Galileo gives an argument to show that the earth is *de facto* a loadstone. The argument involves a comparison which is rhetorically so effective that its persuasive force becomes greater than its purely logical strength:

SALV. Now, Simplicio, suppose that a thousand pieces of different materials were set before you, each one covered and enclosed in cloth under which it was hidden, and that you were asked to find out from external indications the material of each one without uncovering it. If, in attempting to do this, you should hit upon one which plainly showed itself to have all the properties which you had already recognized as residing only in lodestone and not in any other material, what would you judge to be the essence of that material? Would you say that it might be a piece of ebony, or alabaster, or tin? SIMP. There is no question at all that I should say it was a piece of lodestone. SALV. In that case, declare boldly that under this covering or wrapper of earth, stone, metal, water, etc. there is concealed a huge lodestone. [D404; cf. F430—1]

At the end of Salviati's explanation of why an armature greatly increases the amount of iron a loadstone can hold, Sagredo expresses his usual highly favorable evaluation of the argument. He also expresses the feeling that our intellect could desire no tastier food than to have equally clear explanations of the many other magnetic phenomena (F434; cf. D408). Though such language does not add anything to the cogency of the explanation, it conveys very well Galileo's excitement and enthusiasm. The description leaves as strong an impression as that left, two pages later, by Sagredo's application of the label 'magnetic' (F436; cf. D410) to the preceding discussion of magnetism.

In this same context there is a passage of great rhetorical value in which Galileo criticizes the practice of explaining by naming. Simplicio asserts that Peripatetics would account for the effect just explained by Salviati, in terms of sympathy, which is a mutual appetite between similar substances (F436, D410). Sagredo replies that there is sympathy between this manner of philosophizing and the technique of painting by verbal description which was practiced by a friend of his, who would write in various spots on the canvas descriptions of the figures and scenes he wanted there and would then leave to a painter the actual disposition of colors and shapes. The ingeniousness of the comparison and the play on the term 'sympathy' is the source of the rhetorical effect here.

At the beginning of the Fourth Day Galileo mentions several available explanations of the tides. One of these consisted of an account in terms of differences in the depth of the various seas. In the course of giving a brief statement of this cause, Galileo describes it as having recently been 'fished out' (F445; cf. D419) from a certain Aristotelian text, previously not properly interpreted. This is one of the many instances where Galileo is able to use language metaphorically by applying in a critical context terms whose literal meaning refers to the physical phenomena under investigation. The description here seems to have a negative connotation, in spite of the fact that it is given by Simplicio; this is due perhaps to the inappropriateness of the implied image of the philosopher as fisherman, or more probably to the implied image of the Aristotelian texts as an ocean with the same fluidity, formlessness, and susceptibility to taking an almost infinite number of shapes.

On the next page Galileo begins to criticize the various available explanations of the tides. One of these attributed the tides to lunar heat which would warm the water and cause it to expand and thus rise. Besides claiming that this explanation is false, insofar as tidal waters can be observed to have the same density and temperature as ordinary water (F448–9, D422–3), Galileo also gives the following rhetorical refutation: have the people who hold this theory "put a fire under a kettle of water, hold their right hands in this until the heat raises the water a single inch, and then take them out to write about the swelling of the seas" (D420; cf. F446). The persuasive force of this rhetorical criticism is much greater than the logical force of the objective counterevidence also presented; whereas the objective criticism affects only our intellect, the rhetorical one affects our whole being.

Later in the Fourth Day, just before giving his account of the monthly and annual periods, he summarizes some of his criticism of the alternative theories. Here he also makes a new attempt to score a rhetorical point when he says that the various alleged causes of the tides, far from really causing the tides, are instead caused by them, in the sense that the existence of the tides makes the corresponding thoughts arise in frivolous brains (F470, D445). This is a clever linguistic maneuver and may reflect Galileo's real opinion on the matter, but by itself it hardly possesses any rhetorical force. What we have here is a statement of criticism, a *rhetorically interesting* such statement, but the statement contains little if any self-validating force; instead it needs a supporting argument, which of course Galileo gives elsewhere in the Fourth Day. So it is not the present statement by itself which possesses persuasive force.

After Salviati has outlined the general approach he is going to follow in explaining the monthly and annual periods, there is a rather long pause (F472, D446–7) in the dialogue, occasioned by Sagredo's statement that, though he understands what has been said so far, he does not know where this is going to lead. Salviati says that he is not surprised that even Sagredo's usually quick mind fails to see the aim of the discussion, for he himself required a very long time before he was able to master these ideas, so much so that he often despaired of ever being able to do so. Then Salviati makes a reference to Roland from Ariosto's poem *Orlando Furioso,* a reference continued in the response of Sagredo, who says he is glad that Salviati's despair did not have the same effect as Roland's. This passage is striking both because of the length of the pause, when the exposition has just started, and also because of the reference to Roland. I suppose that the rhetorical effect of the pause and of Salviati's confession is to emphasize the difficulty of the topic. In order to understand the reference to Roland, it should be mentioned that in the poem this character, while being in love with a woman, finds clear evidence that she has deceived him with another man; at first Roland refuses to accept the evidence, and then he becomes insane. Obviously Sagredo is glad that Salviati's despair did not lead to madness. What Salviati says is not so obvious; his words are worth translating literally; "despairing of being able to unravel it, as a consolation of myself, I forced myself to be persuaded, like the unhappy Roland, that it might not be true, what however the testimony of so many trustworthy men presented before my eyes" (F472, my translation; cf. D447). The thing that Roland refuses to believe is obviously the infidelity of his paramour, in spite of the evidence. If the analogy is to hold, Salviati's situation must be one where he refuses to believe the falsity of the geokinetic explanation of the monthly and annual periods in spite of some counter-evidence, which should be relatively clear evidence. Galileo discusses and seems to have a plausible answer to the objection that the monthly change in the earth's annual motion presupposed by his theory has not been observed (F479–82, D454–7); however, he does not explicitly discuss or mention the objection that the tides are greater at the equinoxes than at the solstices, instead of the other way around as required by his theory.[9] Therefore, the present analogy to Roland's situation must be a subtle way of recognizing such counter-evidence, for there is no other way of making sense out of the reference. If so, however, the recognition is perhaps too subtle to be rhetorically effective. Of course, it is rhetorically appropriate not to emphasize the weaknesses of one's theory, though at the same time, if the weakness is very obvious, then it is rhetorically *inappropriate* not to mention it at all. So perhaps the subtle recognition does constitute a sufficient qualification after all.

The last rhetorically significant passage is contained in a speech by Sagredo, given after Salviati has expounded the details of the geokinetic explanation of the monthly period. The passage is a simile, and I think that its aesthetic value inclines us favorably toward Galileo's argument, in the sense that the feeling expressed by Sagredo becomes contagious. This is especially true since, besides giving his usual praise, this time Sagredo prosa-ically expresses a difficulty at the end of the same speech, immediately after the poetic image. The simile is one of which Homer himself could be proud:

SAGR. If a very high tower were shown to someone who had no knowledge of any kind of staircase, and he were asked whether he dared to scale such a supreme height, I believe he would surely say no, failing to understand that it could be done in any way except by flying. But being shown a stone no more than half a yard high and asked whether he thought he could climb up on it, he would answer yes, I am sure; nor would he deny that he could easily climb up not once but ten, twenty, or a hundred times. Hence if he were shown the stairs by which one might just as easily arrive at the place he had adjudged impossible to reach, I believe he would laugh at himself and confess his lack of imagination.

You, Salviati, have guided me step by step so gently that I am astonished to find I have arrived with so little effort at a height which I believed impossible to attain. It is certainly true that the staircase was so dark that I was not aware of my approach to or arrival at the summit, until I have come out into the bright open air and discovered a great sea and a broad plain. And just as climbing step by step is no trouble, so one by one your propositions appeared so clear to me, little or nothing new being added, that I thought little or nothing was being gained. So much the more is my wonder at the unexpected outcome of this argument, which has led me to a comprehension of things I believed inexplicable. [D454; cf. F479]

To conclude, we have attempted to define a dimension of Galileo's argu-ment distinct from its purely logical one and from its purely rhetorical one. We have found it in its combined emotional appeal and literary-aesthetic value, and we may label it its rhetorical *force*. This is distinct from the argu-ment's logical structure, studied in the last chapter, insofar as the latter refers to argument and appeal to evidence, whereas rhetorical force involves aesthetic images and appeal to emotions. This rhetorical force is also distinct from rhetorical appearance, studied in an earlier chapter, insofar as the latter refers merely to communicative expressions, whereas the former pertains to persua-sive effectiveness.

We have also found no necessary conflict between reasoning and emotions, or between reasoning and aesthetic expression; for, in taking the form of appeals to emotions, some arguments derive force from the emotions involved, and the logical element is present in connecting the idea being justified to those emotions. Similarly, some arguments derive force from the beauty and pleasantness of the linguistic expressions used and from the aesthetic images involved. As long as one does not confuse an emotional appeal or a brilliant literary expression with an appeal to evidence, there is no problem. Moreover, it is useful to remember that these nonlogical rhetorical devices have their own standards of value; hence, though the pure logician may act as if they did not exist, the concrete logician, or theorist of reasoning, cannot do so.

Finally, our study suggests, as envisaged by Feyerabend, that both literary art and rhetorical persuasion have some role to play in science, for we have seen that in his justification of Copernicanism Galileo displays the expressive ingenuity of the artist and the rhetorical skill of the orator. Of course, this is to say neither that rhetoric and literary art are a guarantee of scientific success, nor even that they are always relevant. As elements of scientific method, they are not universally valid, any more than any other elements, as will be argued exhaustively in later chapters. Nevertheless, they are elements, and both scientists and scientific methodologists need to note the fact.

NOTES

[1] See, for example, W. R. Shea, *Galileo's Intellectual Revolution*, p. 117; and L. Geymonat, *Galileo Galilei*, p. 126.

[2] G. Galilei, *Opere*, edited by Favaro, 7: 44, 51, 53, 79, 96, 190, 248, 275, 302, 303, 363, 372, 373, 380, 382, 388, 431, 477. As usual, subsequent references to this edition of the *Dialogue* will be made in the text by prefixing the page number by 'F'. Similarly, references to Drake's translation will be denoted by 'D'.

[3] See, for example, I. M. Copi, *Introduction to Logic*, pp. 75–6.

[4] Cf. M. Scriven, *Reasoning*, p. xvi; and C. L. Hamblin, *Fallacies*, pp. 41–42.

[5] Cf. the discussion of this argument in Chapters 14 and 16 (Section 6).

[6] This translation is my own. To see that the Salusbury-Santillana translation of this passage, like Drake's, misses some nuances, cf. G. Galilei, *Dialogue on the Great World Systems*, tr. Salusbury-Santillana, pp. 177–178.

[7] See Chapter 16, Section 4.

[8] For more details see Chapter 10, Section on Strauss, Paragraph 6.

[9] Cf. Shea, *Galileo's Intellectual Revolution*, pp. 182–3; and Galilei, *Dialogue*, tr. Drake, pp. 490–491, note to p. 457.

CHAPTER 4

TRUTH AND METHOD:
THE SCIENTIFIC CONTENT OF GALILEO'S *DIALOGUE*

Scientists from Newton to Einstein have found Galileo's *Dialogue on the Two Chief World Systems* to be rich in scientific content. It is well known, for example, that in the *Principia* Newton attributed to Galileo a knowledge of the law of inertia, the law of force, the principle of superposition, the law of squares, and the parabolic path of projectiles;[1] what is not well known is that, as I. B. Cohen has argued, "Newton almost certainly did not read the *Discorsi* − if, indeed, he ever did at all − until some considerable time after he had published the *Principia*",[2] whereas "the evidence is certain and unmistakable that Newton, early in his scientific career, had read the great *Dialogo . . . sopra i due massimi sistemi*".[3] It is also well known that, in his Foreword to Drake's translation of the *Dialogue,* Einstein summarizes its scientific content as being a result about the nonexistence of an abstract center of the universe, with analogies to his own work. According to Einstein, "Galileo opposes the introduction of this 'nothing' (center of the universe) that is yet supposed to act on material bodies; he considers this quite unsatisfactory . . . [because] although it accounts for the spherical shape of the earth it does not explain the spherical shape of the other heavenly bodies".[4] Einstein then suggests

that a close analogy exists between Galileo's rejection of the hypothesis of a center of the universe for the explanation of the fall of heavy bodies, and the rejection of the hypothesis of an inertial system for the explanation of the inertial behavior of matter. (The latter is the basis of the theory of general relativity.) Common to both hypotheses is the introduction of a conceptual object with the following properties:
  (1). It is not assumed to be real, like ponderable matter (or a "field").
  (2). It determines the behavior of real objects, but it is in no way affected by them.
The introduction of such conceptual elements, though not exactly inadmissable from a purely logical point of view, is repugnant to the scientific instinct.[5]

Typical of scientists' attitude is perhaps the judgment expressed by Arthur Schuster in the 1916 Presidential Address to the British Association for the Advancement of Science: "Modern science began not at the date of this or that discovery, but on the day that Galileo decided to publish his *Dialogues* (1632)".[6] The exact reason for this scientific popularity is not clear. In part, it may be that, to use Boris Kouznetsov's eloquent phrase, "an eternal dawn reigns in science",[7] and that therefore scientists feel very much at home in a work characterized by fluid and plastic thought, with which they can identify in some way. Such scientific intuitions, besides having value in themselves, fit very well with the latest biographical evidence, which suggests that the book constitutes Galileo's mature synthesis of physics and astronomy.[8]

SCIENTIFIC RELEVANCE VS HISTORICAL ACCURACY

If scientists' attitudes are so universally positive, the same cannot be said for historians of science. For one thing, many of them tend to pay more attention to the *Two New Sciences,* though in so doing they often engage in a self-defeating exercise insofar as they begin by choosing *it* for its seemingly more scientific character and then end up concluding that its scientific content is rather meager after all. For example, W. Wisan ends her otherwise excellent study of the *Two New Sciences* by saying:

Most of his propositions were of little subsequent interest and only a very few contributed to an advance in scientific knowledge. His mathematics was unoriginal and seems clumsy and obsolete for a work published the year after Descartes' new analytic geometry, and it is difficult to see precisely how the *De motu locali* influenced later treatments of motion. Galileo's laws of motion are, of course, an obvious contribution, and the resolution of projectile motion into its components was an important innovation. But the treatise as a whole seems hopelessly medieval in its separation of 'natural' and 'violent' motions and the failure to give an analysis of circular motion. There is relatively little here that suggests later studies of the motion of rigid bodies.[9]

Other historians are misled by the fact that the *Dialogue* has several other aspects (rhetorical, philosophical, cultural) into thinking that its scientific content is not noteworthy, as if it were impossible for a work to have all these dimensions simultaneously; for example, according to Koyré,

the *Dialogue on the Two Chief World Systems* pretends to be an exposition of two rival astonomical systems. But, in fact, it is not a book of astronomy, nor even of physics. It is above all a book of criticism; a work of polemic and struggle; it is at the same time a pedagogical work, and a philosophical work; lastly, it is a book of history, "the history of Mr. Galileo's mind".[10]

Most historians of science regard the book as a chapter in the history of the struggle for Copernicanism; two attitudes are prevalent here. Some look for a demonstration of the truth of the Copernican system; when they fail to find a scientifically valid one, the scientific worth of the *Dialogue* depreciates in their eyes. For example, Ernan McMullin feels that "in evaluating

Galileo's contributions as a 'man of science' ",[11] one must include the question of whether or to what extent Galileo proved that "the earth is really in motion".[12] McMullin is unimpressed[13] by the proof provided in the *Dialogue,* though he admits that "most scientists among Galileo's readers were persuaded that the Copernican view had been adequately validated by the arguments of the *Dialogo*",[14] and though he is not insensitive to the fact "Galileo's magnificent effort to move the earth has always had and will always have an even greater power to move men".[15] From the point of view of the *scientific* content of the *Dialogue,* and the point of view of the history of *science* in general, one feels like asking rhetorically, "But why should one want to equate science with demonstrative proof?", or like quoting Richard P. Feynman's view:

In learning any subject of a technical nature where mathematics plays a role, one is confronted with the task of understanding and storing away in the memory a huge body of facts and ideas, held together by certain relationships which can be 'proved' or 'shown' to exist between them. It is easy to confuse the proof itself with the relationship it establishes. Clearly, the important thing to learn and to remember is the relationship, not the proof. In any particular circumstance we can either say "it can be shown that" such and such is true, or we can show it. In almost all cases, the particular proof that is used is concocted, first of all, in such form that it can be written quickly and easily on the chalkboard or on paper, and so that it will be as smooth-looking as possible. Consequently, the proof may look deceptively simple, when in fact, the author might have worked for hours trying different ways of calculating the same thing until he has found the neatest way, so as to be able to show that it can be shown in the shortest amount of time! The thing to be remembered, when seeing a proof, is not the proof itself, but rather that it *can be shown* that such and such is true. Of course, if the proof involves some mathematical procedures or 'tricks' that one has not seen before, attention should be given not to the trick exactly, but to the mathematical idea involved.

It is certain that in all the demonstrations that are made in a course such as this, not one has been remembered from the time when the author studied freshman physics. Quite the contrary: he merely remembers that such and such is true, and to explain how it can be shown he invents a demonstration at the moment it is needed. Anyone who has really learned a subject should be able to follow a similar procedure, but it is no use remembering the proofs.[16]

In answer to my rhetorical question, McMullin might answer that it is proper in this context to equate science with demonstrative proof because Galileo himself equated the two.[17] It would not be difficult to show that such a philosophy of science is not at all representative of Galileo's remarks on scientific inquiry (see Chapters 5 and 6); however, for the present purpose, that fact, *if* and to the extent that it is a fact, is not relevant. For we are not examining here the philosophical content of Galileo's *Dialogue,* but rather its scientific content; our own context is here one of history of *science* not history of the *philosophy* of science. If we were doing the latter, then it would be proper first to determine what Galileo's philosophy was, as McMullin does,[18] and then to look in Galileo's works for those elements which can be regarded to constitute science according to that conception of science.[19] It would be clearly necessary to do the latter, as well as the former, for it would be unfair to extract someone's philosophy of science by merely examining his reflections upon science without also looking at his scientific practice. So it may be said as a point in favor of McMullin's account that he does look at both, thus filling with concrete content his account of Galileo's philosophy of science. Nevertheless, such an account is criticizable in its own terms, and moreover, we do not get into science and scientific context by his route.

The second 'Copernican' approach to Galileo's *Dialogue* defines its content to be essentially propaganda for Copernicanism. In recent times, this view has originated from Arthur Koestler,[20] propagandized by Paul Feyerabend,[21] and traces of it can be found in Dudley Shapere,[22] William Shea,[23] and Giorgio de Santillana.[24] The truth behind this interpetation is that the book had considerable practical impact, that, in fact, it is full of rhetorically significant passages, and that practical considerations lurk everywhere in its conception and composition. However, the only responsible way of defining its scientific content in terms of the rhetoric of the earth's motion is in the context of the *science of rhetoric;* for after all the study of the art of persuasion and of achieving practical effects by verbal means is the subject of a discipline which is at least as old as Aristotle's rhetoric.[25] So the only business that a historian of *science* has meddling into the rhetorical analysis of the *Dialogue* is if he is writing the history of the science of rhetoric, not if he is concerned with the history of natural science. Let me make it clear that I think that studies of the rhetorical content of Galileo's *Dialogue* are quite proper; I myself believe that the book is a classic in the art of rhetoric, besides being one in the literature of science. However, these scholars do not provide a serious rhetorical examination of the book, the foundation of such an undertaking being that there can be good as well as bad rhetoric, a point that the above mentioned writers tend to ignore when they speak of propaganda. In fact, given that the context of their own investigations is an ambiguous one between science and rhetoric, when they speak of propaganda, they are contrasting this to valid science, so that the term has for them an implicit negative connotation. Partly to remedy this situation, in previous chapters I undertook the rhetorical analysis of Galileo's book, and I attempted to extract its rhetorical force, structure, and content. However, here I am concerned with its scientific content, scientific in the sense of natural science.

The divergence between scientists' and historians' perception of the scientific content of the *Dialogue* acquires its clearest expression in connection with the interpretations emanating from the medievalist historians. Naturally enough, when these scholars read the *Dialogue* the main thing they perceive is similarities and differences between what's contained therein and the views of various medieval thinkers.[26] Though this kind of exercise informs us about the medieval content, if any, of Galileo's *Dialogue,* it is not clear that it tells us anything about the nature of its scientific content. The less tenable form of the medievalist interpretation would speak of a medievalist origin of, or influence upon, Galileo's book; I believe that this version of the medievalist thesis is presently being rendered of historical interest only (if I may be allowed the pun) through the efforts of Stillman Drake and his followers.[27] Let us then formulate the thesis in the more tenable form according to which the *Dialogue* is claimed to contain views which *de facto* happen to be similar to those found in various medieval texts. If true, this might be very interesting from a number of points of view, which it is not my job to elaborate. For my job here is to ask of what relevance would the alleged medieval aspects of the *Dialogue* be to understanding its scientific content? Here no one, not even the medievalists themselves, would accept the relevant equation (scientific= medieval); so there is no question of confusion, as there is for the identification of science with demonstrative proof, presupposed in one of the other interpretations. But if we avoid the relevant equation, then the relevance can only be defended by the influence thesis; but if we accept the latter, then at least for the present case of Galileo, we draw fire from the guns of Drake and his followers, a fire which is very difficult to resist and which may therefore very well consume us. In the light of the continued resistance by scientific intuition[28] to the infusion of medieval elements into modern science, we may set aside the work of these historians in our attempt to examine the *scientific* content of Galileo's *Dialogue.*

Another common approach to the analysis of the scientific content of this book and of Galileo's work in general, is to examine it from the point of view of Newtonian science. This is of course more satisfactory, but only slightly so. The practice was more justified before the advent of post-classical physics, but nowadays it is no longer proper to equate science with Newton's work. There is no question that the detection of what might be called the Newtonian content of Galileo's book is a highly relevant exercise; of the approaches so far considered it is perhaps the most relevant. Nevertheless the Newtonian approach breaks down completely when the Newtonian content

blinds one to the more-modern scientific content,[29] and it can be shown still
to be inadequate when there is no conflict.[30]

A conflict exists, then, between scientists' and historians' perceptions of
the scientific content of the *Dialogue*. It won't do to point out that historians
can often show that what scientists perceive as being in the *Dialogue* is not
there in some sense, and that by and large scientists find in this work whatever they want to find. For, even when inaccurate, scientists' interpretations
are suggestive of the way the scientific mind works and hence relevant to the
analysis of the book's scientific content. Moreover, though historians tend to
be more accurate, what they sometimes perceive in the *Dialogue* is not there
either, as I show in detail in later chapters for the cases of Koyré, Strauss,
Favaro, Drake, and Clavelin;[31] and, more importantly, one can often show, as
I have done above, that what historians of science perceive in the *Dialogue* is
often not science, but something else. The real problem is, then, that
historians' interpretations *tend* to be accurate but irrelevant, whereas
scientists' interpretations *tend* to be relevant but inaccurate. Is it possible to
combine relevance to science and accuracy to the text? Unfortunately, it may
not be possible because on the one hand the *historical attitude as such* is
bound to introduce irrelevancies, whereas the *scientific attitude as such* is
bound to introduce inaccuracies. Why is this so? I shall first give an abstract
general argument and then a concrete one grounded on the particular
example of Galileo's tidal theory in the Fourth Day.

A historian of science is usually someone who is acquainted with the development of science, and hence with scientific ideas and facts of various time
periods. However great his acquaintance with contemporary science and the
latest textbooks and journals is, this is for him merely the science of the
present period. The minute he holds the latter with some special reverence or
gives it special status, he thereby abandons the historical attitude, which
requires of him a period-free neutrality or objectivity. Of course, his
specialized area of competence is likely to be a particular period, but the
historical sensibility does not allow him to magnify this accident of his biography into a historical or historiographical fact. What happens, then, when a
historian opens a book like Galileo's *Dialogue* and we ask him to tell us about
its scientific content? When the question is asked in the more or less loaded
terminology of 'scientific' content, perhaps he will refuse to answer it, and
confess that he doesn't know what is being asked in asking for the specifically
scientific content of the book. We can make the question less loaded and ask
about the intellectual content of the book. The intellectual elements that he
is going to come up with will be ones involving similarities and differences

with other such elements found in previous and in subsequent periods, thinkers, texts. The relevance of such comparisons and contrasts, the relevance, that is, to science or at least to the interests and concerns of (living) scientists, this relevance is exactly what is hard to see.

Let us look at the scientists' situation now. A natural scientist is one who is making more or less original contributions to the understanding of nature. Science is for him what he himself (and his peers) do. When he opens the *Dialogue* the intellectual elements that are likely to attract his attention are those that have similarity to his own scientific involvements. Because of the growth of scientific knowledge he will necessarily be blind to problems and ideas that may have been central to the book's author but which do not relate to present day scientific research; this relation has to be one identifiable by his scientific intuition, not one demonstrable by historiographical, philosophical, or logical means. I have argued elsewhere that what Joseph Agassi calls inductivism is the typical and inescapable attitude of the working scientist.[32] The growth of scientific knowledge makes it also unlikely, in principle, that the elements detected by the scientist are really in the book; for this would be like saying that Galileo was concerned with problems whose content and substance, as opposed to method and general character, were similar to those of the contemporary scientist; since the latter's problems are usually definable only by reference, at least implicit, to past scientific achievements, it follows that the actual similarity of content and substance is excluded almost in principle. Of course, this becomes less true as one gets closer to Galileo's own period; so that, for example, Newton could accurately claim more similarity of content. However, the real question is whether the extent of this similarity is significant. The fact is that, almost as a matter of definition, the more creative a scientist is, the more his achievements surpass his predecessors (in content, we must remember, not necessarily in approach). It follows that he is less likely to understand them, that is to say understand them in the historical sense of accuracy. Of course, in another sense, the scientific sense, it may be said that he is the only one who really understands his predecessors, since he is the one who superseded them by using their results and building upon them. However, this superior scientific understanding is merely another way of saying that the scientific *relevance* of what he finds in his predecessors is guaranteed, a point I have already conceded; but what I am discussing here is his historical accuracy, which remains necessarily problematic.

TIDES: THE FOURTH DAY

A beautiful example of this tension between scientific relevance and historical accuracy is provided by available accounts of the tidal theory of the Fourth Day. As is well known, in the last 'Day' of the *Dialogue,* Galileo justifies the earth's motion by arguing that only its combined daily axial rotation and annual orbital revolution can explain the existence of tides.

Scientific readers of Galileo's tidal theory, whether their attitude is critical or apologetic, have perceived a number of scientific elements in it. For example, Ernst Mach feels that "the principle of relative motion is a correct feature of this theory, but it is so infelicitously applied that only an extremely illusive theory could result";[33] he also states that "it is noteworthy that Galileo in his theory of the tides treats the first dynamic problem of space without troubling himself about the new system of coordinates. In the most naive manner he considers the fixed stars as the new system of reference".[34] In 1954, V. Nobile gave a mathematical argument designed to show the existence of what he calls a tidal 'Galilean effect', independent of gravitation, and deriving from the combination of the earth's rotation and revolution.[35] In 1962, H. Burstyn[36] argued that the following scientific concepts can be found in the Fourth Day: (1) an imperfect but fundamentally correct theory of the so-called 'tide of reaction';[37] (2) the insight that once the ocean has been set in motion by the primary cause of the tides, it tends to oscillate from its own inertia with a period which may differ from that of this primary cause, since this period depends also on such factors as the size of the basin and the depth of the water;[38] and (3) the fact that "the earth's orbit around the sun is described not by the center of the earth but by the center of mass of the earth-moon system".[39]

The historical accuracy of Burstyn's account has already been questioned. Concerning the point about the earth-moon system, E. J. Aiton has shown that the context of Galileo's argument is such that he is not thinking that the distance between the earth and the sun changes, as the earth-moon system goes around it, but rather only that their orbital speed changes as the sun-moon distance changes due to the moon going around the earth.[40] Though significant, Aiton's argument is merely an expression of what a careful reader of the *Dialogue* readily notices. In his response to Aiton, Burstyn all but concedes the textual incorrectness of his own interpretation when he is willing to admit that his view depends on an anachronism.[41]

Burstyn's second one of the above attributions is indeed correct, and he does deserve great credit for having emphasized this aspect of Galileo's theory, which historians usually neglect. This is one of those things appreciated much more readily by a scientifically minded reader like Burstyn, who is very much aware that the scientific problem of the tides is extremely complex, that Newton's gravitational theory is merely the beginning, that it was not until 1960 that a general numerical solution to the equations of tidal motion was found,[42] and that "in only a very few cases we can carry out the theoretical calculations necessary to approximate closely the tide of a given region of the ocean; in practice we predict the tide for various coastal points by harmonic analysis of records of tidal observations".[43]

As for Burstyn's first point, Aiton also criticizes his interpretation of Galileo's theory as an approximation to the 'tide of reaction'.[44] However, in the light of Burstyn's reply,[45] I find the scientific debate inconclusive. Nevertheless, the textual accuracy of Burstyn's interpretation can be effectively decided by pointing out that Galileo is not considering centrifugal acceleration; he is not adding and subtracting the rotational and the orbital acceleration, but the speeds. This is clear from the text:

in coupling the diurnal motion with the annual, there results an absolute motion of the parts of the surface which is at one time very much accelerated and at another retarded by the same amount. This is evident from considering first the parts around D, whose absolute motion will be very swift, resulting from two motions made in the same diretction; that is, toward the left. The first of these is part of the annual motion, common to all parts of the globe; the other is that of this same point D, carried also to the left by the diurnal whirling, so that in this case the diurnal motion increases and accelerates the annual motion.

It is quite the opposite with the part across from D, at F. This, while the common annual motion is carrying it toward the left together with the whole globe, is carried to the right by the diurnal rotation, so that the diurnal motion detracts from the annual. In this way the absolute motion – the resultant of the composition of these two – is much retarded.[46]

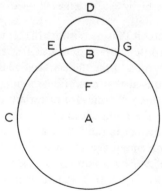

It would be wrong, however, to argue as some do[47] that Galileo was not thinking of centrifugal acceleration because he *could not* have been so thinking; and he could not, because it would have been inconsistent with his own principle of mechanical relativity, according to which all terrestrial phenomena would happen the same way whether or not the earth is moving, so that there is no way of testing its motion with a mechanical experiment within the earth. The reason is that there is evidence that Galileo did not hold the principle of relativity as an absolute truth but rather as a first approximation.[48]

In summary, what one can say about Burstyn's account is that though it is full of brilliant scientific suggestions, it develops a tidal theory which is more correctly regarded as Galilean rather than Galileo's. It is indeed in accordance with knowledge that Galileo shows elsewhere to possess, and it may be regarded as a plausible development and elaboration of what is in his mind. However, it is not *actually* in his mind (in the text); in this sense it is not *Galileo's* theory.

By contrast, the most recent and elaborate account by a historian does not attribute to Galileo beliefs not present in the text, but neither, as we shall see, any beliefs present in the body of scientific knowledge, namely, any scientific content. The account is part of W. R. Shea's *Galileo's Intellectual Revolution,* and it is typical of the rest of the book, which may be regarded as the most extensive historical treatment recently undertaken of Galileo's *Dialogue.*[49] A lack of scientific content of the Fourth Day is being claimed when, after giving a critical summary of Galileo's theory, Shea concludes:

> There can be no doubt that Galileo's theory of the tides opened no new scientific vista to his successors. He neglected to take cognizance of the four well known periods of the tides, he rode roughshod over the discrepancies between his theory and experience, he did not investigate striking observational consequences entailed by his explanation, and he brushed aside contemptuously any appeal to the influence of the moon.[50]

It must be said to Shea's credit that he is aware that it is questionable how his analysis is relevant to contemporary science. In fact, in advocating the 'contextual method' as opposed to the "linear view of the history of science", he is explicit that questions about scientific content (in the sense presently under discussion) are not the only ones for historians of science to deal with. They should also examine the blind alleys of the past if this turns out to be necessary to "assess old theories in terms of the conceptual framework of the scientists who held them, and judge them against the background of the world picture of their age".[52] The condition just mentioned is very important and I shall discuss it shortly. For now, it should be noted that Shea's espousal of the 'contextual method' is tantamount to admitting that he is not analyzing what I have been calling here the scientific content of the *Dialogue*. The fact that Shea is aware of what he is doing does not make it any more relevant to contemporary science.

To be sure, Shea would argue for the *indirect* relevance of his analysis to contemporary science. For the contextual method puts us in touch with the actual thought-processes of past scientists and hence with the actual process of scientific discovery.[53] For example, when we realize that Galileo's discussion of the tides fails to make sense from the scientific point of view,[54] we begin to understand that "he was a natural philosopher who saw beyond the problem of determining the periods of the tides, about which he did not feel strongly, to the great vision of a science in which the real is described by the ideal, the physical by the mathematical, matter by mind"[55]; and this leads to the realization that "Galilean science was not so much an experimental game as a Platonic gamble".[56] In other words, the indirect scientific content will be what it tells us about the nature of science, which we may call the book's methodological content.

I believe that this is the right direction to move toward. However, in trying to extract the methodological content one must use the contextual method quite rigorously. Unfortunately, Shea smuggles in contemporary science, or at least his conception of it, by the back door in his examination when he uses science-dependent evaluations of some of what he attributes to Galileo.

For example, Shea is eager to refer to a criticism of Galileo's theory given by Mach.[57] Shea reports Mach's criticism as amounting to the following:

Galileo makes the error of mixing two different frames of reference. Whereas the motion of the earth is considered relative to the sun, the motion of the water is considered relative to the earth. But relative to the earth, the water can receive no acceleration due to the earth's annual motion, and the water must therefore be at rest relative to the earth.[58]

Unfortunately, though this criticism can be found in Aiton[59] and in Arthur Koestler,[60] I do not find it in Mach.[61] Mach does say that "the principle of relative motion is a correct feature of this theory, but it is so infelicitously applied that only an extremely illusive theory could result"[62]; to show the misapplication Mach describes a thought-experiment in which a rotating sphere is set in uniform motion; for Mach no tides can occur during the uniform progressive motion any more than while it was merely rotating since "the case in question does not differ, according to our view, in any essential respect from the preceding, inasmuch as the progressive motion of the sphere may be conceived to be replaced by a motion in the opposite direction of all surrounding bodies".[63] Second, Mach claims that Galileo's error lies in what he calls a "*negative* conception of the law of inertia"[64]; though obscure, this has nothing to do with mixing reference frames. Mach then goes on to explain that Galileo did not apply the concept of acceleration properly.

Shea's eagerness to misinterpret Mach, or to invoke his authority, represents a violation of the contextual method. Moreover, even when correctly interpreted, such an appeal would be anachronistic. To defend himself against this latter criticism Shea claims that Mach's criticism was expressed by some of Galileo's contemporaries by quoting from a 1633 letter in which the criticism of a group of French physicists is reported.[65] Though it is difficult to read into this letter Mach's criticism, one can find in it traces of Shea's own criticism (from mixing). In the letter it is stated that, in Galileo's theory the various parts of the globe are accelerated relative to the sun but not relative to each other, hence it is difficult to see how the oceans could acquire motion relative to the land. No charge of confusion is argued for or even voiced. Nor does Shea give a justification. Since there is a difference between saying that it is difficult to see how A can follow from B, and saying that A and B are being confused, the contextual nature of Shea's criticism is unsupported.

Next, Shea refers to "Galileo's failure to distinguish centripetal acceleration from linear acceleration".[66] The reality of this failure is undeniable, but so is the uncontextual character of criticism based on it.

Shea's criticism so far has been that Galileo's theory is supported by a fallacious argument, which would be a fair criticism only if it really follows the contextual method, which it does not. His next criticism is that Galileo's theory is inconsistent with the facts insofar as it implies a 12-hour lapse between high and low tide, rather than a 6-hour lapse, which was allegedly well known.[67] What Shea ignores here is the context of Galileo's own discussion[68] which emphasizes that the tides are not due to the primary cause alone (the earth's double motion) but to the interaction between it and various secondary factors (such as water depth and size of basin) which affect the period of oscillation of the water.

In conclusion, if there is no way of analyzing the (substantive) scientific content of the *Dialogue* which is both scientifically relevant and historically accurate, then we will have to be satisfied with analyzing its methodological content as its indirect scientific content. In the latter investigation we will have to use what Shea calls the contextual method, but in a more rigorous manner than he does. For we have seen that he has the tendency to mix contextual evaluations with ones based on contemporary (or at least subsequent) scientific knowledge; and such an impure contextual method would seem even less adequate than the procedure of the so-called linear historian of science, who uses contemporary science primarily as a criterion of choice of topics.

SCIENCE: THE FIRST DAY

Is there really no way of combining textual accuracy and scientific relevance? Should it not be possible to give an account of the *Dialogue* that keeps an eye both on the text and on concepts that are likely to be of interest to the working scientist? To explore further this possibility I have attempted a reconstruction of the scientific content of the First Day. One reason for choosing this section of the *Dialogue* is that it is often alleged to be more philosophical than the other three Days, philosophical in a sense in which this term is contrasted to scientific; hence, our attempt will also be a test of this belief.

The analysis given below has resulted in the First Day being divided into eleven sections, each of which discusses a substantive scientific idea, problem, instrument, fact, observation, or experiment. My interpretation of each section attempts to integrate all its substantive scientific topics around a suitable theme. Whenever a particular claim in my interpretations can be traced directly to a given place of the text, I have indicated the page number in parenthesis. It has turned out that the topic of acceleration predominates in the first third of the 'Day', and that the topic of the moon predominates in the last third; the former fact is somewhat surprising, though the latter obviously is not. The various sections and their scientific topics are as follows:[69]

The importance of acceleration: F33–45.10; D9–21.
The continuity of acceleration: F45.11–47.10 & F51.18–53.13; D21–23 & D27–29.
Acceleration *vs.* speed: F47.10–51.18; D23–27.
An acceleration problem: F53.13–54; D29–30.
The isotropy of space: F54–61.28; D30–37.
The primacy of mechanical properties: F61.28–75.8; D37–50.
Sunspots and the telescope: F75.9–87.14; D50–62.
Amount of visible lunar surface: F87.15–95.30; D62–70.
The moon's mountainousness: F95.31–112.28; D71–87.
The moon's secondary light: F112.29–122.29; D87–97.
The problem of moon spots: F122.30–131; D97–105.

The details of the scientific content of these various sections are, respectively, as follows.

The importance of acceleration (F33–45.10; D9–21). Geometrical (spatial) properties can act as the foundation for cosmological facts, but not for the physics of motion; for the latter one needs to combine spatial properties with temporal ones, such as temporal uniformity; this combination leads to the concept of acceleration.

For example, the three-dimensionality of the world is based on the possibility of constructing three and only three mutually perpendicular straight lines through a given point (F33–8); and the orderliness of the world is based on the existence of circular motions (F43–5). But the geometrical simplicity of straight and circular motions is useless in physics since circularity around one center is as good as circularity around another center (F40), and a straight line off the center is as good as one going through it (F40). Acceleration is the important property in the physics of motion because accelerated straight motion is the natural means of acquiring uniform circular motion (F43–5), which alone retains cosmological orderliness and is in this sense natural (F43–5). And acceleration is a temporal phenomenon in the sense that it takes place in time, i.e. is not instantaneous (F44, 45), and that it depends on the concept of uniformity.

The continuity of acceleration (F45.11–47.10 & F51.18–53.13; D21–23 & D27–29). The continuity of motion implies that a body falling from rest has at some point in its path such a small degree of speed that at this speed it would not have traversed a distance of two hundred *braccia* in a thousand years. This is difficult to understand in the light of the fact that the body can be seen to acquire very quickly a high degree of speed. The problem can be solved as follows.

First, it seems plausible that the degree of speed at any point of the fall is such as to allow the body to go back to the height from which it started to fall. Evidence for this is the plausibility of imagining that a body falling toward the center in a tunneled earth would continue to move past the center for a distance equal to that through which it had gone in falling to the center. Evidence is also provided by (1) the fact that a pendulum removed from the vertical swings back to the other side with equal amplitude, except for the small disturbance of the air and string, and (2) the fact that water seeks its own level. It follows that bodies acquire the same degree of speed when they fall the same amount, regardless of their paths. For example, in falling from C to A by CA the speed at A would be the same as that at B in falling from C to B by CB.

Second, it seems plausible that the less inclined a plane is, the more slowly a body moves in falling by that path. For example, a body will move from D

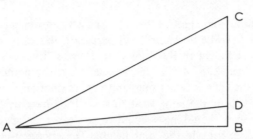

to *A* much more slowly than from *C* to *A*. Since over the horizontal *BA* a body will not move naturally at all, it is easy to imagine an inclined plane, between *DA* and *BA,* that is, whose top is so close to *B*, that the body in falling to *A* over that path will take an arbitrarily long time, for example a thousand years.

Third, it is a fact that a body falling through a given distance, has at the end of that distance a speed which if kept constant would allow it to traverse double that distance in the same time.

It follows that one could drop a body from a point above *B* so close to it that at *B* the body would have a speed as slow as you like. Similarly, dropping the body from *C,* there are points so close to *C,* that at those points the body has a speed as slow as you like.

Acceleration vs speed (F47.10–51.18; D23–27). It is important to distin-

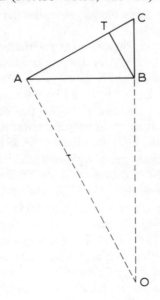

guish acceleration from speed, otherwise one gets into unnecessary problems. For example, let AB be horizontal, CB perpendicular to it, and compare the motion of a ball falling freely from C to B, with its motion along on the inclined plane from C to A. Clearly, in one sense, the perpendicular motion will be faster than the inclined one, and yet, in another sense, the speed of the body at A will be the same as at B, since they have descended the same amount and hence should have acquired a degree of speed which would enable them to go back to the same height. The apparent contradiction can be removed since the quantity which is the same from C to B and from C to A is the speed, whereas the quantity which is greater along CB than along CA is the acceleration. This can be shown as follows.

To say that two speeds are equal means that the distances traversed are proportional to the times elapsed, and to say that one body moves faster than another means that in the same time the first traverses a greater distance than the second, or that in traversing the same distance it takes less time. Now it is easy to visualize that in the time that the body falls from C to B, it would move along CA up to a point T such that $CT < CB$, or in the time that the body falls from C to A, it would have fallen from C past B to some point O such that $CA < CO$. In either case, in the perpendicular motion, greater distances are being covered in the same time. Hence at corresponding times the speeds along the perpendicular are always greater, which is to say that the *speed* is *increasing* faster along the perpendicular than along the incline. That is, the *acceleration* is greater.

On the other hand, in moving from C to A the body takes longer than in falling from C to B. Hence a greater distance is covered in a longer time. All that is required for the speeds to be equal is that the ratio of the time along CA to that of the time along CB should be the same as the ratio of CA to CB. Clearly this *can* happen in spite of the fact that the perpendicular motion (i.e. the increase of speeds along the perpendicular) is faster than the inclined one. For, by the time the body falls from C to B, it would have moved to some point T on the incline, and it is plausible to think that in continuing to move toward A, the excess of the time will equal the excess of CA over CB.

An acceleration problem (F53.13–54; D29–30). Imagine the solar system to have originated in such a way that the planets first moved with straight accelerated motion toward the sun and then, after each had acquired its proper speed, it started to move at this speed in a circular orbit around the sun. The question is whether all planets could have originated from the same place and what was the distance from the sun of this place of origin. To solve the problem one can choose two planets (e.g., Saturn and Jupiter) and from the ratio between their velocities, the distance between their orbits, and the proportion of acceleration for natural motion, one can calculate the distance of their place of origin. Then one checks whether each of the other planets, by falling from this place toward the sun would have acquired the speed and orbit that it actually has. It turns out that the correspondence is amazingly close. The initial calculation involving a pair of planets such as Saturn and Jupiter, would be the following.[70]

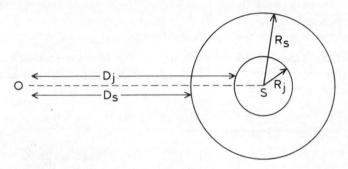

S = sun
O = place of origin of fall
R_j = radius of Jupiter's orbit
R_s = radius of Saturn's orbit
P_j = period of Jupiter's revolution
P_s = period of Saturn's revolution
V_j = orbital speed of Jupiter
V_s = orbital speed of Saturn
D_j = distance fallen by Jupiter
D_s = distance fallen by Saturn
T_j = time of fall for Jupiter
T_s = time of fall for Saturn

(1) $\dfrac{V_j}{V_s} = \dfrac{2\pi R_j/P_j}{2\pi R_s/P_s} = \dfrac{R_j P_s}{R_s P_j}$: this would be known from astronomical data.

(2) $R_s - R_j$ would be known from astronomical data.

(3) $\dfrac{D_j}{D_s} = \dfrac{T_j^2}{T_s^2}$: this would be what Galileo calls "the proportion of acceleration of natural motion".

(4) Since $\dfrac{V_j}{V_s} = \dfrac{T_j}{T_s}$, therefore $\dfrac{D_j}{D_s} = \dfrac{V_j^2}{V_s^2}$, by (3).

(5) Since $R_s - R_j = D_j - D_s$, therefore $\dfrac{D_j}{D_s} = 1 + \dfrac{R_s - R_j}{D_s}$.

(6) From (5), (4), and (1): $1 + \dfrac{R_s - R_j}{D_s} = \dfrac{R_j{}^2 P_s{}^2}{R_s{}^2 P_j{}^2}$.

(7) Therefore, $\dfrac{R_s - R_j}{D_s} = \dfrac{R_j{}^2 P_s{}^2}{R_s{}^2 P_j{}^2} - 1$.

(8) Therefore, $D_s = \dfrac{R_s - R_j}{(R_j{}^2 P_s{}^2 / R_s{}^2 P_j{}^2) - 1}$.

Here, all the quantities on the right side of the equation would be known from astronomical data. The other calculations would involve checking whether for each planet n,

$$\frac{V_n{}^2}{V_s{}^2} = \frac{(2\pi R_n/P_n)^2}{(2\pi R_s/P_s)^2} = \frac{D_s + (R_s - R_n)}{D_s}.$$

The isotropy of space (F54–61.28; D30–37). There is no preferred direction in space. In moving downwards, heavy bodies are moving towards the center of the whole of which they are parts, rather than toward the center of the universe. In the vicinity of other celestial bodies, such as the moon or the sun, their respective parts would fall toward the centers of their respective wholes.

The primacy of mechanical properties (F61.28–75.8; D37–50). The mechanical properties of bodies are the most basic in nature. For example, in the Aristotelian system, the all-pervasive distinction between the terrestrial and the celestial worlds is based on the difference between straight and circular natural motions (F61–3). Or compare the proposition that the earth rotates with the proposition that change derives from contraries; the former is much easier to ascertain and understand (F64). The latter is full of empirical difficulties (F64–5) and of conceptual ones such as: (1) if there exist both changeable and unchangeable bodies in nature, then given that change derives from contraries, the unchangeable bodies should be changeable after all because of their contrariety with unchangeable bodies (F65–9); (2) something (e.g. a whole) can be unchangeable while its parts change (F69–71); (3) if a body has never been observed to change, is it incapable of change or merely unchanged-so-far (F71–5)?

Sunspots and the telescope (F75.9–87.14; D50–62). The telescope makes possible a revolution in astronomy by bringing heavenly objects thirty or forty times closer; for example it led to the discovery of the most significant piece of evidence against the earth/heaven dichotomy, namely sunspots (F80–1). The force of their evidence contrasts with that from comets, which may very well be atmospheric phenomena (F76–7), and with that from the new stars, which might be claimed to constitute changes in the heavens but not changes in some heavenly body (F82–3). The sunspots are significant because they involve what may be regarded as the noblest of the heavenly bodies (F82) and because the changes in their shape and in their speed across the solar disc allow no other interpretation but that they are contiguous to and part of the spherical body of the sun (F77–80). Finally it is incorrect to object (F83–7) that sunspots would be of no use or benefit to man, and hence superfluous, thus making the sun an imperfect body, which cannot be; such anthropocentric teleology is without any justification.

Amount of visible lunar surface (F87.15–95.30; D62–70). A number of facts about the moon are ascertainable relatively easily, for example its spherical shape, opacity, solidity, monthly axial rotation (F90), mutual eclipsing, and mutual showing of phases with the earth. Others require careful reasoning and observation: for example, a little more than half of the lunar surface is visible and the moon's motion is tied to the earth's center (F90). This is shown in the following two figures.[71]

In Figure 1, as the relative position of the moon and a point on the earth's surface S changes through Positions 1, 2, and 3, from the earth's center O one

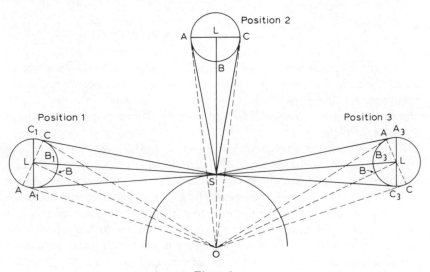

Figure 1.

would always see the same lunar hemisphere *ABC*; in each case the line connecting the centers of the earth and of the moon intersects the lunar surface
at the same point *B*. However from *S* one would see *ABC* only when the
moon is directly overhead (Position 2); when it is in Position 1, one would see
$A_1B_1C_1$; when it is in Position 3, one would see $A_3B_3C_3$. That is, from *S* one
can see CC_1 and AA_3 besides the hemisphere *ABC* (F90–1).

In Figure 2, line *SOS* is in the plane of the ecliptic, *LOL* in the plane of
the moon's orbit. When the moon is north of the ecliptic (left), it becomes
possible to see the region *AY* near the moon's south pole (if the moon and
sun are in opposition); similarly, when the moon is south of the ecliptic
(right) it becomes possible to see region *CZ* near the north pole (F90–1).

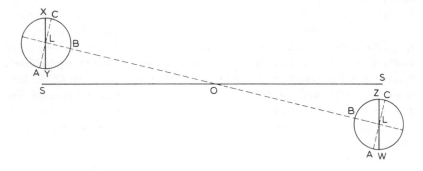

Figure 2.

This can be confirmed by observation. It is possible with the telescope to
see on the lunar disc two diametrically opposite spots, in a northwest, south-
west direction. These spots, which are very near the edge of the disc, can be
observed to vary by a factor of two or three in their apparent distance from
that edge (F90–1).

The moon's mountainousness (F95.31–112.28; D71–78). The mountainousness of the moon can be observed rather directly by the telescope (F112).
The observable details cannot be explained in terms of variations of opacity
such as are found in mother of pearl (F111); nor can mountainousness be
ruled out of existence by the doctrine that heavenly bodies are perfect and
hence perfectly spherical, since shape has nothing to do with perfection
either in fact (F109–10) or even according to the Aristotelian supporting
argument (F109). Moreover, if the moon were perfectly spherical, it would
reflect light in all directions and would transmit only a ray to an observer on

earth, for whom it would be invisible because of the distance; whereas its roughness would allow its being seen the way it is (F95−109).

The moon's secondary light (F112.29−122.29; D87−97). The reflecting power of the earth is something that can be established by a judicious comparison of observations, namely of the moon and of clouds or mountains during the day and of the moon during the day and at night (F112−5). So, the moon's secondary light is to be explained as coming from the earth (F115), and not as being the moon's own light (F115−16), or as deriving from others stars (F116−17), or as coming (directly) from the sun (F117−21).

The problem of moon spots (F122.30−131; D97−105). What is the nature of the moon's large dark spots visible with the naked eye (F124)? Though we can observe with the telescope that the moon surface in those spots is rather flat, we have no way of knowing whether this is sufficient to make them appear darker than the surrounding regions (F125). The effect would result if these dark regions were covered with water (F123−4, 125) or with forests, or with rocks of a naturally darker color (F125). However, there seems to be no water and no life similar to ours on the moon since the lunar day is one month long, the seasons are very insignificant, and no clouds are visible with the telescope (F125−6). Moreover, one should not expect that the nature of these spots is the same as the darker appearances which the oceans would show to someone observing the earth from the moon; for so requires the richness of nature and the omnipotence of God.

INERTIA AND CIRCULARITY

One could proceed with this type of analysis for the rest of the book. However, before doing so we should ask whether it is giving us what we want, in particular whether we are overcoming the tension between relevance and accuracy. Generally speaking I am pretty well satisfied with the accuracy, but I wonder whether it is relevant enough, or at least whether it has the *right kind* of relevance.

Let me begin with accuracy. Part of the justification lies in the page references that I inserted in my account. Such references could not, however, be exhaustive; there simply is not a one-to-one correspondence between every one of my interpretative claims and the sentences in the original text. But there is no reason why there should be, for what is being sought here is not a copy, or a repetition, or even a summary of the original. In an abstract-outline one could provide that sort of one-to-one summary. Here, instead, we are seeking an analysis from a definite point of view, the substantive-scientific. So part of the justification of the above account lies in the nature of the present undertaking.

The accuracy of the above account can also be justified by comparison with other available interpretations. It is not worthwhile to do this for every one of the eleven passages above. Let us select some of the most controversial ones, such as the very first one.

The scientific content of this first passage is usually discussed in terms of the pre-history of the concept of inertia, rather than, as I have done, in terms of the concept of acceleration. This applies both to the negatively critical accounts as well as to the apologetic ones. I will discuss one of each.

By far the most adequate account of the inertial content of Galileo's work has been given by Stillman Drake;[72] his account constitutes an admirable synthesis of textual accuracy and substantive scientific relevance. Drake admits that in this passage Galileo argues explicitly that rectilinear motions cannot be perpetual, only circular ones can; however, he seems to agree that this fact provides some support for the view that in the *Dialogue* the inertial motions of bodies are circular,[73] even though he argues cogently and decisively that the other passages commonly given as evidence for the same view do not in fact support it.[74] The way in which Drake criticizes the alleged scientific import of the present passage is to emphasize its rhetorical content: Drake says that the passage is merely propaganda and that in it Galileo is trying to appear to be more Aristotelian than the Peripatetics, or at least that he begins by accepting some Aristotelian principles so that he can derive un-Aristotelian conclusions therefrom.[75] In short, for Drake "Galileo certainly did not state or believe that the celestial motions would perpetuate themselves merely by being circular in form, but this does not mean that he was averse from letting the philosophers believe that if they wished to".[76]

In criticism of Drake, it may be said that the structure of his argument seems to be the following: the evidence for Galileo's belief in 'circular inertia' based on other passages can be directly refuted;[77] the evidence from the present passage can be criticized only indirectly by arguing that here Galileo is acting not as a physicist but as a propagandist; and Drake's indirect argument (1) implicitly admits that the *scientific* content of this passage is the erroneous view that inertial motions are circular, and (2) requires that one attribute to Galileo the intention of wanting his opponents to reach right conclusions on the basis of wrong reasons. However, to attribute such an intention to Galileo is somewhat far-fetched; its only supporting argument is really that only by such an attribution can one vindicate the scientific content of the present passage; or to be more exact, such an attribution is no better grounded than the one I gave above centering around the concept of acceleration, which is the only thing I wish to show in this context. Moreover, Drake's implicit admission about the specifically scientific content of the passage must be rejected because it lends authority to the following fallacious manner of reasoning: only circular motions can be perpetual, therefore inertial motions are circular. But there is no reason to accept such an argument since it involves an illegitimate equation of perpetual and inertial, or to be more exact, the equation of natural in the sense of ontologically ever-lasting with natural in the sense of logically prior in the conceptual frame-work of classical mechanics.[78] In other words, everyone must, and does, admit that this passage contains the idea that rectilinear motions cannot but circular motions can be perpetual. In my interpretation, I conclude (partly) from this that therefore the passage contains the idea that accelerated motions are important. Koyré and his followers[79] criticized by Drake have concluded that the passage contains the idea that inertial motions are circular; Drake seems to conclude this also, but he destroys the point of this con-clusion by concluding further that therefore the passage is just propaganda. In the light of the additional fact that accelerated straight motion is explicitly discussed in the passage (F43–45), and that acceleration is the topic of several other immediately following passages, the comparative accuracy of my interpretation is thereby vindicated.

Let me justify this further by examining the latest of the negatively critical accounts, namely Dudley Shapere's.[80] Though his book has a number of very serious faults,[81] it does have the merit of discussing explicitly the scientific content of Galileo's work and of explicitly recognizing the danger of reaching simplistic conclusions. So it includes an analysis of the 'inertial' content of Galileo's work[82] and a criticism both of Drake's affirmation, and of Koyré's denial, that Galileo understood the 'essential core' of the inertial concept.[83] Shapere's main conclusion here is that Galileo's relevant views have some similarities and some differences with the concept of inertia of classical physics, but that it is arbitrary to regard any one, or any few ones, of the several features of the concept of inertia as constituting its 'essence'.[84] Part of this claim seems judicious enough and it constitutes a claim about the substantive scientific relevance of Galileo's views as interpreted by Shapere. However, there are other parts to his claim which must be judged in terms of textual accuracy; these are his interpretative claims, to which we now turn.

Shapere considers Galileo's inertial views from several of his works. In his discussion of the *Dialogue*, he reconstructs the 'inertial' content of the latter in terms of five propositions. Of these, only the first four relate to the passage being considered here. They are: (1) *"anything but rest or circular motion is contrary to the orderliness of the universe"*;[85] (2) *"natural circular motion is uniform and perpetual"*;[86] (3) *"the only purpose that might possibly be attributed to straight-line motion is that it could be used to create or restore orderliness"*;[87] and (4) *"if non-circular order-restoring motion does indeed exist, it would be accelerated and rectilinear"*.[88]

Shapere's attribution (2) may be accepted, at least as long as it is interpreted to mean "when motion is both natural and circular, then it is uniform and perpetual", for the evidence is from passages containing arguments about cases of natural circular motion.[89] This is to be contrasted to interpreting Shapere's (2) to mean "natural, i.e. circular, motion is uniform and perpetual", which would be textually groundless. But since Shapere means the former, there is no problem.

The situation is different with his thesis (1). From Shapere's discussion it is clear that this thesis has two parts: the 'anything but' refers to straight motion, which is alleged to be contrary to the orderliness of the universe, while rest and circular motion are alleged not to be contrary to it. From the same discussion, it is also clear that, for Shapere, Galileo equates "contrary to the orderliness of the universe" with 'unnatural', and 'not contrary' with 'natural'. So Shapere is saying that Galileo believes that motion is unnatural when straight, and natural when circular. The problem is with the second part of this thesis, that is with attributing to Galileo a belief in *"the universality of circular motion as natural"*.[90] In fact, though the text does warrant the attribution that straight motion is unnatural,[91] it warrants only the attribution that circular motion *can* be natural.[92] In short, Galileo believes that the only type of natural motion is circular, that *if* motion is natural then it is circular, not that if it is circular then it is natural; and Shapere has simply misinterpreted "not contrary to the orderliness of the universe" as "in accordance with the actual order of the universe".

Shapere's proposition (3) may be accepted, but his (4) is not accurate. His evidence is the passage between the first and the second reference to the allegedly 'Platonic' creation theory.[93] Here he distinguishes two arguments, a general one[94] and a specific one dealing with the continuity of acceleration of falling bodies.[95] Shapere's proposition (4) is a conditional, and he tries to interpret his evidence as two arguments where the consequent clause (acceleration and rectilinearity) is derived from the antecedent clause (non-circular order-restoring motion). That is, for Shapere, the passage he uses is one where the main topic is non-circular order-restoring motion, and the claim being made is that this motion would be accelerated and straight. However, it is really the latter that is the main topic, and the claim being made is that such motion would not be everlasting. In other words, Galileo's point is not that *if* there is non-circular order-restoring motion *then* it would be accelerated and rectilinear but rather that *if* there is accelerated rectilinear motion *then* it is not perpetual. This is so because the straight accelerated motion must have a source-origin of the force field (*"luogo desiderato, cioè dove l' inclinazione naturale lo tira"*[96]) so that when this place is reached the accelerated motion stops or there is a change to circular uniform motion, as in a body rolling down an inclined plane and then moving uniformly along the horizontal.

The accuracy of Shapere's interpretation is prima facie suspect when the reader notices that he supports his attribution of proposition (4) to Galileo by attributing to him arguments that are described by Shapere as being fallacious in several important ways. This is initially suspicious in the sense that if one cannot interpret some passages as providing actual argumentative support for a certain conclusion, then how is one going to support one's interpretative claim that the author held that conclusion? Even if one could find an explicit assertion by the author stating that a certain proposition is his conclusion, why should one attach more weight to this evidence than to the evidence of the actual content of the passage? It would seem that the actual content of a passage is what is really contained therein, not what the author may have said that it contains. At any rate, since in this passage Galileo nowhere makes the needed explicit assertion, there would seem to be no reason at all for attributing to him Shapere's proposition (4).

In fact, when one examines Shapere's evidence more closely one finds that it is based on several textual misunderstandings. For example, in his discussion of Galileo's general argument[97] Shapere claims that

the argument does *not* prove that a body moving in natural straight-line motion will continually accelerate; all it shows (assuming the Principle of Continuity, of course) is that, *if* the body is to arrive at any particular finite non-zero velocity, it will have to pass successively through all lesser velocities; it does not show that when any given speed is reached, the body will continue to increase (rather than maintain or decrease) its speed beyond that until it arrives at its goal.[98]

The passage Shapere refers to is the following:

Every body constituted in a state of rest but naturally capable of motion will move when set at liberty only if it has a natural tendency toward some particular place; for if it were indifferent to all places it would remain at rest, having no more cause to move one way than another. Having such a tendency, it naturally follows that in its motion it will be continually accelerating. Beginning with the slowest motion, it will never acquire any degree of speed (*velocità*) without first having passed through all the gradations of lesser speed – or should I say of greater slowness? For, leaving a state of rest, which is the infinite degree of slowness, there is no way whatever for it to enter into a definite degree of speed before having entered into a lesser, and another still less before that. It seems much more reasonable for it to pass first through those degrees nearest to that from which it set out, and from this to those farther on.[99]

Shapere apparently has taken the phrase 'continually accelerating' to mean 'everlastingly accelerating' rather than 'nondiscontinuously accelerating'. Though in the abstract these are possible meanings of the adverb 'continually', in the context in which the above quoted argument occurs[100] Galileo's main claim is that straight accelerated motion is both nondiscontinuous and non-perpetual, but rather transitory and instrumental for achieving a given 'determinate velocity'[101] in uniform, circular motion which alone can last forever. Moreover, if one checks the original Italian text, one finds that the phrase used is *'continuamente accelerando'*,[102] which is much less ambiguous than 'continually accelerating' and closer to 'continuously accelerating'.

METHODOLOGY

On the basis of the critical comparison just given, and of others that could be made for the other passages, I feel that the accuracy of my account of the scientific content of the First Day is vindicated. As I suggested earlier, what I am dissatisfied with is its relevance. Let us examine in turn the various passages as I have reconstructed them. The first (F23–45)[103] stresses the importance of acceleration; however, though it is indeed a fact of contemporary science that acceleration is a phenomenon and concept of crucial importance, it is clear that the Galilean justification given in the passage is of merely historical interest; that is not the way a contemporary scientist would justify its importance. Similar remarks apply to the second (F45–7, 51–3) and third passages (F47–51): though it is true that acceleration is continuous, and that it must be distinguished from speed, the contemporary scientist would not appreciate the justifications given by Galileo. It is different with the next passage (F53–4) dealing with the acceleration problem of whether it is possible to originate the solar system by letting the planets fall toward the sun and then changing their accelerated, straight motion into a uniform, circular one: the problem is one that would still interest a contemporary scientist, at least in the sense that he would regard it as a useful exercise for beginning students; moreover, Galileo's method of solution would be regarded as acceptable, but several important details of the substance of his solution would be found incorrect. The next two passages, on the isotropy of space (F54–61) and on the primacy of mechanical properties (F61–75), would be found as being correct in their main claim, but uninteresting as regards the justification of those claims. For the next passage on the telescope and sunspots (F75–87), a scientist would agree with the historical significance of this instrument and of this phenomenon, and he could even appreciate Galileo's justification of this significance; however, for him nowadays, though telescopes are still used, and though sunspots are still studied, their present significance is not as great as what it was in the seventeenth century. The next passage (F87–95) emphasizes the fact that slightly more than 50% of the lunar surface is visible from the earth; I believe that both the argument and the conclusion would be of some interest to a contemporary scientist. Similar remarks apply to the next two passages on the moon's mountainousness (F95–112) and on its secondary light (F112–22). The problem of moon spots (F122–31) would be regarded by a scientist as being exactly the right *type* of problem studied in science; Galileo's remarks on the solution would be appreciated more for their tentativeness and judiciousness, than for the correctness of their details.

In summary, the scientific content of the First Day consists of (1) the at least implicit assertion of a number of *facts* that would be still accepted as such and used by a contemporary scientist, (2) the explicit formulation of a number of *problems* recognizably scientific in their character due to their subject matter or the concepts involved, (3) the basic correctness and scientific character of a number of *arguments* given in support of scientifically accepted facts, and (4) the basic correctness and scientific character of a number of *methods* or approaches used or suggested for the solution of recognizable scientific problems. It should be noted that not all of these four kinds of scientific content are present in *each* passage, but rather that each passage contains *one or more* of them. Of course, it would be no problem for the textual analyst to point out *portions* within most of these eleven passages which are devoid of scientific content, but this fact may be judged as irrelevant. The important fact is that every plausibly self-contained discussion in that First Day has scientific content. This is important because it helps to explain the persistent appeal that the *Dialogue* has for scientists. For what is true of the First Day is even more so of the other three: we have already seen above that the allegedly ill-fated Fourth Day addresses itself to scientific problems that may not be completely resolved even today; the wealth of scientific content of the Second Day is usually admitted; the Third Day is equally rich when examined from the point of view of the theory of parallax and the practice of the evaluation of parallax measurements.[104]

Generalizing, we may say that the scientific content of the *Dialogue* consists of the specific facts, problems, arguments, and methods mentioned therein which have some substantive correspondence with the facts, problems, arguments, and methods found in the latest science textbooks or contemporary scientific practice. This has interesting implications for the theory of science and the question, What is science? For what it means is that science is a collection of facts, problems, arguments, and methods which have substantive inter-historical similarities. To say this involves denying that science is a purely formal enterprise in the sense that it would be possible for a theory to be scientific if it was held for valid (though factually incorrect, or somewhat incomplete) reasons or had been arrived at in a proper manner. This is a very plausible and widespread view; it corresponds to what Joseph Agassi calls 'conventionalism',[105] and one version is Thomas Kuhn's philosophy of science;[106] nevertheless, it does not receive support from our examination of Galileo's *Dialogue* coupled with scientists' reaction to this book.

Another consequence of our analysis is a denial of the view that the essence of science is its factual truth-claims, so that those people and theories are scientific whose claims happen to correspond (perhaps approximately) to the substantive content of present-day scientific knowledge. This is the philosophy of science that Agassi calls 'inductivism'.[107] According to this view, what matters most is that one's results be right, independently of the rationale or methods underlying them. Against this view our analysis suggests that truth-content must be combined with proper method and/or rationale.

Third, this study of the scientific content of Galileo's *Dialogue* suggests certain refinements to the distinction between the context of discovery and the context of justification. I have discussed elsewhere[108] the contextual though nonarbitrary character of this distinction. What needs to be noted here is that it would be a mistake to regard a book as belonging solely to the context of justification. For it has been shown above that, besides the justifications found in a book, what can make an impression on scientific judgment is its problems or its mere claims. Moreover, even when the context is obviously one of justification, the most striking aspect of such a justification may be the fact that certain ideas were formulated in the course of the justification. In other words, in the course of proving a proposition p one may find himself formulating some other proposition q such that though p is false, and the proof incorrect, nevertheless the conception of q is scientifically significant. Or to give a concrete application, suppose that the whole *Dialogue* is categorized as an attempt to prove Copernicanism, and hence placed in the context of justification; in the course of such an attempted proof one may find himself formulating the principle of mechanical relativity, or of conservation of motion. Then the same book constitutes context of discovery, from the point of view of these principles. Or, suppose that we regard the Fourth Day as a justification of the proposition that the earth's (double) motion is the (primary) cause of the tides; even if one regards such a proposition false, and the justification fallacious, one can be struck by the discovery inherent therein of the necessary role of secondary factors relating to the natural period of the local sea basin and to the differential effects of water depth. In other words, from the point of view of the distinction between the context of discovery and the context of justification, a book like the *Dialogue* belongs at least as much in the former as in the latter. This is so because the book though not unconcerned with justification, is full of that fluidity and plasticity of thought that characterizes the context of discovery.

Having said this much, my dissatisfaction with the scientific relevance of the scientific content so far considered lingers on. It is not the case that such substantive scientific content is not relevant, but rather that there is another point of view from which the book's scientific relevance is so much greater that one sees little point in continuing to elaborate this first more direct kind of scientific content; that point of view is that of methodology. That is, the book is full of methodological discussions about scientific inquiry. Such methodological content might be regarded as a second, indirect, higher-level kind of scientific content. So far we have examined only the book's scientific content at the level of substantive scientific concepts, problems, and ideas. However, scientists are even more interested in extracting methodological lessons about the proper scientific procedure and the nature of scientific knowledge. In fact, as I shall soon show, even when a scientist refers to the substantive scientific content, he usually does so for the purpose of making a methodological point. Since the book's methodological content would be of interest to philosophers as well as scientists, it seems advisable to concentrate on it. In fact, since discussions about the nature of knowledge and the means for acquiring it are also in the domain of what is ordinarily called 'philosophy', we may also say that the book has great philosophical content and that this content is its methodology. Thus, the book's true (i.e. more important) scientific content is its methodological content, which coincides with its (only type of) philosophical content. Let us not forget, however, that it is a very important fact that the book possesses great scientific content of the first kind. For it is this kind of content that makes it possible to regard its methodological content as dealing with *scientific* methodology. Before examining the scientific-methodological-philosophical content (next chapter), let us however show that even for a scientist the substantive scientific content often has primarily a methodological significance. I shall take the examples of Newton and Einstein with which this chapter began.

In the *Principia*, Newton's reference to Galileo occurs in the scholium that follows the statement of the three laws of motion and six corollaries;[109] the scholium begins by saying, "Hitherto I have laid down such principles as have been received by mathematicians, and are confirmed by abundance of experiments. By the first two Laws and the first two Corollaries, *Galileo* discovered . . . ."[110] The section referring to Galileo ends as follows: " . . . On the same Laws and Corollaries depend those things which have been demonstrated concerning the times of the vibration of pendulums, and are confirmed by the daily experiments of pendulum clocks".[111] Then Newton goes on to say, "By the same, together with Law III, Sir Christofer Wren, Dr. Wallis, and Mr. Huygens, the greatest geometers of our times, did severally determine . . . ."[112] The first two laws are the law of inertia and the law of force; the first corollary is the principle of the composition of velocities, the second is the principle of the composition and resolution of forces. I suggest that the function of Newton's reference to Galileo is to serve as a methodological argument in support of the first two laws and corollaries. The structure of Newton's argument is the following: In his work on the law of squares, the parabolic trajectory of projectiles, and the period of vibration of pendulums, Galileo used and accepted the first two laws and corollaries. Therefore, in dealing with the problems discussed in this book, we should accept the first two laws and corollaries. The argument, of course, presupposes two things: that Galileo is an appropriate scientific model, and that the situation he was in is appropriately similar to Newton's. That is, the argument is simultaneously an argument from authority and from analogy. The truth of Newton's premise here is the sort of thing that historians have disputed. At any rate, there is no question that historical investigations can be of great service to a scientist for the establishment of such propositions on which such arguments can be grounded. The plausibility of the *inference* of course depends on the two presuppositions. In Newton's time the appropriateness of using Galileo as a scientific model may have been a matter of some controversy, nowadays it no longer is; whereas the appropriateness of the analogy appealed to by Newton neither was nor is disputable, since the two situations involve obviously similar problems about the motion of bodies. However, for other such arguments that a scientist may want to give, the analogy between the two situations may be subject to question.

Turning now to Einstein's Foreword to Drake's translation of the *Dialogue*, Einstein's argument quoted above[113] may be reconstructed as follows:

(E1) In the *Dialogue*, Galileo rejected the hypothesis of a center of the universe for the explanation of the fall of heavy bodies.

(E2) The hypothesis of an inertial system for the explanation of the inertial behavior of matter is analogous to the one being rejected in Galileo's *Dialogue* (because both hypotheses introduce a conceptual object which (1) does not have the same kind of reality that matter or fields do, and (2) affects, without being affected by, the behavior of real objects).

(E3) Therefore, this hypothesis about an inertial system should be rejected (insofar as it is as unscientific as the one rejected by Galileo).

This argument is very explicit in what Einstein says, in fact he even gives a subargument to justify the appropriateness of the analogy (E2). He takes for granted the propriety of Galileo as a model, and his historical claim (E1) seems to contain no obvious textual inaccuracies.

Such methodological arguments are common in scientific research, and they have an obvious 'relevance'. In examining the methodological content of a book like the *Dialogue* one cannot, however, provide specific instances of such methodological arguments; but one can provide the groundwork for them. That is, one can describe or reconstruct the various scientific discussions and simultaneously formulate in general (though judiciously qualified) terms a methodological claim which each scientific situation illustrates. The scientist can then use such a series of methodological reconstructions by determining which one (or which ones) of them applies (or apply) to the situation or problem he himself faces. In such a determination he can be assisted by deciding whether his own case is an instance of the general claim formulated in the methodologist's reconstruction. A book like the *Dialogue* lends itself very well to this kind of methodological analysis because it is full of explicit methodological and epistemological remarks; hence we have two ways of checking the reliability of our reconstructions, namely whether they do justice to Galileo's explicit methodological remarks, and whether they do justice to the concrete scientific discussions in the course of which such remarks occur. Such a methodological reconstruction for everything discussed in the *Dialogue* is provided in the next chapter.

NOTES

[1] I. Newton, *Mathematical Principles of Natural Philosophy*, pp. 21–22.
[2] I. B. Cohen, 'Newton's Attribution of the First Two Laws of Motion to Galileo', p. XXVI.
[3] *Ibid.*, p. XXVIII.
[4] A. Einstein, 'Foreword', in G. Galilei, *Dialogue Concerning the Two Chief World Systems*, tr. Drake, pp. xi–xiii.
[5] *Ibid.*, p. xiii.
[6] Quoted in S. M. Uzdilek, 'Galileo Galilei, The Founder of Experimental Philosophy and . . . ', p. 230.
[7] B. Kouznetsov, 'Galilée et Einstein. Prologue et Épilogue de la Science Classique', p. 63.
[8] S. Drake, editor and tr., *Galileo Against the Philosophers*, p. xi: "since it can now be shown beyond reasonable doubt that Galileo's important work on terrestrial physics was essentially completed before he ever applied the telescope to astronomy, it would follow that . . . not an unreasonable zeal for a system incapable of scientific demonstration at the time, but patient study of observational data, first in physics and then in astronomy, led Galileo to effect his synthesis of a new physics and the new astronomy of his famous *Dialogue* of 1632." For the exhaustive evidence, see Drake's *Galileo at Work*.
[9] W. L. Wisan, 'The New Science of Motion: A Study of Galileo's *De motu locali*', p. 298.
[10] A. Koyré, *Études Galiléennes*, p. 212.
[11] E. McMullin, 'Introduction: Galileo, Man of Science', p. 3 .
[12] *Ibid.*
[13] *Ibid.*, pp. 35–42.
[14] *Ibid.*, p. 35.
[15] *Ibid.*, p. 43.
[16] R. P. Feynman, R. B. Leighton, and M. Sands, *The Feynman Lectures on Physics*, Vol. I, p. 14–1.
[17] McMullin, 'Galileo, Man of Science', pp. 31–35.
[18] *Ibid.*
[19] *Ibid.*, pp. 35–42.
[20] A. Koestler, *The Sleepwalkers*, pp. 473–479.
[21] P. Feyerabend, *Against Method*.
[22] D. Shapere, *Galileo: A Philosophical Study*, p. 105.
[23] W. R. Shea, *Galileo's Intellectual Revolution*, p. 117.
[24] G. de Santillana, 'Introduction', in G. Galilei, *Dialogue on the Great World Systems*, tr. Salusbury-Santillana, p. XXX; cf. p. 349, n. 34. See also his *Crime of Galileo*, pp. 174–182.
[25] For recent statements see C. Perelman and L. Olbrechts-Tyteca, *The New Rhetoric*; also Benedetto Croce, *Philosophy of the Practical*; idem, *La Poesia*; also M. Natanson and H. W. Johnstone, Jr. (Eds.), *Philosophy, Rhetoric, and Argumentation*.
[26] The classic locus of such an approach is the works of Pierre Duhem. The best recent elaboration, partly grounded on new textual evidence, is the work of W. A. Wallace, such as 'Mechanics from Bradwardine to Galileo'; 'Galileo and Reasoning *Ex Suppositione*'; and 'Galileo and the *Doctores Parisienses*'.

[27] See, for example, S. Drake's 'Galileo's Discovery of the Law of Free Fall'; 'Galileo's Experimental Confirmation of Horizontal Inertia: Unpublished Manuscripts'; 'The Uniform Motion Equivalent to a Uniformly Accelerated Motion from Rest'; 'Galileo's New Science of Motion'; *Galileo Studies,* and *Galileo at Work.*

[28] See, for example, R. J. Seeger, *Men of Physics: Galileo Galilei, His Life and Works,* pp. 36–37; and idem, 'On the Role of Galileo in Physics', p. 36.

[29] Examples of attempts to examine the post-classical physics scientific content of Galileo's work are B. G. Kouznetsov's 'L'idée d'homogénéité de l'espace dans le *Dialogo* de Galilée'; 'Galilée et Einstein. Prologue et Epilogue de la Science Classique'; 'Style et Contenu de la Science'; and 'Le soleil comme centre du monde, et l'homogénéité de l'espace chez Galilée'. Also, B. M. Kedrov and B. H. Kouznetsov, 'La logique de Galilée et la logique de la physique actuelle'. Also E. Agazzi, 'Fisica galileiana e fisica contemporanea'. The passage on semicircular fall in the *Dialogue* (pp. 162–67 in Drake's translation) is a good example of how concern with classical physics can blind one to its content from the point of view of modern physics. One can very easily read into it the kind of force-free geometrical physics that is characteristic of kinematical general relativity, whereas it is commonly taken to indicate that Galileo's view of 'inertia' did not coincide with Newton's.

[30] For example, at the beginning of the First Day Galileo suggests that the solar system may have originated by letting the planets fall with uniform acceleration from their place of creation toward the sun until they reached their respective orbits, at which time and place the motion they had acquired must have been changed from straight and accelerated to circular and uniform. In this case there is no conflict between classical and modern physics, and the suggestion is in some sense wrong. However, in trying to understand what Galileo has in mind in this passage, Newton-minded scholars uncritically use Kepler's third law in reconstructing Galileo's reasoning. For example, in 'Galileo, Newton, and the divine order of the solar system', I. B. Cohen considers no fewer than five different reconstructions each of which, however, presupposes Kepler's third law (pp. 212–218). I believe Cohen's justification speaks for itself: "In this presentation I have, of course, introduced Kepler's third or harmonic law, which seems not to have been known to Galileo. This anachronistic procedure is wholly justifiable, however, since the third law is a quite accurate representation of the relation among planetary data as known to Galileo and other Copernicans" (p. 213)! The same approach is taken by S. Nakayama, in 'Galileo and Newton's Problem of World-Formation'. By contrast, in 'Galileo's "Platonic Cosmogony" and Kepler's *Prodromus*', S. Drake, with more adequacy, interprets Galileo's cosmogony as actually involving a *groping toward* Kepler's law (pp. 185–187); Drake's article is primarily an analysis of previously unavailable documents. However, there is no documentary excuse for the Newtonian readings of the passage in the *Dialogue* since my reconstruction below shows that by merely considering this passage it is possible to give an interpretation which may be said to correspond to the tip of the iceberg uncovered by Drake.

[31] See Chapter 9 for a critique of Koyré, and Chapter 10 for the others.

[32] In my *History of Science as Explanation,* especially pp. 131–141, and 229–238. Cf. J. Agassi, *Towards an Historiography of Science,* esp. pp. 1–28. I might add that what Agassi calls 'conventionalism' (pp. 28–48) is analogous to the historical attitude as I have just described it; cf. my *History of Science as Explanation,* pp. 131–141.

[33] E. Mach, *The Science of Mechanics,* p. 263.

[34] *Ibid.*, p. 264.

[35] V. Nobile, 'Sull' argomento galileiano della quarta giornata dei *Dialoghi* e sue attinenze col problema fondamentale della geodesia', especially p. 432.

[36] H. L. Burstyn, 'Galileo's Attempt to Prove that the Earth Moves'.

[37] *Ibid.*, p. 182.

[38] *Ibid.*, p. 164.

[39] *Ibid.*, p. 167.

[40] E. J. Aiton, 'On Galileo and the Earth-Moon System', pp. 265–266.

[41] H. L. Burstyn, 'Galileo and the Earth-Moon System: Reply to Dr. Aiton', p. 401.

[42] H. L. Burstyn, 'Galileo's Attempt to Prove that the Earth Moves', pp. 168, 181.

[43] *Ibid.*, p. 174.

[44] E. J. Aiton, 'Galileo and the Theory of the Tides: Comments by E. J. Aiton'.

[45] H. L. Burstyn, 'Galileo and the Theory of the Tides: Reply by Harold L. Burstyn'.

[46] G. Galilei, *Dialogue Concerning the Two Chief World Systems,* tr. S. Drake, p. 427. Cf. Galilei, edited by A. Favaro, *Opere* 7, 453.

[47] E. J. Aiton, 'Galileo's Theory of the Tides', pp. 57–58.

[48] Cf. H. L. Burstyn, 'Galileo's Attempt to Prove that the Earth Moves', p. 169.

[49] W. R. Shea, *Galileo's Intellectual Revolution.* About one-half of the book deals with the *Dialogue.* More recently, a more adequate and original account of the tidal argument has been given by H. I. Brown's 'Galileo, the Elements, and the Tides'.

[50] Shea, *Galileo's Intellectual Revolution,* p. 184.

[51] *Ibid.*, p. vii.

[52] *Ibid.*

[53] *Ibid.*

[54] *Ibid.*, p. 185.

[55] *Ibid.*

[56] *Ibid.*, p. 186.

[57] *Ibid.*, p. 175.

[58] *Ibid.*

[59] E. J. Aiton, 'Galileo's Theory of the Tides', p. 46.

[60] A. Koestler, *The Sleepwalkers,* pp. 464–466.

[61] Shea refers to E. Mach, *The Science of Mechanics,* pp. 262–264, which corresponds to Chapter II, Part IV, Section II of the ninth German edition.

[62] Mach, *Science of Mechanics,* p. 263.

[63] *Ibid.*

[64] *Ibid.*, p. 264.

[65] Shea, *Galileo's Intellectual Revolution,* pp. 175–176. The letter is the one of September 5, 1633 found in G. Galilei, edited by Favaro, *Opere* 15, 251–252.

[66] Shea, p. 176.

[67] *Ibid.*, pp. 177–178.

[68] G. Galilei, *Dialogue,* tr. Drake, pp. 428–436; idem, edited by Favaro, *Opere* 7, 457–462.

[69] Page numbers within the text of my interpretations refer only to Vol. 7 of Favaro's edition of Galileo's *Opere,* whether or not the numerals are prefixed by 'F', which is short for 'Favaro'. References for the main subdivisions are given immediately after my chosen titles for these sections; note that for these section titles two sets of page numbers are given: the ones denoted by 'F' refer to Galileo's *Opere,* Vol. 7, and contain

line references as well; the ones denoted by 'D' refer to Drake's translation. Thus, for example, the references to the third section (on acceleration *vs* speed) is 'F47.10–51.18; D23–27' and means that the section goes from p. 47, line 10 to p. 51, line 18 in Favaro's edition, and from p. 23 to p. 27 in Drake's translation.

[70] The figure and the computation are not, of course, explicitly in the text.

[71] The figures and the arguments are, of course, not explicitly in the text.

[72] S. Drake, 'Galileo and the Law of Inertia'; idem, 'Semi-circular Fall in the *Dialogue*'; idem, *Galileo Studies*, pp. 240–278.

[73] Drake, *Galileo Studies*, pp. 252–53.

[74] *Ibid.*, pp. 253, 257–278.

[75] *Ibid.*, pp. 253–254.

[76] *Ibid.*, p. 254.

[77] *Ibid.*, pp. 256–278.

[78] My view here is similar to the one held by Libero Sosio when he stresses the importance of distinguishing between 'cosmological-architectonic' and dynamical considerations; see 'Galileo e la cosmologia' in G. Galilei, *Dialogo sopra i due massimi sistemi*, edited by L. Sosio, pp. IX-LXXXVII, especially pp. L-L1; J. A. Coffa is making essentially the same point when he states that "the heavy task facing the circularist trying to rest his case on the present passage is to show that it has inertial implications at all"; see his 'Galileo's Concept of Inertia', p. 281.

[79] A. Koyré, *Etudes Galiléennes*, pp. 161–341, esp. 205–211. W. R. Shea, *Galileo's Intellectual Revolution*, pp. 116–121. Though Shea does not really discuss the scientific content of the passage, but rather approaches it from the point of view of the history of Platonist anti-Aristotelianism, he does not give up the opportunity of stating in the course of his discussion that "the triumph of circularity is complete, but it is purchased at the price of excluding the possibility of rectilinear inertia" (p. 118).

[80] D. Shapere, *Galileo*, pp. 87–125, esp. pp. 87–98.

[81] See M. S. Mahoney, 'Galileo's Thought'; C. S. Schmitt, 'Review of Shapere's *Galileo*'; P. Strømholm, 'Galileo and the Scientific Revolution'; and my 'Philosophizing About Galileo'. Strømholm goes so far as saying that Shapere's book deserves 'condemnation'. My own analysis, while severely critical is simultaneously an explanation and implicitly a defense from certain kinds of criticism.

[82] Shapere, *Galileo*, pp. 87–121.

[83] *Ibid.*, pp. 121–125.

[84] *Ibid.*

[85] *Ibid.*, p. 87; italics in original.

[86] *Ibid.*, p. 89; italics in original.

[87] *Ibid.*, p. 91; italics in the original.

[88] *Ibid.*, p. 93; italics in the original.

[89] Galilei, *Dialogue*, tr. Drake, pp. 31–32.

[90] Shapere, *Galileo*, p. 89; italics in the original.

[91] Galilei, *Dialogue*, tr. Drake, pp. 19–21, and 31–32.

[92] *Ibid.*, pp. 31–32.

[93] Galilei, *Dialogue*, tr. Drake, pp. 20–28.

[94] Shapere, *Galileo*, pp. 93–95; cf. Galilei, *Dialogue*, tr. Drake, pp. 20–21.

[95] Shapere, pp. 95–97; *Dialogue*, tr. Drake, pp. 22–28.

[96] Galilei, edited by Favaro, *Opere*, 7, 44.

[97] *Dialogue,* tr. Drake, pp. 20–21.

[98] Shapere, p. 94.

[99] Shapere, pp. 93, 94; *Dialogue,* tr. Drake, p. 20.

[100] *Dialogue,* tr. Drake, pp. 19–21.

[101] Galilei, edited by Favaro, *Opere,* 7, 45; the Italian phrase is *determinata velocità.*

[102] *Ibid.,* p. 44.

[103] In acdordance with the conventions of footnote 69 above, references consisting of page number preceded by 'F' denote pages from Galilei, edited by A. Favaro, *Opere,* 7.

[104] Concerning the central section of the Third Day (F369ff.) W. Hartner says that "this long and very clear discussion of parallax is a masterpiece". See his 'Galileo's Contribution to Astronomy', p. 190.

[105] J. Agassi, *Towards an Historiography of Science,* pp. 1–28; cf. my *History of Science as Explanation,* p. 131–141.

[106] T. S. Kuhn, *The Structure of Scientific Revolutions.*

[107] Agassi, *Historiography;* cf. my *History of Science as Explanation,* pp. 131–141.

[108] See my *History of Science as Explanation,* pp. 229–238.

[109] I. Newton, *Mathematical Principles of Natural Philosophy,* pp. 21–22. I have suggested above that since Newton almost certainly had read the *Dialogue* but not the *Two New Sciences,* he may be taken to have 'read' these things into the former.

[110] Newton, p. 21.

[111] *Ibid.,* p. 22.

[112] *Ibid.*

[113] Galilei, *Dialogue,* tr. Drake, p. xiii.

CHAPTER 5

THEORY AND PRACTICE:
THE METHODOLOGICAL CONTENT OF
GALILEO'S SCIENCE

Galileo's *Dialogue* is full of explicit remarks about topics such as: (1) the role played in physical inquiry by experience, authority, mathematics, logical analysis, conceptualization, human interests and capacities, causal investigation, explanations, comprehensibility, criticism, simplicity, quantitative and qualitative considerations, open-mindedness, and ignorance; (2) the nature of, and interrelationships among, these things and others, such as the Socratic method, unconscious knowledge, and astronomical investigation; and (3) the distinction between method and results, between discovery and justification, and between theory and practice. Remarks about such matters are universally regarded to be methodological remarks, and when a book contains them with the frequency and intensity that the *Dialogue* does, one has to conclude that such a book is a work on method, whatever its title may be and whatever other dimensions it may possess.

How is one to understand the methodological content of the *Dialogue*? A systematic attempt to understand must be distinguished from the attempt to systematize this methodological content. The former is prior to the latter since systematization is a refinement in understanding, and before a refinement is possible one must have a basic understanding of the fundamentals. Moreover, in order to systematize a set of methodological claims one must first be in possession of them; hence their detection and extraction is prior. Finally, just as it is obvious that the *Dialogue* contains a wealth of methodological remarks, it is equally obvious that no systematization of them is present in the text; hence such a systematization would not be the explanation of the methodological content present therein, but rather the theoretical elaboration of it.

A systematic attempt to understand the book's methodological content should be first an attempt to understand *every* methodological remark made therein; second one should try to understand the remark in the context in which it is made. It seems redundant to state such a requirement except that the context is almost always a nonphilosophical but a scientific one, namely the discussion of a concrete physical problem, and because of this the context has seldom been taken into account by interpreters. Such a contextual analysis has the double advantage that sometimes one can use a very explicit methodological remark to throw light on a concrete scientific discussion or series of discussions, and sometimes one can use the clarity of a particular scientific discussion to make sense of a number of relatively unexplicit, but present, methodological remarks.

Next, a systematic analysis of methodological content must not be confused with the analysis of the philosophical significance. Such significance can be present without methodological awareness, e.g., in the work and accomplishment of such people as Jesus or Euclid, or in the occurrence of such historical events as the French or industrial revolution. In other words, by methodological content I do not mean the philosophy *implicit* in the book, but rather the philosophy explicitly discussed in it, that is, the philosophy which is both more or less explicitly *stated and* illustrated with examples of scientific discussions. In speaking of the greater or lesser degree of explicitness of statement, I am admitting that Galileo's methodological remarks are not always equally explicit; I would also admit the existence of borderline cases difficult to decide. Nevertheless, an analysis of methodological content should always have some evidence from Galileo's words as a reason for attributing to him a methodological claim.

To explain further this distinction between methodological content and philosophical significance, i.e., between explicit and implicit philosophy, I will call attention to one feature of the investigation undertaken below. This investigation consists of an analysis of the entire *Dialogue* into 55 contiguous passages each of which can be reconstructed as the illustration by scientific example(s) of some methodological statement(s). Each passage is evidence that Galileo achieved a very judicious combination of scientific practice and philosophical reflection. Such a science-philosophy synthesis is, to repeat, substantiated 55 times in the book, every section of which can be integrated into it. Thus we have overwhelming evidence of the implicit presence of the philosophy that one of the highest values is to synthesize theory and practice, that is, to enlighten one's experience and to act on one's thoughts. Such a philosophy is a central doctrine of both Socrates's moral philosophy and of Benedetto Croce's purified pragmatism. This means that it helps us to understand Galileo's *Dialogue* by regarding it partly Socratic and partly Crocean. However, in Galileo such a philosophy is totally unexplicit; it would be completely wrong to attribute to him such a doctrine. Unlike this doctrine, all the methodological theses attributed to him in the analysis below are held by him with some awareness. This is the sense in which this analysis provides the methodological content of the *Dialogue*.

Outline

Pages[1]	Methodological Topic
FIRST DAY	
33–8/9–14	Mathematical reasoning and number mysticism
38–57/14–32	Conceptual frameworks
57–9/32–5	Sense experience
59–71/35–47	Aristotle's authority as a logician
71–82/47–57	The role of authority in natural philosophy
82–95/57–70	Teleology and anthropocentrism
95–112/71–87	The necessity of experiments
112–7/87–91	Cognitive awareness
117–24/91–98	Different categories of error
124–31/98–105	The powers of human understanding
SECOND DAY	
132–9/106–14	Independent-mindedness and the misuse of authority
139–50/114–24	Simplicity and probability
150–9/124–33	Open-mindedness and rational-mindedness
159–69/133–43	Logical analysis
169–80/143–54	Reasoning vs experience, and the Socratic method
180–8/154–62	The Socratic method: understanding vs information-knowledge
188–93/162–7	Fanciful vs relevant digressions
193–6/167–71	Unconscious knowledge
196–206/171–180	Hypothetical and circular reasoning
206–9/180–3	Critical experimentalism
209–14/183–8	The proper time for experiments
214–7/188–90	Criticizing and understanding
217–23/190–7	Knowledge and recollection
223–9/197–203	Geometry and philosophy
229–37/203–10	Mathematics and physical reality
237–44/210–8	Causal investigations
244–60/218–33	Physico-mathematical synthesis
260–72/233–47	Comprehensibility and truth
272–81/247–56	The deception of the senses
281–8/256–64	Inconceivability-claims and facts
289–90/264–6	The primacy of nature, over man
290–3/266–9	Logical distinctions
293–8/269–75	The superficiality of abstract answers

METHODOLOGICAL ANALYSIS AND RECONSTRUCTION[2]

33–8: *Mathematical reasoning and number mysticism*

Mathematical reasoning should not be confused with number mysticism any more than the latter should be confused with Pythagorean or Platonic mathematicism; hence mathematical reasoning should not be excluded in principle from the study of nature. An example of number mysticism is the Aristotelian justification of the three-dimensionality of the world based on the perfection of the number three. An example of genuine mathematical reasoning is the proof that the world has three dimensions because there can be at most three mutually perpendicular lines through a given point. The intuitive clarity (F36) of the latter example shows that it is not impossible to use mathematical arguments in natural philosophy.

38–57: *Conceptual frameworks*

Conceptual frameworks (*'precetti d'architettura'* [F40, F42–3]) should be formulated independently of empirical considerations. Aristotle's theory of simple (natural) motion (F38–43) is a conceptual framework because it is the conceptual foundation of the earth-heaven distinction. It depends on empirical considerations insofar as it equates (1) simple lines with straight and circular thus excluding the cylindrical helix; (2) straight with up or down; (3) circular with around the center of the universe; (4) different simple motions with natural motions of different simple bodies; and (5) bodies having simple natural straight motion with earth and fire. A more correct conceptual framework is the theory of universal circular motion (F43–57) according to which, in a well-ordered universe, all movable bodies must move with circular motion. The basis of this theory is the following conceptual distinction: straight and circular motions are two different stages of natural motion; straight motion can be acquired naturally but cannot naturally continue forever, whereas circular motion can naturally continue forever but cannot be acquired naturally without straight motion.

57–9: *Sense experience*

Sense experience cannot be dismissed, but it should be treated critically, in the sense that it is subject to different interpretations and one should not assume that his own interpretation is the only possible one. Hence it would be naive to reject the theory of universal circular motion on the grounds that it conflicts with the natural up and down motion of terrestrial bodies and thus violates the principle that sense experience should prevail over any reasoning. In fact, sense experience tells us merely that this up and down motion appears to be straight; to interpret it as really being straight involves the uncritical acceptance of the principle that apparent vertical motion implies actual vertical motion; and this implication does not hold as it will be shown later. In other words, naive empiricism is to be rejected in favor of critical empiricism, which involves logical analysis and logical synthesis (reasoning, logical practice) in an essential way.

59–71: *Aristotle's authority as a logician*

Aristotle's authority as a logician is not sacrosanct and at any rate pertains primarily to the domain of logical theory rather than logical practice. His authority as a logical practitioner cannot be grounded on his undisputed authority as a logical theorist, but only on the soundness of his concrete reasoning. The latter can only be tested by the logical analysis of his actual arguments, which analysis one should therefore be free to carry out. If you do this you will discover that his logical practice leaves much to be desired. For example, consider Aristotle's argument that the point toward which heavy bodies fall is the center of the universe. Logical analysis reveals that this argument begs the question because it assumes that, in moving away from the earth's circumference, light bodies are moving either (1) toward a greater circle concentric with the center of the universe, or (2) along lines which extended in the opposite direction pass through the center of the universe. Or, consider Aristotle's argument for the immutability of the heavens based on the natural circular motion of heavenly bodies; the analysis of the conceptual status of natural and of circular motions (which conceptual analysis is a kind of logical analysis) has revealed that all movable bodies must move with circular motions; hence the only conclusion that follows in this context is *either* that the earth like heavenly bodies is unchangeable *or* that change has nothing to do with difference in motion. Or consider the argument from contrariety; logical analysis reveals several faults with this argument: (1) it uses an assumption (the premise asserting a connection between change and contrariety) which is more problematic and questionable than its main intended consequence (namely, that the earth stands still); (2) it may be self-contradictory because it suggests a parallel argument concluding that heavenly bodies are changeable as well; (3) it is groundless to the extent that the contrariety of rarity and density exists in the heavens; and (4) it is ambiguous because the various mentioned "bodies" sometimes refer to *whole* bodies and sometimes to *parts* of bodies.

71–82: *The role of authority in natural philosophy*

Aristotle's authority in natural philosophy has little if any function because acceptance of it implies its rejection in favor of direct sense experience. Both his example and his own words show this. Consider the question of celestial unchangeability. Aristotle's *a priori* arguments in its support have already been examined and found to be inconclusive (F76); moreover, they probably reflect his method of justification rather than his method of discovery, which is probably reflected in his *a posteriori* argument. So consider this argument. The argument is logically invalid (F71–5); but Aristotle's procedure ('modo di filosofare', F75) is such that, if followed today, it leads one to believe that the heavens are changeable. For, his procedure is to conclude that the heavens are unchangeable on the basis of seeing no changes in the heavens; hence, the same procedure would force one to conclude that the heavens are changeable on the basis of seeing some changes; now it so happens that today one can observe changes in the heavens (F76–80); hence acceptance of Aristotle's example in matters of procedure leads one to reject his authority concerning the substantive claim of celestial inalterability, and to rely on sense experience rather than on an authority. As for his words, one of his most important principles is that sense experience should prevail over any reasoning, and the principle is more fundamental than the claim that the heavens are unchangeable. Now, applied to the sense experience possible today and to Aristotle's own reasoning, this principle implies that it is more in conformity with Aristotle to argue "the sky is changeable because so shows us sense experience" than to argue "the sky is unchangeable because reasoning so persuaded Aristotle". Hence acceptance of Aristotle's words in methodological matters leads one to let sense experience prevail over Aristotle's own substantive conclusions, and hence to the rejection of his authority in substantive matters.

It may also be concluded that no authority has any methodological role in natural philosophy, however understandable the security of relying on it is (F81), since Aristotle's procedure in his *a posteriori* argument for celestial immutability is not an isolated example, but rather *a posteriori* arguments are methodologically prior to *a priori* ones, since the former reflect the method of discovery and the latter the method of justification (F75–6).

In other words, the concrete example of Aristotle and the inalterability of the sky shows that if we take the methodological point of view, if we distinguish between an author's methodological claims and his substantive claims, if we distinguish between methodological theory and methodological practice and do not neglect the example of the latter besides the use of the former, if we distinguish between methods of discovery and methods of justification within methodological practice, and if we believe that *a posteriori* reasoning is the typical method of discovery and *a priori* reasoning the typical method of justification, then it is impossible for an authority to have a methodological function.

82–95: *Teleology and anthropocentrism*

Teleological anthropocentrism is the thesis that the whole universe exists for the sake and benefit of man (F84, 85). A good example of reasoning which is both teleological and anthropocentric is the following Aristotelian argument (F83–85; cf. F109): (1) changes would make celestial bodies imperfect, since such changes would not be of any use or benefit to man, and hence they would be superfluous; therefore, (2) for celestial bodies, being unchangeable would be their way of being perfect; (3) but celestial bodies are unchangeable, because celestial changes would be vain and useless, and hence there cannot be any; it follows that (4) celestial bodies are perfect, that is have all perfections. Here, teleological anthropocentrism is presupposed in steps (1) and (3). The doctrine is objectionable for at least two reasons. First, one valid reason for teleological anthropocentrism would be if the place where man lives — the earth — were appropriately special and different from the other bodies in the universe. But, using the nearest celestial body as an example — the moon — it can be shown that it and the earth are rather similar (F87–95): in fact they are both spherical, opaque, lacking their own light, solid, and irregular in their surface; and each illuminates and eclipses the other and shows phases and uneven brightness to it. Second, teleological anthropocentrism leads one to conclusions, such as the perfection of celestial bodies mentioned above, which can be shown to be false. For the moon has a rough rather than perfectly smooth surface, because of the fact that it is visible at all and because of its mountainous appearance through the telescope (F87–8; cf. F95–112).

95–112: *The necessity of experiments*

Experiments are sometimes necessary (F100, 101, 108) in natural philosophy. One such situation occurs when the facts involved are counterintuitive (F96), as in the question of the manner the moon reflects sunlight. For example, (1) if you hang a mirror on a wall and compare the way the mirror and the wall reflect the light, you will see that the wall looks brighter than the mirror from all places except one, where the mirror's reflection goes (F96–8). Or, (2) compare the reflection of sunlight from a flat and from a spherical mirror by comparing their effect on the brightness of a neighboring wall; you will see that the spherical mirror has no noticeable effect on the illumination of neighboring objects (F100–1). Or, (3) fold a sheet of paper and let one part receive light perpendicularly, and the other part obliquely; you will see the latter to be less bright than the former (F105). Or, (4) fold a sheet of paper into two unequal parts, let some light shine perpendicularly on the smaller part and obliquely on the larger, and look at the paper from an angle such that the two unequal parts appear of equal size; you will notice that the obliquely-lit part still looks less bright than the other (F108–9). From the first two experiments one can argue that if the moon's surface were not rough, it would be invisible (F96–105); from the latter two, that if its surface were not rough, a full moon would appear brighter at the center and darker at the edges (F105–9).

Another example of the necessity of experiments in natural philosophy occurs when the relevant facts need to be ascertained with some precision, or when a man-made model of the natural situation is possible, as in trying to explain the uneven appearance of the lunar surface to the telescope by differences of opacity and transparency. Such an explanation cannot be faulted by means of general considerations, but only by the specific details of the telescopic appearances, all of which are reproducible with a solid opaque ball subject to varying illumination, though none with a solid ball of nonuniform transparency (F111): you can see (F112) that the various lighted bulges cast shadows; that the length of these shadows varies depending on the distance of the bulge from the boundary of light and darkness; that this boundary is very irregular; that in the dark region beyond this boundary many bulges are already lit; that the length of the various shadows of the lighted region decreases until there are no more shadows when the whole hemisphere is lit; and that then as the darkness starts growing the same bulges can be recognized and cast shadows whose length increases.

112–7: *Cognitive awareness*

The intellectual contents of the human mind are not always open to introspection or known best to the person involved; this applies both to the real reasons underlying one's beliefs and to propositions that one may know without knowing that he knows them. For example, someone may *say* that the reason why he thinks that the earth cannot reflect sunlight as well as the moon is that the earth is so dark and opaque (F112–3); however, cross-examination reveals that his real reason is that whereas the earth can be seen lit only during the day, the moon can be seen lit also during the night, for from this he is led to the error of comparing the moonlight as seen from a dark place (night) with the earth-light as seen from a lighted place (day), and he naturally concludes that the moon reflects light much more easily that the earth. An example of knowledge without awareness is the proposition that the earth can reflect light as well as the moon (F115); someone may think that he believes the opposite, but cross-examination can show that he knows all the facts from which the correct conclusion easily follows: that the moon appears more shiny at night than during the day, that during the day it appears about as shiny as clouds, and that the earth-light cannot be seen at night. Another example of unconscious knowledge is the proposition that the moon's secondary light is light reflected by the earth onto the moon (F115–6).

These examples also implicitly illustrate the importance of the rationale underlying beliefs for the resolution of disagreements. For when that rationale is seen to be wrong, as in the first example above, it is often an easy matter to see what is the proper conclusion to be drawn, as in the second example above.

117–24: *Different categories of error*

Different categories of errors exist; they are not all equally culpable. The acceptance of an erroneous theory by its originator or at a time when no better theory exists is a relatively minor methodological transgression. However, if a follower defends a grossly erroneous theory at a time when a better one has been published, that can only be an appeal to the ignorant masses. An example of such a methodologically objectionable appeal is the recently resuscitated theory that the moon's secondary light is sunlight which penetrates the somewhat transparent lunar body. Glaring falsehoods asserted by this author are that the secondary light is greater at the edges than in the middle, and that in solar eclipses the portion of the lunar disc outside the solar disc is darker than the rest.

124–31: *The powers of human understanding*

The powers of human understanding are likely to disappoint both the optimists and the pessimists. On the one hand, the human mind is not a measure of what can occur in nature since if you ever try to understand fully some one thing you realize how little you know (as the example of Socrates shows). Moreover, the human mind is infinitely inferior to the divine mind both in number of propositions known and in manner of knowing them; for God keeps infinitely many propositions simultaneously in His mind without having to use a step-by-step knowing process. On the other hand, the human mind can know a few propositions (such as those of mathematics) with an 'objective certainty' (F129) which equals that which God has; and moreover, one cannot but marvel at the various artistic and scientific achievements of man. An example of this limited but real power of the human mind is our knowledge of the moon: though we can think of a number of conditions that might cause the big dark spots on the moon, we have no way of knowing which one it is; and though we can have no idea of what kind of life there is, we know that it cannot be anything like on earth because the moon has very long days and nights (one month combined), no significant seasons, and no rain.

132–9: *Independent-mindedness and the misuse of authority*

The misuse of Aristotle's authority by his followers is laughable and ridiculous.

They often oppose to the evidence of sense experience, Aristotle's mere authority or mere words, rather than other experiences or Aristotle's *reasons;* an example would be for someone who has been shown experimentally that the nerves originate from the brain, to say that he would have to accept it as true if Aristotle had not said that the nerves originate from the heart (F133—4). They also often claim that, and act as if, all knowledge is contained in his books: an example is the Peripatetic who tried to show that the telescope was known to Aristotle in its fundamentals by suitably interpreting, *ex post facto,* the passage where he describes the effect of looking at stars from the bottom of a well (F134—6).

Aristotle's authority results only partly from the strength of his demonstrations, but primarily from lack of independent-mindedness, as well as academic opportunism and the failure to appreciate what might be called epistemological modesty. This last attitude means the failure to see that not only is there nothing wrong with claiming ignorance about certain things, but it is actually better to do so since if one does this then he will be taken more seriously (at least by serious philosophers) concerning the things that he does profess to know (F137). An example of academic opportunism is the Peripatetic who had written a book on the soul in which he denied that Aristotle regarded it as immortal on the basis of certain less well-known passages; when he was told that it would have been difficult for him to get permission to publish, he decided to use certain other Aristotelian passages favorable to immortality (F137—8). It is primarily lack of independent-mindedness which generates the weight of Aristotle's authority because, no matter how strong his arguments, it is we who must evaluate this strength; it follows that in philosophy at least, authorities play no *essential* role (F138—9). It also follows that it is imperative to examine carefully his arguments concerning the question of the earth's motion.

139—50: *Simplicity and probability*

Arguments based on the principle of simplicity are neither worthless nor conclusive, but rather probable. They are not worthless because, in the absence of conclusive arguments, the simplest idea is the most acceptable; they are not conclusive because a single piece of counterevidence or conclusive counterargument is sufficient to refute an idea based on simplicity, and they are probable because nature tends to act by means of the fewest possible operations. For example, the arguments in favor of the earth's diurnal motion are based on the principle of simplicity insofar as they amount to saying that it is simpler (1) to let the earth rotate on its axis at a relatively low speed rather than to let innumerable celestial bodies revolve around the earth at incredibly higher speeds; (2) to let each planet nave only one motion rather than two contrary ones; (3) to let the law of periods of revolution be universally valid, rather than to make an exception for the case of the motion of the uppermost sphere; (4) to let all the fixed stars do one thing − stand still − rather than to let them revolve over unequal orbits; (5) to let the earth change its axis of rotation than to let all fixed stars change their orbits in unison; (6) to let the earth rotate than to coordinate the motions of all fixed stars in such a way as to keep their relative positions unchanged; (7) to prevent the earth's rotation from being transferred to the rest of the universe, rather than to prevent the diurnal motion of the heavens which is transferred through all the spheres from being transferred to the earth; and (8) to let the earth rather than an *ad hoc* highest sphere be the *primum mobile*. These arguments are obviously inconclusive but they do lend some initial probability to the earth's diurnal motion (F148). For this probability to remain, one must show that the apparent counterevidence does not really exist and that the apparently conclusive objections are in fact unsound.

150−9: *Open-mindedness and rational-mindedness*

It is very important to know and understand the arguments, reasons, and alleged evidence against one's view (F153), since its justification is strengthened if undertaken in the light of the awareness of counterarguments (F154−5), and since truth emerges from the conflict of opposing views if this conflict takes the form of the assessment and evaluation of the logical strength of the relevant arguments (F155−8). Let us call open-mindedness the willingness and capacity to know and understand the arguments against one's view, and rational-mindedness the willingness and capacity to accept the views which are supported by the best arguments; and let us call logical analysis the (art of) understanding the structure of arguments and evaluating and comparing their strength. Then we may say that open-mindedness and rational-mindedness are essential in the search of truth, and logical analysis is in turn essential for both of these traits.

The motion of the earth provides a very good example of all this. First the methodological superiority of the proponents as compared to the opponents is obvious insofar as the former are usually knowledgeable of the counterarguments, whereas the opponents are usually not knowledgeable about the favorable arguments (F154−5); moreover, the proponents have even improved the clarity and cogency of the counterarguments (F153, F159).

Second, these counterarguments against the earth's motion are so numerous and so undeniably convincing and conclusive in appearance that it will be a very good test of the power and usefulness of logical analysis to see what results it yields when applied to them.

159–69: *Logical analysis*

The logical analysis of the arguments against the earth's motion yields the following results: (1) (F159–62) Aristotle's first objection is an argument from violent motion and has the following structure: the earth cannot move circularly because such motion would be violent, and hence not perpetual; it would be violent because if it were natural, the parts of the earth would also move circularly by nature, which is impossible since the natural motion of its parts is straight downwards. This argument is a fallacy of equivocation since the clause "the parts of the earth would also move circularly" can mean either that these parts would move around their own centers or that they would move around the earth's center. In the first case, the corresponding proposition, namely that "if the earth's motion were natural, then its parts would also move circularly by nature", would be obviously false, and hence the argument would be unsound by reasons of falsehood of one premise. In the second case, the argument would be unsound by reason of invalidity, since it would contain two incompatible steps; for one step claims that the earth's circular motion would not be perpetual because it would not be natural, thus equating the natural with the perpetual, whereas another step would claim that the earth's parts do not naturally move around its center because their natural motion is straight downwards, and this is so because their spontaneous motion when free to move is straight downwards, thus equating the natural with the spontaneous under certain supposedly natural conditions. The invalidity derives from the fact that this second notion of natural is both inconsistent with the first, and implausible because the conditions under which the spontaneous motion is straight downwards are not really natural but rather contrived artificially or seldom occuring in nature. (2) (F162–4) Aristotle's second argument is another fallacy of equivocation because its conclusion could mean either a denial of the diurnal motion or of the annual motion. (3) Aristotle's third argument is the argument from natural motion and has the following structure: the natural motion of the earth's parts is toward the center of the universe, and therefore the whole earth stands still therein. This argument is groundless since the empirical argument that the earth's parts move toward the center of the universe is circular (as it is shown in F58-61), and since the *a priori* justification of the same premise presupposes an untenable, confused, or inconsistent concept of what it is for motion to be natural (as it is argued in F33–62 and F159–62). (4) (F164–9) Aristotle's fourth argument is the argument from vertical fall, and it is circular when stated in terms of *actual* vertical fall, and invalid when stated in terms of *apparent* vertical fall. The former version has the following structure: The earth must stand still because bodies fall vertically, which they could not do if it moved; and bodies fall vertically, since they are seen to fall vertically and since apparent vertical fall implies actual vertical fall. Now, it turns out that this implication holds if and only if the earth stands still; hence one would have to assume that the earth stands still to justify the corresponding premise in the argument, namely the proposition that if bodies are seen to fall vertically, they really do fall vertically.

The apparent vertical fall version of the argument has the following structure: The earth stands still because bodies are seen to fall vertically; now this could not happen if the earth were moving since if the earth were moving and bodies were seen to fall vertically, then bodies would be moving with mixed motion, around and toward the center; but such mixed motion is impossible since bodies dropped from the mast of a moving ship fall behind. The invalid step in this argument is the last one, which presupposes that the nonoccurrence of mixed motion on a moving ship would imply physical impossibility (or nonoccurrence on a moving earth). Such an implication is invalid because the motion of the ship is violent whereas the earth's diurnal motion would be natural, and because the relevant portion of air does not move along with the ship but would do so (near the earth's surface) if it moved.

169–80: *Reasoning vs experience, and the Socratic method*

Experiments are sometimes unnecessary to ascertain the results of a test, for sometimes it can be argued on the basis of known or more easily ascertainable facts, what these results must be. For example, consider dropping a rock from the mast of a moving ship; the Aristotelians believe that the rock will be left behind, and from this, by analogy, argue that the vertical fall of a rock dropped from a tower implies that the earth stands still. It is doubtful whether they ever made the experiment; but at any rate, it can be shown that on the moving ship the rock will fall at the foot of the mast. Here, the more easily ascertainable facts are that (a) the undisturbed downward motion of bodies on an inclined plane is accelerated, (b) their undisturbed motion up an inclined plane is decelerated (F171–3), and (c) the cause of the motion of projectiles is not the motion of the air surrounding them (F175–80).

From (a) and (b) one may conclude that (1) the motion of bodies on an horizontal plane is conserved if undisturbed, and consequently that (2) the horizontal motion of the rock on the moving ship, even after being dropped, continues if undisturbed. Now from (c) one can conlude that (3) the cause of the motion of projectiles is the 'virtue' impressed on them by the projector, and consequently that (4) the cause of the horizontal motion of the rock, after it has been dropped, would be the horizontal 'virtue' impressed on it by the hand holding it before dropping. Since there is no way in which this horizontal impressed virtue could be disturbed by the vertically downward tendency due to weight, it follows that (5) the horizontal motion of the dropped rock is undisturbed, and hence by (2) that (6) the horizontal motion of the dropped rock will continue, and therefore that (7) the rock will end up at the foot of the mast. It should be noted that such a justification can be effectively carried out by means of the Socratic method of questioning (F171, 176–7).

180–8: *The Socratic method: understanding vs information-knowledge*

The Socratic method (F183) is very effective in making explicit, knowledge which is implicit in the mind. This is not teaching in the ordinary sense, that is in the sense of imparting or conveying new information, but rather in the sense of deriving what was previously not accepted from propositions that are accepted. Alternatively we may say that the knowledge in question is not information or mere knowledge, but understanding.

For example, someone may be skeptical about whether certain projectiles (1) can move faster after they hit the ground than in the air and (2) can move faster over a later than over an earlier segment of their whole path. (1) can be made simultaneously certain and comprehensible by appropriately questioning the skeptic and making him realize(F186) that when the projectile is something like a hoop, that can spin, then the spinning does not add to its forward motion as long as it is in the air, but it does so when it hits the ground if the spinning motion is appropriately in the same direction as the forward motion. (2) can be demonstrated to the skeptic by making him realize that, on falling to the ground, a hoop may hit a rock slanting in the direction of the forward motion and thus may acquire more spinning which is added to the forward motion. Thus from the doctrine of the conservation of motion one can understand the apparently paradoxical consequence that the motion of certain projectiles can sometimes increase, when a basic part of their motion is conserved and other amounts are added by various secondary factors.

188–93: *Fanciful vs relevant digressions*

There are two kinds of digressions: some are justified by their relevance to the main argument and by the logic of the discussion, which is a complicated one where several people are involved. Others are justified by the whim of the writer or persons involved or by the intrinsic beauty rather than by the reasoned tenability of the ideas involved (F188). The digressions up to this point have been typically of the first type; an example of the latter type of digression is the following fanciful (F192) speculation concerning the actual path of a falling body on a rotating earth: that the path is an arc of a circle having as a diameter the line going from the earth's center to the point from which the body was dropped. This is a path that is almost too good to be true (F192) insofar as it means that whether or not a body is dropped or held still on a rotating earth, it would be moving (a) circularly, (b) along distances of the same length, and (c) with the same uniform motion.

193–6: *Unconscious knowledge*

Some knowledge is unconscious (F194) in the sense that it is possible to arrive at it by reasoning (F196), though experience may be necessary for those who are unwilling or incapable (F196). For example, by means of Socratic questioning and hypothetical reasoning involving an analogy with arrows shot from a crossbow mounted on a moving carriage, it is possible to show to someone who initially believes otherwise, that on a rotating earth the range of eastward and of westward gunshots would not differ (F194–6).

196–206: *Hypothetical and circular reasoning*

Hypothetical reasoning is a very valuable skill, but a difficult one to master; deficiency in this skill often results in circular reasoning or arguments that beg the question. For example, when people argue that east and west gunshots would have unequal ranges if the earth were rotating, they are reasoning as if the state of the cannon ball before firing would be one of rest even on a rotating earth; this amounts to a failure to take seriously such rotation when examining the consequences of such rotation. The same is true for the arguments from vertical fall, from vertical gunshots, from north-south gunshots, and from point-blank gunshots. In all these cases, the opponents of the earth's motion pretend to be engaging in hypothetical reasoning examining the   consequences of the hypothesis that the earth moves, and yet there is one crucial step where they disregard this assumption; it is the step where the initial state of the projectile or falling body is taken to be rest.

206–9: *Critical experimentalism*

The uncritical acceptance of alleged experimental results is to be avoided. One way of being critical is to watch out for the possibility that the results have been merely hypothesized in conformity with some theory and for the possibility that the effect is so small that one cannot check its presence experimentally. For example, consider the question of whether point-blank gunshots hit high toward the east and low toward the west. The Aristotelians claim that experience shows this does not happen (and since they think it would have to happen if the earth rotated, they conclude that the earth stands still). Now, by means of a rough calculation based on certain (otherwise objectionable) Aristotelian principles, one can show that the magnitude of the effect would be about 1/25 of a 'braccio', which is experimentally undetectable.

209–14: *The proper time for experiments*

The proper time to make an experiment is when reasoning is incapable of deciding whether a phenomenon would happen, or why it does happen, or how it can happen. For example, consider the phenomenon of the flying of birds on a rotating earth. Even someone who understands very well the motion of projectiles on a rotating earth (cf. Sagredo's superiority over Salviati in F196–206, especially F202 and F204-5), cannot help but think that flying birds are *essentially* different, for just as a (live) bird does not do the same thing a rock does in the case of the natural motion of falling, so the bird would not do the same thing as inanimate bodies in the case of the natural circular motion along with a rotating earth (F212). Experiment can show that, for the latter type of motion, there is in reality no essential difference between an inanimate body and a bird, and that it is an irrelevant difference for them to have different sources of motion, an external projector for the projectile and an internal principle for the bird. The experiment is made under deck on a ship: one observes the behavior of flies, fish, butterflies, dripping water, rising incense, and projectiles when the ship is standing still and when it is moving. One observes no difference deriving from the state of rest or of uniform motion of the ship, and that the flies, fish, and butterflies share the common motion of the moving ship just as easily, naturally, and effortlessly as the inanimate bodies. The experiment is also very effective in answering the geostatic arguments from falling bodies and projectiles, which are criticizable in other ways as well.

214–7: *Criticizing and understanding*

Destructive criticism is more effective when preceded by constructive elaboration which strengthens the idea being criticized. For example, consider the objection to the earth's motion which claims that if the earth rotated, then all bodies on its surface would fly off into the heavens, because turning gives a body an impetus away from the center of the whirling. Before answering this objection it is better to make a logical clarification and a substantive addition. The elucidation is that the objection should properly be stated by saying that if the earth rotated, then the unattached bodies or creatures on its surface could not have come into existence on its surface; this corrects the usual version that if the earth rotated then the unattached bodies would *move away from it*, which is a virtually universally shared formulation, and which is a misstatement since it would show at best that the earth cannot *begin* rotating. The substantive addition is the evidence of a pail full of water, tied at one end of a rope, and swirled from the other end; the water will not fall out of the pail regardless of the angle of the plane of its circular motion; moreover, if a hole is made in its bottom, the water will come out through the hole in all directions, but always away from the center of the swirl.

217–23: *Knowledge and recollection*

Plato's doctrine that knowledge is a kind of recollection is in a sense true. The sense in which it is true is that knowledge is sometimes acquired by reasoning based on facts ascertainable by reflection upon certain chosen aspects of one's sense experience. For example, one can easily conclude that bodies on the surface of a rotating earth would not be extruded into the heavens after having been made, by questioning, to reflect that: (1) when a body which has been made to move along a circle is let go, it acquires a motion tangential to the point of release; (2) as the body starts moving along the tangent, it moves away from the point of tangency much more than away from the circumference, so that to prevent the body from moving away along the tangent all that is needed is for it to move a very small amount toward the circumference; and (3) bodies, however light, do have a tendency to move toward the surface (circumference) of the earth due to their weight.

223–9: *Geometry and philosophy*

Geometry is indispensable in natural philosophy. For example, only by means of geometrical considerations is it possible to understand how and why the falling tendency of a terrestrial body, however small one chooses to make it, is always sufficient to overcome any centrifugal tendency on a rotating earth and thus to keep it on the surface. The main geometrical consideration is that, given an arbitrarily large ratio, one can find a point outside a circle such that the tangent from that point bears that ratio to the portion of the secant outside the circle; other reinforcing considerations are that (1) the geometry of the situation is such that the tangential tendency can be overcome by an arbitrarily small weight, even though the tangential tendency were diminishable to infinity in only one way, whereas the downward tendency were diminishable to infinity in two ways arising from the decrease in weight and from the approach to the point of tangency; and (2) the decrease of the downward tendency due to decreasing weight is not only not faster than linear but much slower.

229–37: *Mathematics and physical reality*

Mathematical truths are about abstract entities in the sense that they are statements about the necessary consequences of certain definitions and axioms. For example, the proposition that a sphere touches a plane in only one point is about abstract spheres and planes in that it is a necessary consequence of the definition of a sphere and the principle that a straight line is the shortest distance between two points. Mathematical truths are also about physical reality, though only conditionally; that is, a mathematical proposition is physically true if the abstract entities it is about happen to exist as material entities in physical reality. For example, the proposition that a sphere touches a plane in only one point is true of physical reality in the sense that *if* there happen to be material spheres and planes, *then* they touch in only one point. Third, mathematical truths are *applicable* to physical reality because and insofar as material entities approach or approximate abstract ones; for when material entities do not approximate one type of abstract entity, they approximate another. For example, if and to the extent that material spheres and planes touch in more than one point, they instantiate abstract spheres and planes that are imperfect, and of these it is equally true in mathematics that they touch in more than one point. Finally, the real problem is to use the proper type of abstract entity in terms of which to interpret physical entities and processes, for though one is sure that the latter must correspond to *some* type of abstract entity or situation which is treatable mathematically, one cannot be sure of which one. It is along these lines that one could properly question the relevance of the geometrical proof given in the previous section.

237–44: *Causal investigations*

Causal investigations are important in natural philosophy. Sometimes one may want to investigate how the cause of a certain effect varies. For example, it is important to realize that in circular motion, the *cause* of extrusion or projection varies directly as the circumference speed when the radius is constant (F238 begin.) and inversely as the radius when the circumference speed is constant (F239, 243). Sometimes one is certain about a fact but may want to know the causal explanation of it. For example, there is no doubt that equal circumference speeds produce a greater tendency to extrusion when the radius is smaller; but it is important to know 'the cause why' (F238) or reason for it (F239); this is to be found by considering that the force needed to keep a given body from escaping along the tangent will be greater when greater is the speed it would need to move from the tangent back onto the circumference; and such speed is greater for the smaller circle since the circumference speeds, and hence the relevant times, are equal. Finally, sometimes a cause can be discovered by comparing its effect to a second effect whose cause is known, and using the way in which the given effect differs from the second one as a clue to how the unknown cause of the given effect differs from the known cause of the second effect. For example, in a steelyard, why does a lighter weight balance a heavier one? The cause is that the lighter weight moves more than the other. For compare the steelyard and the equal-armed balance; in the latter equal weights balance each other because they are moving equally; in the steelyard it is *unequal* weights that are balancing each other; so this must happen because they *move* unequally.

244–60: *Physico-mathematical synthesis*

It is not enough that the questions treated be important and worthwhile; it is essential that the method of treatment be worthwhile (F246). To be worthwhile the method must be both mindful of physical facts and sophisticated mathematically. For example, the problem of a body falling from the moon to the earth is a very interesting one. In dealing with it, however, a recent author has not treated it properly: first, in computing the time of fall he has arbitrarily assumed that its speed of fall would be equal to the speed of diurnal rotation at the moon, instead of determining the actual speed with which bodies actually fall (F246); second, his answer that the time would be six days seems to involve the mathematical absurdity that the radius of a circle is six times longer than its circumference (F247, 258–9). A proper treatment of the question would begin by noting the fact that falling bodies accelerate (F248), showing the mathematical sensitivity of asking by what proportion the velocity increases, ascertaining that the distances from rest vary as the square of the times, that a body falls 100 *braccia* in 5 seconds, and then using these values together with the law of squares to compute the time of fall from the moon as being about 3½ hours (F248–52).

A second example (F252–56) to illustrate the proper combination of physical and mathematical considerations is the double-distance law, which states that a body would move double the distance through which it has fallen from rest if it were allowed to move uniformly at the acquired speed for a time equal to that of fall. The physical considerations involve both a pendulum and a body oscillating about the earth's center in a tunnel going from a point on the surface to the point diametrically opposite to it; here, observation in the case of the pendulum, and physical intuition in the case of the tunneled earth, tell us that the impetus acquired by the body while it undergoes acceleration is such as to enable it to move an equal distance while being *decelerated* by the same amount. The mathematical considerations involve summing up all the speeds during both acceleration and deceleration and equating this sum to the sum-total of constant speeds during an imaginary second half of the motion if it had continued uniformly instead of decelerating; another mathematical consideration involves summing up all the speeds during a given accelerated motion and comparing this sum (measured by a right triangle) to the sum-total for a uniform motion at a speed equal to that acquired during acceleration (measured by a rectangle twice the area of the triangle).

A third example (F256–8) of physical-mathematical synthesis involves the question of whether a pendulum would come to rest even if air resistance is disregarded. The answer is that it would because any physical pendulum is actually a series of pendulums of different lengths, due to the matter in the string, and all these pendulums are trying to vibrate at different rates and hence interfere with each other, with the slower continually slowing down the faster. The physical observation is that pendulums of different lengths vibrate at different rates, the rate being faster the shorter the pendulum; the mathematical consideration is the analysis of a single material pendulum into a series of pendulums of different lengths.

260–72: *Comprehensibility and truth*

The aim of natural philosophy is understanding, not mere knowledge; not merely truth, but comprehensible truth. Hence the discussion of the comprehensibility of an idea is always relevant and important; this is especially true when it is not known whether the idea corresponds to reality, for at this stage incomprehensibility constitutes a strong reason for rejection; when the idea is known to be true, incomprehensibility is merely a reason for doing further work. In other words, comprehensibility and truth are distinct: if an idea is comprehensible, that does not make it true, nor if it is incomprehensible, does this make it false; nevertheless natural philosophy aims at both.

A good illustration of all this is the idea of the earth's rotation and the difficulties that have been raised concerning its cause. First (F260), it would be invalid to argue that the earth cannot move because there would be no way to explain the rotation in terms of either an internal or an external principle; the reason is that ignorance of the explanation does not imply non-existence of the effect; in other words, inexplicability does not imply false-hood.

Second (F260–1), a rotating earth would be no more inexplicable than the fall of heavy bodies, the motion of projectiles, or the alleged diurnal motion of heavenly bodies; for it is no explanation to call their respective causes 'gravity', 'impressed virtue', and 'informing or assisting intelligence'; in other words, knowledge of the name does not imply knowledge of the essence of a cause, or, naming is not the same as understanding.

Third (F262–72), it cannot be argued that an explanation of the earth's rotation would be inconceivable, i.e., that this rotation would be inexplicable-in-principle; an explanation could always be given in terms of an internal principle, by attributing the rotation to nature and eliminating the need for an external cause. The reason is that the idea of the earth's rotation is at least as comprehensible as the diurnal motion of the heavens (which is regarded as comprehensible); moreover, there are no conceptual difficulties with the earth's rotation; what are usually mentioned as difficulties turn out to be nothing but unproblematic consequences of it. In other words, causal in-explicability, even in principle, does not imply incomprehensibility, for intrinsic comprehensibility is possible.

272–81: *The deception of the senses*

The deception of the senses can be instructive rather than a cause for despair. For many alleged deceptions are not deceptions at all, while others are deceptions of reason rather than of the senses. The main lesson to be learned is to be critical toward the senses, i.e., to realize that there is a difference between appearance and reality and that it is by reason that we go from sensory appearances to claims about reality (F280–1).

For example (F272–8), it would be no deception of the senses if on a rotating earth we perceived bodies falling vertically and failed to perceive any circular component in their motion; for motion exists only relative to things that do not share it, hence motion shared by an observer and an observed object does not exist for him and is imperceptible to him; on a rotating earth, the circular component of the motion of the falling body would be shared by the observer and hence would not be there to be perceived. Second (F278–9), the failure to perceive a wind due to the annual motion would not be a deception of the senses because wind is perceivable only when air moves relative to the observer. Third (F279–80), the failure to *feel* any motion on a rotating earth would by itself be merely a sensory appearance; it would generate a deception only if one went on to say that therefore the earth stands still since if it moved one would be able to feel its motion; but this would be a deception of reasoning, not of perception, since the error would be in assuming that one can feel mere motion, whereas experience with navigation shows that one can only feel *changes* of motion.

281–8: *Inconceivability-claims and facts*

It is wrong to argue that a doctrine is false because it has conceptual difficulties, that is difficulties concerning the way in which it *could* conceivably be true (F287); in other words, it is wrong to say that something is not the case because there is no way to conceive *how* it *could* be the case. It is wrong to draw factual conclusions from inconceivability claims. The problem is that inconceivability claims frequently end up being falsified by factual discoveries. For example, it is wrong to argue that the earth does not move on the grounds that there is no way to conceive how a simple body like the earth could have the three distinct motions that it is supposed to have according to Copernicus. For the only factual content of this inconceivability claim is that there is no way for the earth to have its different motions the way that an animal can move in various ways; on the other hand, the earth could move with more than one motion the way that Saturn, the sun, Jupiter's satellites, and loadstones have been discovered to do.

289–90: *The primacy of nature over man*

It is wrong to argue that something is the case because it fits well our understanding, for nature is prior to the human mind. For example, it is wrong to argue that different substances must behave differently otherwise we could not come to know them; what's wrong with this is that it is empty of factual content; insofar as it does refer to natural reality the principle must be qualified to specify the difference in behavior. In other words, though we know a priori that different substances must behave differently in *some* way (in order to be different), it does not follow from this that they must behave differently in a particular given manner. For example, there is no necessity for water and air to have to behave differently with respect to their circular diurnal motion on a rotating earth; for even on a rotating earth they would retain enough differences.

290—3: *Logical distinctions*

One should not confuse (F293) rhetorical appeals with arguments any more than sound with unsound arguments, or the correctness of an argument-form with the truth of premises. Examples of mere rhetorical appeals are the following: to say, as the Aristotelians do, that it is more fitting to have the earth at the center of the universe (rather than between Mars and Venus) because that way there is a better separation of the pure (heavenly bodies) from the impure (earth); or to say, more plausibly, that it is worse to have the (impure) earth at the center because this would be like having a lazaretto in the heart of a city; or to say, with Copernicus, that it is more fitting to have the sun at the center because this would be like having a great lamp in the middle of a temple so that it can best illuminate all places. An example of an unsound argument is to say that the sun and stars move because they are more like the planets than the earth is, and this is so because the sun, stars, and planets are all incorruptible and unchangeable, while the earth is not; the Aristotelians beg the question here since they base their last premise (the corruptible/incorruptible distinction) on the difference with respect to motion of the earth on the one hand and all the other bodies on the other. An example of an argument with valid form but false premise is to say that the sun and stars move because they are more similar to the planets than the earth is, and this is so because the sun, stars, and planets are all *luminous,* whereas the earth is not; here the last premise is not true; the planets are not luminous. Thus the following has both valid argument-form and true premises: the planets and the earth are alike in that they all lack their own light, and they all differ from the sun and stars, which have their own light; since the planets obviously move, the earth probably does also, but the sun and stars do not.

293—8: *The superficiality of abstract answers*

It can be a sign of superficiality (F295) to ignore the empirical-factual aspect of a question and deal only with its conceptual-verbal aspect in cases when the two approaches suggest conflicting answers. For example, consider the question of which is more probable and credible: the Copernican commitment to immense stellar distances, or the Ptolemaic commitment to immense stellar speeds of diurnal motion; and consider Kepler's argument that the Copernican view is more probable because it is harder to conceive a property beyond the model of the subject than to increase the subject without the property. It is true but irrelevant to object that Ptolemy's immense diurnal stellar speeds are not inappropriate, in the sense that they are just the right ones needed to complete merely one revolution per day at their distance from the center. It is much more important that those speeds are inappropriate in the sense that they go against the norm, observable in the revolutions of the planets and of Jupiter's satellites, namely that the speed of revolution decreases and the period increases as the distance from the center increases.

299–337: *Quantitative data and qualitative conclusions in astronomy*

Quantitative astronomical data are so problematic that at best they yield only qualitative conclusions. For example, consider the observational data available for the 1572 nova; when the distance is computed from the difference between the nova and the polar parallaxes due to the change in observer position, the data yield physically impossible distances in most cases and widely different distances in the rest. The physically impossible distances are those implied by negative parallax difference; the other distances vary from 1/48 to at least 716 earth radii. This may lead one to reject all the data as unreliable, or else to correct them appropriately. The failure to do either was Chiaramonti's main mistake. One way to correct the observational data would be from the point of view of the qualitative question whether the nova is sublunary or superlunary; the observational data could be divided into those that yield sublunary distances and into those yielding superlunary distances; one could then try to make consistent the sublunary set of data on the one hand, and the superlunary set on the other, by changing the observational values. When this is done, the corrections needed to harmonize the super-lunary data are very much smaller than those needed for the sublunary data, from which one might conclude that the nova was probably superlunary. Another way to correct the observational data would be from the point of view of the range of distance suggested by most observations; then only the data yielding impossible distances would be changed; since almost all of the data yield impossibly great distances and constitute a majority of the data, the great majority of the observational data may be said to suggest a super-lunary distance.

337–46: *The reliability of astronomical measurements*

Quantitative astronomical data are not equally reliable; what's needed is a critical evaluation of the methods for collecting them. For example, the data concerning the elevations of the 1572 nova on positions off the meridian are practically worthless because such positions are much more difficult to measure than its meridian positions, whose measurements have been shown to be quantitatively worthless and usable only for qualitative conclusions. The most reliable data are from the measurements of distance between the nova and the pole or some neighboring star; such measurements are the simplest ones to make and neither the effects of refraction nor the use of the sextant present any real problem.

346–9: *Procedure vs results, and Aristotle's authority*

Aristotle's authority is misused by the Peripatetics when they follow him in a specific conclusion he reached rather than in the manner whereby he arrived at it. The advantage of following his procedure is that one can apply it to a novel situation to reach novel conclusions when conditions have changed. For example, since Aristotle argues that the earth is at the center of the world because it is the center of celestial revolutions, it is reasonable to conclude that, if it could be shown that the sun and not the earth is at the center of celestial revolutions, then he would conclude that the sun and not the earth is at the center of the world; the novel situation here would be one where the available evidence made it more likely that the sun rather than the earth was at the center of celestial revolutions. In other words, the follower of the spirit of Aristotle, as opposed to the blind follower of his letter, should be able to see that the proposition that the center of the world is the center of celestial revolutions is more basic than the proposition that the earth is the center of the world, and from this to prefer the former in cases of conflict.

349–68: *Simplicity, insight, and sense experience*

Insight is admirable, but the only reliable procedure is to take sense experience seriously and combine it with reason (F355–6, 363), that is, to infer truth from rational simplicity only where there is no disagreement with sense experience. For example, one cannot but marvel at the insight of Aristarchus and Copernicus who became convinced of the truth of the heliocentric system, because of its rational simplicity, despite the insuperable empirical difficulty that the planets did not show the necessary changes in size and shape (F355, 362–3, 367, 368); the Copernican attitude was insight rather than mere intellectualistic stubbornness because the telescope has nowadays made possible the observation of the necessary changes, which constitute direct evidence from sense experience of the truth of the heliocentric system.

However, though admirable, Copernicus's *procedure* is not rationally justifiable because he was inferring truth from mere simplicity and disregarding
clear empirical counterevidence (F354–368). Nevertheless, Copernicus's
conclusions are the most probable ones *nowadays* (F349–354) when we can
observe with the telescope phases in Venus and great changes in the apparent
size of the planets.

368–72: *Computational vs philosophical astronomy*

There are two distinct attitudes possible in astronomy, a merely computational attitude and a philosophical attitude. The merely computational
astronomer tries to give the reasons for celestial appearances in terms of
devices from which those appearances can be calculated, without worrying
about the physical reality of these devices; the philosophical astronomer tries
to give the reasons for celestial appearances only in terms of devices that can
be physically real. Ptolemaic astronomy is a good example of merely computational astronomy; its physically questionable devices are: (1) circular
motions which are uniform with respect not to the center but to a point off
the center; (2) circular motions which are simultaneously from east to west
and from west to east; (3) planetary motions which are retrograde; and (4)
epicycles which are so large that they cross the orbits of other bodies.
Copernican astronomy, on the other hand, is a good example of philosophical
astronomy; its primary assumption is the physical reality of the earth's annual
motion, which gives such a simple explanation of retrograde planetary motion
that this alone would be conclusive to an unbiased mind.

372–83: *Logic vs methodology in theoretical explanation*

Logically speaking, whenever certain observable facts are shown to follow
from a theory, this does not constitute a conclusive argument for the truth of
the theory, unless one shows that there is no other way of explaining those
facts (F377–80). Nevertheless, methodologically speaking, given two different theories each of which explains the relevant facts, the theory which
has greater explanatory power and which is less *ad hoc* is the better and more
probable one. For example, consider the intricate way in which sunspots can
be observed to move across the solar disc, namely along curved and slanting
paths whose curvature and slant are continuously changing in such a way that
they curve downwards and slant upwards for half a year, they curve upwards
and slant downwards for the other half a year, they are straight twice a year
when the slant is greatest, and their slant is absent twice a year when their
curvature is greatest. This motion of sunspots can be shown to follow very

easily by assuming that the earth has the annual and the diurnal motion and that the sun rotates around an axis which is inclined to the plane of the ecliptic. The same observations can also be explained by assuming that, in addition to the diurnal and annual motions, and in addition to an axial rotation around an axis inclined on the ecliptic, the sun has a fourth motion, namely that its inclined axis of rotation itself rotates once a year around the axis of the ecliptic. Now aside from the obvious greater simplicity, the geokinetic theory has greater explanatory power insofar as it can explain the yearly period of the motion of sunspots, whereas there is no reason why this should be so in the geostatic theory (F382).

Moreover, and more importantly, the geostatic theory is completely *ad hoc* (F382); the sun's fourth motion is added merely out of the necessity of accounting for the observations under consideration. By contrast, the geokinetic theory not only does not have any such *ad hoc* element, but it made possible or at least facilitated Galileo's discovery of the relevant facts; the geokinetic theory is freed of an *ad hoc* character by the mere fact that the discovery of the full details of the motion of sunspots *could* have been made by predicting (F379) these on the basis of it (F372–80). The geokinetic theory is therefore better and more probable.

383–5: *The Bible*

It is in bad taste and disrespectful to inject Biblical references in a scientific discussion. That is what the author of the "booklet of conclusions" does when he says that, according to the Copernicans, Christ *ascended* into hell and *descended* into heaven, and when Joshua ordered the sun to stand still it was the earth that stood still.

385–99: *Objectivity and the concept of size*

The concept of size is in some ways subjective and relative, but it is not teleologically anthropocentric.

Our ideas about the size of fixed stars are subjective in the sense of being dependent on our perception of apparent stellar diameters (F387–91); now, *under ordinary conditions* apparent stellar diameters are perceived much greater than they are by factors ranging from 24 (F389) to more than one hundred (F388), due to the irradiation of the eye which makes us see them with a halo; it is possible, however, to reduce or eliminate this observational error by observing them with a telescope, or during the day, or with the help of a fine string behind which they can be made to hide. Moreover (F393–4), our ideas about *very large* sizes are unclear and confused in the sense that our imagination cannot distinguish between one such size and another; an example would be the size of the universe, as measured by stellar distances. Finally (F396), the concepts of 'large' and 'small' are relational in the sense that to say that something is large (or small) makes sense only relative to a reference class; hence, to say that the stellar sphere of the Copernican system is too large means that it is too large as compared to other celestial spheres; but this is not so because if we compare the Copernican stellar sphere to the lunar sphere, the ratio is much smaller than that obtained from a comparison of an elephant to an ant or of a whale to a gudgeon. Though the concept of size is subjective and relative in these ways, it is not, however, teleologically anthropocentric (F394–96, 397–98); that is, it is wrong to think that the size of things fulfills a purpose definable in terms of human interests; to think so would be as absurd and arrogant as it would be for a grape to think that, just because the sun acts on it in a way which is perfectly suited to cause its ripening, the sun exists and acts only for its benefit; hence it is wrong to argue that the universe cannot be as large as required in the Copernican system because the space between Saturn and the fixed stars would be superfluous and useless; this argument involves several nonsequiturs: to ground nonexistence on superfluousness, superfluousness on lack of purpose for man, and lack of purpose for man on ignorance of the purpose for man.

399–406: *Ad Hominem arguments*

Ad hominem arguments must be used with care if they are to have a methodological function, in addition to their rhetorical one. That is, they are easily effective for the purpose of discrediting the person whose beliefs are used as premises, but such arguments are insufficient to resolve the issue of the factual truth of these beliefs (F399). For an *ad hominem* argument is one in which the arguer derives a conclusion not acceptable to an opponent from premises which are accepted by him (the opponent), but which may not be too well grounded; hence, such an argument does not establish its conclusion, but only that one cannot both accept the premises and deny the conclusion.

For example (F399–400), consider the Tychonic argument that if the earth has the annual motion, since there is no annual change in the stellar appearances, then the fixed stars are at such an immense distance from the earth that, in order to be visible as they are, their size must be of the order of planetary orbits. This argument is *ad hominem* against Copernicus in the sense that the proposition that there is no annual change in stellar appearances, though accepted by him, has not been supported by evidence or argument. In fact, no one has tried to observe such changes in stellar appearances, or even inquired into exactly how stellar appearances should change; instead there are a lot of misconceptions about the matter. One misconception (F400–3) is that the elevation of the celestial pole should change, which is absurd because if the earth had the annual motion then the celestial pole would be defined by the terrestrial one, whose elevation can change only by moving on the earth's surface; Tycho himself (F400) seems to have made this error since, in discussing the matter, he claims that no change in polar elevation is in fact observed and fails to note that the point is misconceived. Another misconception (F403–4) is that the annual changes in the elevation of the fixed stars should be comparable to those resulting by moving on the earth's surface, which is wrong because the latter is highly curved, whereas the annual motion takes place on a plane. A third misconception (F405–6) is that the annual changes for fixed stars would be comparable to those observable in the sun, which though more difficult to clear up is also wrong, as it will be shown later (F406–16, and F416–23).

406–9: *The value of criticism*

It is advisable to test one's arguments and theories against the criticism of men of sound judgment and sharp intellect; for if they survive the criticism, we can have great confidence in their correctness (F408). For example, consider the argument that, for fixed stars lying in the plane of the ecliptic, the earth's annual motion would not cause an annual change in their apparent position, since their elevation relative to the ecliptic would stay constant regardless of how the earth moved in that plane. If one presents this argument to an intelligent person like Sagredo, his responses lead to the following qualifications. That is, the conclusion would be right if all fixed stars were equidistant from the center; but if their distances vary, as they probably do, then some changes should appear in the relative positions of those lying in the ecliptic plane.

409–14: *Theories and observations*

When one succeeds in detecting certain specific observational consequences of an hypothesis, and there is no other way to explain these observations, then one must accept the hypothesis; but when one fails to detect these consequences, and there is another way to explain this failure, then one need not reject the hypothesis (F413–14).

For example, the observational consequences of the earth's annual motion, as regards the appearance of the fixed stars, are the following (F409–12): annual changes in the apparent magnitude of the stars, the change being greatest for those in the plane of the ecliptic and decreasing to nothing for those near the pole of the ecliptic; annual changes in elevation above the ecliptic, the change being greatest for stars near its pole and decreasing to nothing for those in its plane; differences in both these changes (in magnitude and in elevation) due to the distance of the stars, the changes being greater for nearer stars. Now, if one were able to detect these variations in stellar appearances, since there would be no way to explain them other than by the earth's annual motion, then we would have to conclude that the earth had this motion. But the fact that no such variations are generally known or have generally been observed does not force us to conclude that the earth lacks the annual motion, because such general ignorance is explicable by the fact that no one has searched for them seriously or systematically enough, or with the appropriate instruments, or with the necessary skill (F413–14). Even if we had seriously searched and failed to observe these variations, we still would not have to reject the earth's annual motion, since our failure could be explained by stellar distances so great, and hence changes so small, as to be undetectable with our available instruments (F413).

414–6: *The inexactness of astronomical instruments*

Available astronomical instruments cannot be trusted much for the observation of very small effects, such as stellar parallax, even though some instruments are better than others. In fact, it is common for astronomers to disagree by several minutes in the values they get for the elevation of fixed stars, and even of the pole, let alone for the comets or new stars. The fact that Tycho's instruments are the best does not mean that they are good enough for the very exact observations needed to observe annual parallaxes. What is needed is instruments whose sides are miles long, so that differences of seconds in celestial elevation correspond to distances of the order of *braccia* along the instrument. For this purpose one can use topographic features on the earth's surface such as mountains. An example of such an instrument to try to observe stellar parallaxes, would be to observe the changes in the way that a particular star would be hidden by a beam on top of a building, at the top of a mountain, when one carried out the observations from the valley below, at different times of the year.

416–25: *The complexity of simplicity*

The concept of simplicity is not simple; it has at least two aspects, mathematical and physical. For example, consider the Copernican explanation of the seasonal variations in solar elevations and in length of nights and days

(F416–22). Someone (Salviati, F422) may be impressed by the fact that this explanation uses two terrestrial motions which are (1) both in the same direction, namely west to east, and (2) performed in periods corresponding to their magnitudes. He may then (F422–3) contrast this to the fact that the geostatic explanation requires celestial motions which (1) violate the *symmetric* character of the relationship between speeds and size of the thing moved, by having the very large highest sphere move at a very high diurnal speed, and the very much smaller planetary spheres move at a very much slower characteristic speed; (2) are performed in opposite directions; and (3) are performed against one's individual inclination, in the case of the participation by the planetary spheres in the diurnal motion. He may conclude (F423) from this that the Copernican explanation is superior in simplicity, and hence more probable. Another person (Sagredo, F423) may agree with the conclusion but give different reasons for the greater simplicity, namely its being more in accordance with the widely accepted philosophical principles that nature (a) does not multiply entities without necessity, (b) uses the simplest and easiest means in producing its effects, and (c) does nothing in vain. A third person (Simplicio, F423) might object to the first person's concept of simplicity as being excessively mathematical in its stress on the correspondence between periods of revolution and magnitude of orbit, or symmetry between speed and size of body moved, which involve quantitative relationships. Instead he may be impressed (F424) by the fact that the geostatic explanation attributes a single natural motion to each simple body, whereas the Copernican explanation violates this principle; and from this he may conclude that the Copernican explanation is *inferior* in simplicity. The difference in their concepts of simplicity would remain even if it were shown to him that, in a sense, the Copernican explanation does not violate the principle that a simple body must have only one natural motion (F424–5), for the three or four terrestrial motions supposedly required can be reduced to one as follows: the straight-downwards motion of terrestrial bodies is a motion that belongs to the parts of the earth, not to the whole earth; the earth's diurnal and annual motions since they are in the same direction, are two only in a special sense, a sense related to that in which a ball rolling down an inclined plane has two motions; and finally, Copernicus's "third motion" does not exist, the phenomenon is really a kind of rest.

425–31: *Open-mindedness, intellectual cowardice, and curiosity*

Curiosity and awareness of the infinitesimal extent of human knowledge tend to make one willing to pay attention to new ideas, whereas intellectual cowardice hampers open-mindedness (F426). For example, consider Gilbert's idea that the primary substance of the terrestrial globe is loadstone. Peripatetics have generally dismissed this idea as chimerical and insane, even though neither Aristotle nor anyone else has refuted it. Such dismissal is as cowardly as being afraid of one's shadow (F426–7). For it is easy to show that the idea is initially as plausible as other theories about the primary substance of the earth (F427–9). In particular, to think that this primary substance is earth in the sense of topsoil is implausible, since digging into the ground reveals hard substances like rocks, since weight and pressure would harden underground substances even if they were like soil, and since this theory is probably the result of the linguistic accident that the word 'earth' denotes both the terrestrial globe and the substance of topsoil. Moreover, without actually repeating Gilbert's proof, a description of his procedure ought to be sufficient to awaken one's curiosity and take his idea seriously (F429) (which incidentally illustrates the fact that the *method* of arriving at an idea as well as its *content* may motivate the open-minded individual). Gilbert's procedure is to infer the nature of the substance from knowledge of its properties (F429–30) and may be compared to the following: suppose (F430) you were given samples of a large number of substances all covered by a cloth so that you could not see what they were; suppose further that your problem was to guess what the substances were; finally, suppose that one of these covered samples exhibited all the properties known to be present only in loadstones; then, clearly, you could infer that the sample in question was loadstone. Similarly, Gilbert reaches his conclusion (F430–31) after finding that the earth shows all the properties of a loadstone: for example, a magnetic needle tilts from the horizontal as it is moved away from the equator; and the poles of a magnet behave as if they were in the presence of a bigger magnet, namely they have equal strength only at the equator, while its south pole is stronger in the northern hemisphere, and the north in the southern.

431–6: *Causes: role, discovery, and justification*

The development of a new field by the conception of new ideas, by the observation of new effects, and by the initial investigations of causes requires uncommon originality and so is much more admirable than the subsequent perfecting of the field; for example, Gilbert is to be envied for his magnetic philosophy (F432–3). Nevertheless, the explanation of the causes of puzzling and curious known effects is an obvious second step as well as one of the most intellectually rewarding activities (F432; F434, Sagredo); one such effect is the fact that an armatured loadstone can sustain much more weight (in some cases 80 times more) than without the armature. Now, in causal investigations, one thing to avoid is explaining by naming, for example, to explain magnetic phenomena in terms of the 'sympathy' between qualitatively similar substances; to do so would be like painting by giving verbal descriptions (F436). Another thing to avoid is excessive readiness to accept a causal explanation as conclusively proved; practice with mathematical demonstrations makes one less willing to do so (F432). Many of Gilbert's explanations show this excess (F432), whereas the explanation here to be given for the greater weights sustained by armatured loadstones is an example where the conclusiveness is greater than any other to be found in all magnetic philosophy (F434); it is, in fact, almost as great as the conclusiveness of mathematical demonstrations (F434, F435). What can give a causal explanation a certainty comparable to the mathematical one is the direct confirmation of the presence of the cause, after one has been lucky enough to think of it from a knowledge of the effect (F434—5, Salviati). A quasi-method to acquire this luck is to compare the effect to be explained with another significantly similar but appropriately different effect whose cause is known (or contextually unproblematic); then check whether an appropriate variation in the known cause is responsible for a corresponding variation in the effect equivalent to the difference between the two effects being compared; if not, check what new condition is present in the effect to be explained, which is not present in the effect whose cause is known; this new condition must be the cause we are looking for (F433). For example, let us compare the loadstone with the armature to the loadstone without the armature; let us check whether an increase in the force of the loadstone is responsible for the increase in the weight that the armatured loadstone can sustain; we find that this is not so, since the armatured loadstone does not act at a greater distance, and since it does not attract more strongly through an interposed paper or goldleaf; next we notice that the only variation introduced by the armature is the difference in the contacts with the sustained weights; so we conclude that this difference is the cause of the increase in the weight sustained with the armature (F433). This difference in contacts amounts to a greater contact area between iron and armature than between iron and loadstone, that is a greater number of lines of force (so to speak, *"filamenti . . . che collegano i due ferri"* [F434]) uniting the two surfaces. Fortunately, the actual presence of this cause can be confirmed by a number of facts and experiments (and this is what gives the explanation almost mathematical certainty): the texture of iron is much finer than that of loadstone (F433—4); loadstones contain many impurities, and one can observe that a needle is not attracted by them but slips onto the surrounding loadstone substance (F434); the sharp point of a needle is not attracted any more strongly by the armature than by the bare loadstone (F434); if you let a needle be suspended between a bare loadstone and a nail, with one end touching the loadstone and the other the nail, when you separate the two the needle will remain attached to the one which it was touching with its eye (F434); finally, if you take a loadstone and smoothen and polish it on one side, you will clearly see on the resulting surface many bright spots on a dark background, and you can observe that iron filings are not attracted to these spots but to the surrounding surface only (F435–6).

436–41: *Parts vs wholes*

The principle that "the same is true of the whole as of the parts" is often misapplied and misunderstood (F440). Sometimes it is ignored when it should be applied. For example (F437–9), consider what Peripatetics are likely to say when confronted with the fact that loadstones have at least three natural motions. They would say that this does not refute the doctrine that a simple body like the earth cannot have the several natural motions attributed to it in the Copernican system, because loadstones are not simple bodies; however, if loadstones are not simple, since they are parts of the earth, then the principle in question could be *properly* applied, and it would follow that the earth is not simple either, thus allowing it to have its Copernican motions without violation of the doctrine (of one natural motion per simple body). Moreover, they would try to account for the multiple motions of loadstones in terms of the different motions of their simple constituent parts; however, if these parts are elementary bodies, then their natural motions would presumably be straight (up or down), but unfortunately you cannot mix straight motions and get a circular motion for the result (and here too we would have a proper application of the principle being discussed); hence Peripatetics would be led to the absurd consequence of allowing celestial substances as component parts of loadstones.

Sometimes, the principle in question is applied when it shouldn't. For example, consider (F439–40) Gilbert's own assertion that if a small loadstone sphere were balanced properly and free to move, it would turn on its axis; however, there would be no reason for this to happen even on the assumption that the earth is a loadstone and has the diurnal motion, for such a spherical loadstone would be doing the same as the whole earth by turning around the center of the earth rather than around its own center; there would be a reason for it to happen only if the small sphere were prevented from turning around the center of the whole. Another example of a misapplication is (F440–1) Sacrobosco's argument that since small drops of water can be seen to arrange themselves in a round shape, that shows that the whole element water on the globe forms a sphere; by such reasoning one would have to conclude the absurdity that any body of water forms a sphere; whereas the relevant sense in which the same is true of the whole as of the parts concerns the tendency of water to form a sphere around the center of gravity of the whole (the earth's center).

442–5: Choosing the effect in a causal investigation

Knowledge of the effects is helpful and necessary for the investigation and discovery of causes, but such knowledge is always limited both in extent and reliability; hence the best way to proceed is to start from those features of the effects that are certain and principal, and to try to arrive at a cause which is primary and which would serve as a pointer for the direction to follow in extending and deepening our knowledge in a consistent manner (F443–4).

For example (F444–5), in the investigation of the cause of the tides, the principal and best known effects are the fact that the time interval between high and low tide is six hours, and the fact that the motion of the water is merely upwards and downwards in some places, merely eastwards and westwards in others, and a combination in still others. The primary cause that will be proposed is the motions of the earth (F445), to which the necessary additions and articulation will later be made, by taking the fluid properties of water as secondary causes (F454–62), and by explaining the monthly and annual periods as due, respectively, to actual variations in the earth's annual motion and to merely effective variations in its unchanging diurnal one.

445–62: The artificial reproduction of cause and effect

In causal investigation, a proposed cause should be rejected if it does not allow an artificial reproduction of the effect, and it should be accepted if such an artificial reproduction is possible (F447, 456; cf. 471–2). For example, the only cause of the tides that passes this test is the motion of the earth. The failure to pass this test is the real reason why one finds implausible the theories of lunar attraction, lunar heat, differences in water depth, and seabed caves; none of these conditions would produce up and down, or back and forth motion in only some places of a container of water (F447–9). On the other hand, by accelerating or retarding a container of water one can make the water rise and fall at its extremities, and move back and forth at the middle, as one can observe on boats that carry water (F450–1); and the

motions of the earth would produce a daily acceleration and a daily retardation at every point, as the diurnal motion component of this point is respectively added to or subtracted from the annual motion component (F452–4);
moreover, the various fluid properties of water, which can be investigated by
means of artificial models (F456), are bound to act as secondary causes
interacting with the primary one (F454–6), to produce the complexity of
effects observable in the motion of sea water (F457–62), some of which
effects are observable in artificial models (F457, F459).

462–70: *Reasoning ex suppositione*

Reasoning *ex suppositione* involves two steps: a known effect is explained
by making a supposition, and then one takes the effect as evidence and confirmation for the supposition; such reasoning is pointless if the supposition is
false, and invalid if there are in the context reasons to reject the supposition
(F462). For example, consider the argument for the earth's motion based on
the tides (F462, postil). This argument is *ex suppositione* in the above sense,
and hence must be defended from any contextual objections to the supposition that the earth moves, which would render the argument invalid. One
such objection (F363–4) is that if the earth's motion causes the tides then it
would cause a perpetual westward wind, which however does not exist. The
answer to this is twofold: first, the earth's motion could cause the tides
without causing the perpetual wind because air unlike water does not have
the propensity of retaining acquired motion, which is the property that
makes water move back and forth when accelerated or retarded; second,
prevailing westward winds do exist, especially in equatorial regions. Another
objection (F467–8) would be that since prevailing westward winds do exist,
and since they could be explained as caused by the diurnal motion of the
lunar orb, and since this explanation would be more economical and less
innovating (F467) than a geokinetic explanation, therefore we ought to reject
the earth's motion. The problem with this objection is that (F468–70) the
prevailing winds could not be produced by the turning of the lunar orb since
it is supposed to have a very smooth surface, and hence its motion could not
be transmitted to the element fire; moreover, the turning of this element
could not produce the turning of the denser element air. On the other hand
(F465–6), these winds could be easily produced by the earth's diurnal
motion, as the air lying over its larger smoother sections would be less easily
carried along than that trapped closely among mountains, which would
constitute a further reason favoring the supposition in question.

470–1: *Confessed ignorance and occult qualities*

The confession of ignorance is sometimes preferable to the claim to know, for such claims are sometimes a mere figment of the imagination (F470, F446) and the result of vanity or verbosity (F470). For example, it would be better to say that you don't know the cause of the tides, rather than explaining them in terms of attraction by occult qualities, which is intellectually repugnant (F470), or in terms of lunar or solar heat and light, which is empirically ridiculous and absurd (F470–1, F445–7).

471–84: *The method of concomitant variations*

Whenever there is a regular variation in the effect, there must be a corresponding variation in the cause (F471). Or to be more exact, whenever an effect varies in a regular way, it is best to explain this variation in terms of the same primary cause, by means of corresponding variations in it (F471–2), rather than in terms of a novel cause (F474), which would run the risk of being *ad hoc*. For example, consider the monthly and annual variations in the magnitude of the daily tides. It would be best to account for such monthly and annual periods in terms of variations in the accelerations and retardations that cause the diurnal period, namely variations in the speed of the earth's annual or diurnal motions or both (F471–4). Now, there are independent reasons to believe that the earth's annual motion would undergo monthly variations: the evidence comes from what may be regarded as a highly confirmed law of revolution, namely that whenever the cause of circular motion remains unchanged, such motion is slower along larger circles and faster along smaller ones (F474–8); for the earth-moon system would go around the sun in a circle whose effective radius would undergo monthly variations as a result of the moon being now in opposition and now in conjunction with the sun (F477–8). The fact that there is such *independent* evidence for variations in the earth's annual motion is more important than the fact there is no *direct* evidence for them from astronomical observations; for such a lack may be due to the failure to search for the variations, or to their very small magnitude (F479–82); hence we are entitled to conclude that the monthly period of the tides is caused by these lunar variations in the earth's annual motion (F479). Regarding the annual period, the inclination of the earth's axis would cause annual variations in the *effective* speed of the diurnal motion of points on the surface of the earth, the effective speed being the speed projected on the plane of the ecliptic (F482–4). Hence, by the same principle, we may conclude that the annual period is caused by such annual variations in the earth's actually unchanging diurnal motion.

484–9: *Searching vs results: the supremacy of method*

An idea may need refinement (F484–5), and yet it may be very valuable if it is essentially correct and points in the right direction (F485–6); conversely, a theory may be highly supported (F487), and yet it may be false (487–8), if it is possible for God to have created a world not in accordance with it (F488); this is so because the mental exercise required in the investigation is very valuable and probably the most important thing (F489). In other words, the important thing in natural philosophy is that the approach or method be right.

The geokinetic theory of the tides is a good example of this. It provides only a general account of the basic diurnal, monthly, and annual periods, and only a sketch of the complicated and somewhat irregular details, not all of which are well known anyway (F484–5); yet it represents a breakthrough, especially insofar as it suggests a mechanical approach (F485–6). It constitutes a very strong argument in favor of the earth's motion (F487), yet it could be mistaken (F487–8), because it is possible for God to cause the motion of tides in some other way.

NOTES

[1] Two sets of page numbers are provided here: those before the slash refer to G. Galilei, *Opere,* edited by Favaro, Vol. 7; those after the slash refer to G. Galilei, *Dialogue,* tr. S. Drake.
[2] The page references here are all to Favaro's edition, both those in the various subheadings, and those in parenthesis in the text; the latter numbers are prefixed by an 'F', which is short for 'Favaro', as a reminder that his edition is being referred to.

PART II

LOGICAL AND METHODOLOGICAL CRITIQUES

CHAPTER 6

CONCRETENESS AND JUDGMENT: THE DIALECTICAL NATURE OF GALILEO'S METHODOLOGY

IMPLICIT PHILOSOPHY

The first thing to note about the methodological content of Galileo's *Dialogue* is that it is more or less evenly spread throughout the book. As Table I (below) indicates, the First Day contains on the average a distinct methodological discussion about every 10 pages,[1] the Second Day about every 7¼ pages, the Third Day every 9 pages, and the Fourth every 8 pages. In other words, the methodological intensity of the book does not vary much from beginning to end.

TABLE I

Frequency of philosophical discussions

DAY	Number of Subdivisions	Text pages		Commentary pages	
		Total	Average	Total	Average
I	10	99	10	6.3	0.63
II	23	167	7 1/4	14.4	0.63
III	16	143	9	11.0	0.69
IV	6	48	8	3.3	0.55

Second, one cannot help but be impressed by the quantity and variety of philosophical topics which are discussed. Though the number of philosophical passages distinguished in my reconstruction is 55, and though many of them overlap, the number of individual philosophical topics touched upon is at least 92, as an index of these topics shows (Table II).

This large number of individual topics cluster around a much smaller number of themes (15), as Table III shows. Nevertheless, there is no one philosophical thesis that predominates, though some are discussed more frequently than others. It turns out that there are three topics which are much more frequently discussed than others. The topic of sense experience, and the topic of causes, explanation, and understanding occur each in 11 different passages; logic and reasoning in 10. The topics of unconscious

TABLE II

Index of individual philosophical topics

At least 92 individual topics are mentioned as indicated in this table.

Abstract entities, 229–37
Anthropocentrism, 82–95, 289–90, 385–99
A priori vs a posteriori, 71–82
Ad hominem arguments, 399–406
Aristotle's authority, 59–71, 71–82, 132–9, 346–9
Astronomical instruments, 414–6
Astronomy, 299–337, 368–72
Authority, 59–71, 71–82, 132–9, 346–9
Bible, 383–5
Causal investigation, 436–41, 471–84
Causes, 237–44, 260–72, 431–6, 445–62
Certainty, 431–6
Conceptual frameworks, 38–57
Cognitive awareness, 112–7
Comprehensibility, 260–72, 281–88
Concepts and facts, 293–8
Concrete entities, 229–37
Cowardice, intellectual, 425–31
Criticism, 206–9, 214–7, 406–9
Curiosity, 425–31
Data, 337–46, 299–337
Deception of the senses, 272–81
Digressions, 188–93
Discovery vs justification, 71–82
Empiricism, 57–9, 206–9
Errors, 117–24
Experience, 169–80
Experimentalism, 206–9
Experiments, 95–112, 169–80, 206–9, 209–14
Explanation, 260–72
Explanation, causal, 237–44, 431–6
Explanation, theoretical, 372–83
Geometry, and philosophy, 223–9
Human understanding, 124–31
Ignorance, 132–9, 470–1
Inconceivability, 281–8
Independent-mindedness, 132–9

Insight, 349–68
Introspection, 112–7
Knowledge, and information, 180–8
Knowledge, and recollection, 217–23
Knowledge, limitations of, 425–31
Knowledge, unconscious, 193–6
Logic, 57–9, 59–71
Logical analysis, 150–9, 159–69, 290–3
Logical distinctions, 290–3
Logic vs methodology, 372–83
Mathematical practice, 431–6
Mathematical reasoning, 33–8
Mathematics, 33–8, 229–37
Mathematics and physics, 244–60
Mathematics, applicability of, 229–37
Measurement, astronomical, 337–46
Methodological point of view, 71–82
Method of concomitant variations, 237–44, 471–84
Method, importance of, 244–60, 484–9
Naming, as a way of explaining, 260–72, 431–6
Nature, primacy over man, 289–90
Number mysticism, 33–8
Objectivity, 385–99
Observations and theories, 409–14
Occult qualities, 470–1
Open-mindedness, 150–9, 425–31
Opportunism, academic, 132–9
Optimism, 124–31
Originality, 431–6
Parts vs wholes, 436–41
Physico-mathematical synthesis, 244–60
Physical reality, and mathematics, 229–37
Platonic mathematicism, 33–8
Practice vs theory, 59–71, 71–82
Probability, 139–50
Probable arguments, 139–50
Procedure vs results, 346–9
Pythagorean mathematicism, 33–8

TABLE II (Con't)

TABLE III

Index of main philosophical topics

The large number of individual philosophical topics cluster around a much smaller number of themes.

TABLE III (Con't)

Open-mindedness: 150−9; curiosity and intellectual cowardice, 425−31
Practice vs theory, 59−71, 71-82
Simplicity: 139−50, 349−68; different concepts of, 416−25
Socratic method and unconscious knowledge: cognitive awareness, 112−7; Socratic
    method, 169−80, 180−8, 193−6; unconscious knowledge, 193−6; knowledge and
    recollection, 217−23

---

knowledge and the Socratic method, of authority, of mathematics, and of
method occur each 5 times; the doctrine of anthropocentrism, the nature of
astronomical investigation, the activity of criticism, and the topic of ignorance
are each discussed in 4 different passages; simplicity and the activity of conceptualization each 3 times; open-mindedness (in a sense distinct from but
related to independent-mindedness) and the distinction between practice and
theory are each discussed in two different passages.

The next interesting question to ask is whether the primary topics tend to
predominate in certain parts of the book. Table IV shows that, except for the
extremely short Fourth Day, the range of topics discussed in each day is
about the same. In other words, except for the preponderance of causal
discussions in the Fourth Day, no one major topic predominates for very long
in the rest of the book. It is true, however, that most major topics have places
of high concentration, with two or more consecutive passages devoted to the
same general topics; for example, astronomy on pages 299−346, authority on
pages 59−82, cause on pages 442−62 and 470−84, experience on pages
38−59 and 206−14, logic on pages 57−71 and 139−80, mathematics on
pages 223−37, the practice/theory distinction on pages 59−82, and the
Socratic method on pages 169−188.

One could go on with such attempts at "data analysis." They are legitimate
insofar as, in part, they attempt to discern the structure of, or patterns inherent in, the methodological content, and in part they attempt to systematize this content. In other words, after having analyzed the (relatively)
explicit methodological content, a next step would be to study the implicit
philosophy. Now, this implicit philosophy can be of two kinds, namely the
relatively systematized philosophical doctrine *implied by* the relatively un-
systematized methodological claims extracted so far, and the one (or the few)
pattern(s) *exhibited by and inherent in the* methodological content extracted
so far. Though these two types of implicit philosophy are abstractly possible,
it so happens that for the *Dialogue,* the philosophy clearly inherent in the
structure of the philosophical content precludes an implied philosophical
system. This can be seen as follows.

TABLE IV

Distribution of philosophical topics

Day:	I: 33–131	II: 132–298	III: 299–441	IV: 442–89
Anthropocentrism	82–95, 124–31	289–90	385–99	
Astronomy			299–337, 337–46, 368–72, 414–6	
Authority	59–71, 71–82	132–9	346–9, 383–5	
Cause		180–8, 214–7 237–44, 260–72, 281–8	372–83, 431–6	442–5, 445–62 470–1, 471–84
Conceptualization	38–57	281–8, 293–8		
Criticism	117–24	206–9, 214–7	406–9	
Experience	38–57, 57–9, 95–112	169–80, 206–9, 209–14, 244–60 272–81	349–68, 372–83 409–14	
Ignorance	124–31	132–9	425–31	470–1
Logic	57–9, 59–71	139–50, 150–9, 159–69, 169–80, 196–206, 290–3	399–406	462–70
Mathematics	33–8	223–9, 229–37, 244–60	431–6	
Method	71–82	244–60	346–9, 372–83	484–9
Open-mindedness		150–9	425–31	
Practice/theory	59–71, 71–82			
Simplicity		139–50	349–68, 416–25	
Socratic method	112–7	169–80, 180–8, 193–6, 217–23		

The overwhelmingly obvious structural feature of the *Dialogue* was already mentioned at the beginning of last chapter; it is the synthesis of practice and theory, or experience and self-awareness, or to be more specific, scientific practice and philosophizing about science. Every single passage in the book supports such skillful combination by Galileo. Such characteristic may be given the single label of 'concreteness'. We may say then that the philosophy exemplified by the *Dialogue* is a concrete philosophy of science. The second general feature of its explicit methodological content is really a consequence of the first, but it deserves special mention; it is judgment. That is, we find Galileo emphasizing on various occasions different and in some cases opposite things; for example, sometimes the need, and sometimes the superfluousness of experiments; sometimes quantitative considerations, sometimes qualitative considerations; sometimes antiverbalistic mathematical analysis, sometimes verbal-oriented logical analysis; sometimes causal explanation, other times phenomenological description. There is no inconsistency here, except when the methodological claim is improperly generalized, something I have always tried to avoid in my reconstructions above.

If concreteness and judgment are the two essential features of the book's methodological content, then it would be un-Galilean to try to arrive at a philosophical system implied by it. The reason is that the systematization would involve two things, namely generalizing some of the methodological claims, and analyzing their interrelations in the abstract, that is in abstraction of the scientific context; but the generalizing would conflict with the judgmental feature, whereas the abstract interrelating would conflict with the concreteness. This character of the philosophy implicit in the *Dialogue* is not only clearly present, but also helps to understand two otherwise rather puzzling facts, namely (1) the discrepancy between the place of Galileo in the actual history of philosophy and in history-of-philosophy books, and (2) the uniqueness of his role in the philosophy of science.

GALILEO AND THE HISTORY OF PHILOSOPHY

It is rather obvious that the kind of history of philosophy one writes depends very much on what one conceives philosophy to be. Let me give a personal anecdote to provide a very simple but incisive illustration of my point. A few years ago I happened to have some discussions with a philosopher, now deceased, who believed that the history of philosophy was almost a desert between Aristotle and Frege, whom he regarded as the greatest philosopher since Aristotle. It is easy to see what his concept of philosophy was: he equated philosophy with formal logic. From this point of view his historical interpretation makes sense, just as it perhaps makes sense to regard him as "the most brilliant American philosopher of this era", a claim made by the publisher of his posthumous book entitled *Formal Philosophy*.[2]

Let me now inject a little more substance as well as history into such an illustration and step back to the year 1917, the year of publication of Benedetto Croce's *Theory and History of Historiography*.[3] The theoretical part of this treatise concludes with a chapter entitled 'Philosophy and Methodology'. In it Croce formulates the concept of philosophy as methodology, according to which philosophical inquiry is the methodological aspect of historical understanding, in the broad sense of the term 'historical' so that historical understanding is any kind of concrete analysis, or applied knowledge, or inquiry into specific problems. The implications for the historiography of philosophy were expressed by Croce as follows:

Even the *history itself of philosophy* has hitherto been renewed only to a small extent, in conformity with the new conception of philosophy. This new conception invites us to direct our attention to thoughts and thinkers, long neglected or placed in the second rank and not considered to be truly philosophers because they did not treat directly the 'fundamental problem' of philosophy or the great *peut-être*, but were occupied with 'particular problems'. These particular problems, however, were destined to produce eventually a change of view as regards the 'general problem', which emerged itself reduced to the rank of a 'particular' problem. It is simply the result of prejudice to look upon a Machiavelli, who posited the conception of the modern state, a Baltasar Gracián, who examined the question of acuteness in practical matters, a Pascal, who criticized the spirit of Jesuitry, a Vico, who renewed all the sciences of the spirit, or a Hamann, with his keen sense of the value of tradition, as minor philosophers, I do not say in comparison with some metaphysician of little originality, but even when compared with a Descartes or a Spinoza, who dealt with *other* but not *superior* problems. A schematic and bloodless history of philosophy corresponded in fact with the philosophy of the 'fundamental problem'. A far richer, more varied and pliant [history of] philosophy should correspond with philosophy as methodology, which holds to be philosophy not only what appertains to the problems of immanency, of transcendency, of this world and the next, but everything that has been of avail in increasing the patrimony of guiding conceptions, the understanding of actual history, and the formation of the reality of thought in which we live.[4]

To be sure, Croce did not practice what he here preached, and though his works on Hegel, Vico, and Marx assure him a significant place in the history of the history of philosophy, he did not interpret them as being essentially methodologists; instead he gave *critical explanations* centered around, respectively, the concept of knowledge as concrete-universal, the concept of art as expression, and the philosophy of the practical. Nor has a history of philosophy as methodology been written by any of his followers, who do not even interpret Croce himself as a methodologist, though in the abstract they pay lip service to the doctrine; instead, they interpret him, in effect, as a metaphysician of mental activity, when the methodological interpretation would be to regard him as a methodologist of criticism. I should mention that on one occasion Croce did take seriously his methodological conception of philosophy; this occured when he interpreted Galileo as a philosopher,[5] insofar as he was a methodologist, that is as a philosopher-pure-and-simple, which is not to say a pure or a simple philosopher.

However, if one looks at history-of-philosophy books, the most striking fact about Galileo is the neglect he receives. Ueberweg devotes to him four lines[6], Bréhier one page[7], Copleston about three[8]. W. T. Jones has probably the longest and most detailed account[9] to be found in a general, comprehensive history of philosophy; yet his 11 pages should be contrasted with the 15 on Bacon, and with the complete chapters on Hobbes (25 pages), Descartes (27 pages), Spinoza (22 pages), Leibniz (12 pages) and Locke (34 pages).

This neglect would create no problem if the place of Galileo in the actual history of philosophy were as insignificant as it is in history-of-philosophy books. But this is far from being the case, and one could document this fact in a number of ways: by exhibiting the wealth of philosophical content of Galileo's books (as I have done above); by discussing the philosophical sensation that he caused among 17th century philosophers; by tracing and defining the nature of his influence on important subsequent philosophers such as Hume[10], Kant[11], Husserl,[12] and Ortega;[13] by authoritative quotations from more knowledgeable historians, such as the following from Sir David Brewster:

Had Bacon never lived, the student of nature would have found in the writings and works of Galileo not only the principles of inductive philosophy, but also its practical application to the noblest efforts of invention and discovery;[14]

or the following from E. A. Burtt:

It is difficult indeed to leave Galileo without pausing a moment to reflect on the simply stupendous achievements of the man ... and then, as if these accomplishments were not enough, we must turn to him likewise as the philosopher who sufficiently perceived the larger implications of his postulates and methods to present in outline a new metaphysics – a mathematical interpretation of the universe – to furnish the final justification of the onward march of mechanical knowledge. ... In view of these manifold and radical performances Galileo must be regarded as one of the massive intellects of all time.[15]

However such documentation would really be unnecessary because I regard it as uncontroversial that Galileo has a very significant place in the actual history of philosophy, for even those philosophers who neglect him in their own accounts pay lip service to his philosophical greatness. For example, Ueberweg's four lines on Galileo belie his own practice:

Galileo Galilei (1564–1641) [*sic*]) acquired by his investigation of the laws of falling bodies a lasting title to esteem not only as a physicist, but also as a speculative philosopher. Worthy of note are his maxims of method: independence of authority in matters of science, doubt, and the founding of inferences on observations and experiments.[16]

Thus the neglect is not justified as being a deserved one; nor could it be justified by saying that Galileo's place is in the history of science and not in the history of philosophy. This would amount to equating science with natural philosophy and then excluding the latter from philosophy; now, to do this, whatever its theoretical or present day justification, would be unhistorical since neither the exclusion nor the equation characterize the relevant historical period. I don't think the neglect is justifiable at all.

Let me therefore try to find an explanation for it. This will give us some insight into the nature of the study of the history of philosophy, or at least into the nature, however perverse, of general histories of philosophy. The *raison d'être* of such books is the attempt to find common topics and themes so that one can discern some kind of development in human thought about such topics and themes. A thinker whose philosophical ideas cannot be easily related to the others has no place in such a development. In other words, the principles underlying such books are the following: (1) history of philosophy is the study of the development of philosophical ideas; (2) philosophical ideas are ideas about such topics as God, the soul, immortality, free will, goodness, justice, truth, knowledge, and beauty; and (3) philosophical ideas originate from other philosophical ideas in the double sense that (a) ideas about one philosophical topic originate partly from ideas about other philosophical topics, and (b) the philosophical ideas of one philosopher originate from the philosophical ideas of earlier philosophers. (3a) requires that an immense amount of integration be present in order to qualify as philosophy, and (3b) requires that one have the appropriate connections to qualify as a philosopher. Such a concept of philosophy, or philosophy of the history of philosophy, excludes those thinkers who, though they had significant ideas about *some* philosophical topics, acquired these ideas from their experience, and hence were unconcerned about developing a *system* of philosophical ideas, and about justifying them on the basis of other philosophical ideas; that is, their philosophical ideas are unsystematic and justified primarily on the basis of practical considerations, where 'unsystematic' means relatively unintegrated at an abstract level, and where 'practical' refers to any general human activity upon which one may reflect from a metalevel. As we have seen, Galileo was such an unsystematic, practice-oriented philosopher.

The way to criticize such a concept of philosophy presupposed by general histories is partly by showing that it leads to the neglect of philosophers who ought not to be neglected (as in the case of Galileo), and partly by showing the inaccuracies of specific historical claims made about the origin of the ideas of the various philosophers that are discussed in such books. For example, I believe it could be shown that Hume derived many of his philosophical ideas from Galileo, in the following indirect but precise sense, namely by taking Galileo as an example, that is by trying to use in his own area of concern the approach that Galileo followed in his; here a comparison of the *Dialogues Concerning Natural Religion* and of Galileo's *Dialogue Concerning the Two Chief World Systems* would be a very relevant exercise, though beyond the scope of the present investigation. A third way to criticize the above mentioned philosophy of the history of philosophy would be to oppose to it a different one and to show the novel insights to which it leads; one of these alternatives would be to adopt Croce's concept of philosophy as methodology and to begin writing the history of philosophy accordingly. This I propose to illustrate by arguing that there is a very interesting methodological similarity between Galileo and Socrates, that is a similarity at the level of the method followed by each in their respective and very different fields. This discussion will also illustrate the kind of integration and unity that such a 'Crocean' history of philosophy would have; it would be unity at the level of method followed. The developments in such a history would be restricted to the following types: application of the same basic method to a new domain, and/or refinement of the method as applied in an old domain.[17] For the purpose of my argument I shall accept Vlastos's general interpretation of Socrates's philosophy.[18]

According to Vlastos, Socrates's greatest accomplishment, "among the greatest achievements of humanity",[19] is his method, i.e., his method of moral inquiry; only this interpretation makes sense of what Vlastos calls the Socratic paradox, which is the central fact of his life. In other words, Socrates was first and foremost a methodologist, a moral methodologist (Vlastos's own word being 'searcher'); virtually all of Socrates's activities can be understood as instances of preaching or practicing this method.

The 'Socratic paradox' consists of two groups of apparent inconsistencies.

First, on the one hand Socrates claims that the improvement of your soul is the highest value; on the other he does not seem to care about improving the soul of his fellow men since he is such a tireless critic, such a 'despotic logician' (to use Nietzsche's phrase), and so unwilling to give the answers and solutions. Second, on the one hand he explicitly asserts that knowledge is virtue, implicitly exhibits in his behavior supreme confidence in being right, and in general feels that he hás a mission to fulfill; on the other hand, he also explicitly asserts that he is not wise and implicitly exhibits in concrete discussions ignorance of the answers and solutions.

The methodological interpretation solves the paradox as follows. His tireless criticism, despotic logic, and unwillingness to give the answers derive from the nature of his method of moral inquiry and from his conviction that the improvement of the soul consists in the mastery of the method, or to be more exact, in the improvement of your ability to use the method, which to some extent is a universal human patrimony.[20] Here, the relevant essential characteristics of the method are the following: it is critical in the sense that the role of the teacher or expert is to offer criticism; it is logical or rational in the sense that it makes essential use of logic and reasoning and rigorously follows the argument wherever it leads; and the method is maieutic, or "Socratic" in the usual more restricted sense, insofar as it aims to make the person involved arrive at the solution himself. This is so in part because the Socratic teacher, according to Vlastos, does not really know the answer in advance, does not possess knowledge of the result, but only or primarily knowledge of the method of discovering it. And this brings us to the solution of the second element of the paradox: when Socrates declares that virtue is knowledge, this virtuous knowledge is knowledge of the method of inquiry and not knowledge of the answers or results, of which he is genuinely ignorant until the inquiry is completed, i.e., temporarily completed; for such virtue-knowledge is open-ended and always ready and willing to reexamine itself. In other words, Socratic knowledge is really methodological insight, not knowledge in the usual sense.

What are the essential features of the Socratic method of moral inquiry? I have already mentioned that it is critical, logical-rational, maieutic, and open-ended. Vlastos also stresses[21] that it is *difficult* in the sense that it requires sincerity, humility, and courage; that is, it requires commitment and involvement on the part of the inquirer. Other terms that come to my mind are seriousness, intellectual honesty, or 'concreteness' in a Crocean-Hegelian sense. What this really amounts to is that the method requires a give-and-take between action and thought, or experience and reflection, or theory and practice (in a Crocean sense). The answers that the inquirer is to test must be ones that really inhere in his experience. Finally, and perhaps most importantly, the method is judgmental; it is not a method in the sense in which some people conceive of method, namely as an infallible rule. Instead, 'it offers no guarantee',[22] but it is what the teacher aims at improving.[23]

In speaking of Socrates's method of moral inquiry, Vlastos distinguishes it from other specific Socratic achievements or doctrines, e.g., that it is always wrong to repay evil with evil, or that piety could amount to nothing but the improvement of your soul, or that no harm can come to a good man. These maxims alone would rank Socrates as a great moralist, but it is important to make the distinction between them and the method.

Finally, a correction should be made to one of Vlastos's critiques of Socrates, namely of the principle that virtue is knowledge. He argues plausibly that, *contra* Socrates, knowledge is neither a necessary nor a sufficient condition for virtue. However, this argument uses a concept of knowledge which is both un-Socratic and different from the one that Vlastos himself has attributed to Socrates, in order to solve the Socratic paradox. Socratic knowledge is methodological insight and not information stored in one's head and immediately available to recall; and it is concrete, i.e., serious, i.e., sincere-humble-courageous; it is not divorced or abstracted from one's experience or practice; hence it is to be found as much in what one explicitly says, as in what inheres implicitly in action. Therefore, I do not accept Vlastos's counter-example of the courageous person who cannot give a general explanation of the concept, or the one of the person who fears death in spite of his ignorance about it; that is, I do not accept them as proper examples involving the concept of knowledge that he himself is articulating and suggesting as being the Socratic one.

In summary, Socrates's greatest achievement is his method of moral inquiry, which is distinct from his specific moral maxims, and which can be characterized as being critical, logical-rational, maieutic, open-ended, concrete, and judgmental.

In the case of Galileo, it is a commonplace that his greatest achievement is his method; it is also obvious that this is a method of physical investigation. Moreover, it is easy to see that his method is critical, logical-rational, maieutic, and open-ended. The novelty is to suggest that Galileo's method of physical inquiry is concrete and judgmental; yet I believe I have shown this already, though the proof was neither short nor easy. So let me repeat what it is that was shown. Concreteness is my Crocean label to refer to what Vlastos calls sincerity-humility-courage in the case of Socrates. This involves primarily using one's own experience to test one's abstract ideas, and in turn using the latter to enlighten one's own actual experience. It is a synthesis of two aspects of human activity: the involvement in a particular situation, and the reflection on this specific involvement; I speak of *synthesis* in such a way as to oppose it to an eclectic mixture of the two; and moreover I speak of two *aspects* of human activity to indicate that they are not normally separate or separable in reality, but rather merely distinguishable in the mind. In the case of Galileo the two relevant aspects are scientific practice and philosophizing. Scientific practice is the element where Galileo is involved in the discussion and solution of some scientific problem or topic, for example concerning the motion of the earth; philosophizing is the element where Galileo remarks about such things as the relationship between theorizing and observation, the role of mathematics and of authority in physical inquiry, the origin, justification, and limitations of physical knowledge, etc.

The judgmental character of Galileo's method refers to the fact that Galileo's work is characterized by a proper balance (1) between physical inquiry and philosophical awareness, and (2) among the following: observation and speculation, mathematical (quantitative) analysis and qualitative considerations, causal investigation and positivism, negative criticism and constructive articulation, anti-authoritarianism and respect for original thinkers, anti-anthropocentrism and humanism, and anti-verbalism and logical analysis. The balancing of all these requirements is the most fundamental feature of Galileo's work in the sense that it alone does justice to the complexity and wealth of philosophical remarks that we find in his writings. This is not to say that I have portrayed Galileo as an eclectic, or that I have followed an eclectic approach in analyzing his method. The reason for this is that the balancing of the numerous activities mentioned in (2) above is a consequence of the theory-practice synthesis, mentioned in (1) above, which therefore must be regarded as an essential feature of Galileo's method, especially since it is merely another aspect of the concreteness mentioned earlier. Finally, such philosophical-scientific synthesis would alone insure Galileo's originality in the history of thought, which originality may otherwise be difficult to see in the light of the existence of precursors among previous students of nature and among previous epistemologists and theorists of method, none of whom combined the two types or levels of inquiry in a nontrivial way.

GALILEO AND THE PHILOSOPHY OF SCIENCE

Galileo occupies a unique position in the philosophy of science. He has a

significant place both in the history of the philosophy of science and in the
history of science, that is, both in the theory of scientific practice and in the
practice of which philosophy of science is the theory. Moreover, virtually
every philosopher of science has felt or feels the need to come to grips with
Galileo, in the sense that he either derives his theories from his analysis of
Galileo, or he tests the theories he has otherwise formulated by applying
them to the case of Galileo.[24] Finally, it has turned out to be possible for
philosophers of science of almost any persuasion to use Galileo for their purposes, in the sense of finding evidence from Galileo to support their theories.
It is this last fact to which I wish to call attention. I believe it has no counterpart with any other scientist. There are indeed plenty of scientists who are
more highly regarded by some one school of philosophers; for example,
Newton by inductivists and positivists, Einstein by hypothetico-deductivists
and explanationists. There is no one scientist, however, in whom almost any
philosopher, no matter how idiosyncratic, can with some semblance of
plausibility find his own predilections or prejudices. This fact, which I shall
document below, constitutes both a problem and a singular characteristic of
Galileo. It is a problem from the point of view of the current state of the
philosophy of science, since the existence of these conflicting interpretations
represents an unsatisfactory state of affairs. At the same time the existence of
such a spectrum can be regarded as a uniquely important and importantly
unique feature of Galileo's work, indeed its most basic and general aspect.
For there is no other scientist for whom such a spectrum is possible.

Let us begin our list of interpretations with what is perhaps the most unexpected one. According to Feyerabend, "Galileo violates important rules of
scientific method which were invented by Aristotle, improved by Grosseteste
(among others) and canonized by logical positivists such as Popper. Galileo
succeeds because he does not invariably follow these rules".[25] This interpretation is used by him to support what may be called a counterinductivist
methodology of science, which asserts that "in order to progress, we frequently have to step back from the evidence, reduce the degree of empirical
adequacy of our theories, abandon what we have already achieved, and start
afresh".[26]

Another philosopher has recently found evidence that Galileo was a
progressive, somewhat eclectic, scholastic, Thomistic Aristotelian[27] using the
methodology of demonstration *ex suppositione*. This is reasoning of the form
"p; if p, then q; $\therefore q$" where "p stands for a result that is attained in nature
regularly or for the most part, whereas q states an antecedent cause or condition necessary to produce that result".[28]

The Platonist interpretation of Galileo also continues to receive support, and the latest word on it is perhaps that of Howard Stein, who has given a new twist to the interpretation. He claims that "the notion of 'inquiry' suggested by Galileo's exposition is one which it is not far-fetched at all to compare with Plato's notion of 'dialectic'; and the movement, through inquiry, from fallacies and paradoxes to science, invites comparison (for instance) with the 'divided line' of the *Republic*".[29]

Empiricists, too, continue to find evidence that Galileo was a sophisticated experimentalist; it must be admitted, in fact, that they have effectively refuted Koyré's *original* claims to the contrary.[30] But that is not to say that the 'rationalist' interpretation has not unearthed its own evidence. In fact, recent studies have added a new twist to the interpretation by suggesting that, though Galileo did depend on actual experiments, and though he usually had actual experimental evidence supporting his theories, he often concealed this empirical underpinning, out of rationalist commitments.[31]

So Galileo has been variously portrayed as counterinductivist, Aristotelian, Platonist, empiricist, and rationalist. Some of these differences derive from, and would be resolvable in terms of, the different kinds of evidence examined; the main distinction here is between published works and unpublished manuscripts and documents. Some of the differences are due to emphasizing different *aspects* of Galileo's work; two distinctions would be relevant here: context of discovery vs context of justification, and scientific practice vs reflections on this practice. However, I shall follow another approach.

One may find useful in this context the cliché that each interpretation is right in what it asserts and wrong in what it denies. In other words, I think it is true that there is evidence that Galileo was a counterinductivist, *and* an Aristotelian, *and* a Platonist, *and* an empiricist, *and* a rationalist. What is wrong is to attribute to Galileo one of these characteristics and to deny the presence or significance of the others. This tendency toward exclusiveness and intolerance can be illustrated by reference to the Feyerabend counterinductivist interpretation. Actually, it is clear from the above quotation that Feyerabend is making three distinct claims. The first is the historical thesis that Galileo's work displays the feature of counterinductivism, i.e., violation of important methodological rules such as to take the evidence seriously, and to accept the most empirically adequate theory. Feyerabend's second claim is a causal explanation of Galileo's success, namely that it was because he was such a counterinductivist that he succeeded. Feyerabend's third point is the philosophical claim that the counterinductivist methodological rule is sound.

Hence a plausible reconstruction of Feyerabend's position is that he is grounding his philosophical claims on his causal explanation, and his causal explanation on his historical thesis. Now I believe that the philosophical claim is a plausible conclusion from the causal explanation, in the sense that the additional premises needed to complete the inference are all rather plausible. It is also obvious that the philosophical claim does not follow from the historical thesis alone. Let us then examine the step from the historical thesis to the causal explanation, that is the following argument: Galileo practiced a counterinductivist methodology; therefore, his success is due to this method-ology. This is a common type of historical inference, much studied by philo-sophers of the historical and social sciences, though rarely in a context like the present one. The full details of the argument would be the following:

(1) Galileo was a successful scientist.

(2) There must be a cause for this success, since every event has a cause.

(3) Galileo practiced a counterinductivist methodology.

(4) His counterinductivist methodology is a *possible* cause for his success. (Or, pedantically stated: the fact that (3) is a possible cause for the fact that (1).)

(5) There is no other feature of Galileo's work which could have been the cause of his success.

(6) Therefore, the counterinductivist methodology was the cause of his success.

(1) and (2) raise no relevant problems. (3) is Feyerabend's historical thesis, which I am accepting here for the purpose of this discussion.[32] (4), the possible-cause claim, is something that I am inclined to accept as obvious from the context in which the historical thesis (3) is discussed and supported; in other words, Feyerabend's discussions of (3) are designed to support (4) as well. (6) is the causal explanation and does follow from (1) through (5). The problem lies in (5), the denial of alternative causes. The existence and history of the other interpretations shows that this denial, so necessary to the argument, is indefensible. In fact, for each of the other interpretations, one can formulate a parallel argument, whose only problem would be the denial of alternatives. Thus I conclude that Feyerabend's causal explanation of Galileo's success is not justified, even if his historical thesis were; and since his historical argument in support of his philosophical claim depends on the causal explanation, this claim is historically groundless.

Now, what has been said of the counterinductivist interpretation, and what has been easy to do in this case because of Feyerabend's explicitness, could be said and done for the other interpretations. Each is subject to the same problem, and hence their philosophical relevance is problematic, though their historical basis is unquestioned.

What is one to do? What one can do is to follow a combinatory or synthetic approach. This is both an obvious line to try in such a situation, and also in accordance with another traditional interpretation of Galileo, so far unmentioned. Shapere's book on Galileo,[33] whatever faults it may possess,[34] does have the merit of recognizing that a synthetic approach is the correct one, since that is what the book attempts. It is true, however, that the book fails in the attempt, and instead of a synthesis it ends up with what may be called inconclusive eclecticism. I have discussed elsewhere[35] the methodological value of Shapere's book (methodological in the sense of philosophical methodology, that is at the level of the methodology for the history and philosophy of science). There I did not mention the synthetic approach but stressed another feature of Shapere's method which is valuable, namely the method of logical analysis which I emended and qualified for other purposes. Be that as it may; I believe that we ought to accept the synthetic, multifaceted approach and the logical-analytical approach suggested or inspired by Shapere's book, for there is no reason why synthesis should degenerate into inconclusive eclecticism, or why logical analysis should degenerate into scholasticism.[36] However, for substance we must look elsewhere.

Substance is what one finds in Clavelin's *Natural Philosophy of Galileo,* perhaps the best recent example of the synthetic tradition,[37] now under consideration. His view is that

the reason, therefore, why no scientific problem was the same again as it had been before Galileo tackled it lay largely in his redefinition of scientific intelligibility and in the means by which he achieved it: only a new explanatory ideal and an unprecedented skill in combining reason with observation could have changed natural philosophy in so radical a way.[38]

The new explanatory ideal is the following:

To explain, according to Galileo, meant above all to proceed from a certain number of principles and concepts and with the help of a model to an intelligible reproduction of the phenomena under investigation. This definition highlights two characteristic traits of Galileo's rationalism, above all, the almost complete reduction of physical to rational necessity: once the causal ideal of Aristotelian physics had been abandoned, explanation had the sole task of establishing an implicative relationship between the facts, as derived from a model, and the guiding principles of reason. At the same time, simplicity became an important physical criterion: nature follows the simplest path, that is, the one which permits, on the side of reason, the simplest deductions. However, explanation does not merely tend to provide a rational reconstruction of phenomena; its true aim is to turn every physical problem into a mathematical one and thereby to [utilize] for its analysis the existing mathematical science.[39]

Some comments and some analysis are needed here. Two basic features are being attributed to Galileo, the first at the level of philosophy, or theory of scientific practice, the second at the level of scientific practice or method. Philosophically, Galileo formulated a new theory of scientific knowledge; scientifically, he practiced a new skillful combination of thinking and sense-experience. Clavelin's account of Galileo's scientific method or practice is difficult to excel; it is the latest and best discussion of Galileo's synthesis of reason and observation. Occupying the second half[40] of the chapter on "Reason and Reality," Clavelin begins by criticizing the Tannery-Koyré apriorist thesis about the role of experience in Galileo's scientific practice;[41] he then analyzes Galileo's work on the cohesion of matter to illustrate his consciousness of the limitations of mathematical reason and necessity of experience;[42] he concludes with an analysis of the simplifications made by Galileo in his geometrization of the motion of heavy bodies to show his awareness of the complexity of, and his judiciousness regarding, the relationship between reason and reality.

It is important to note that there is something very special about this methodological feature described as "proper combination of reason and observation." It is not as specific as counterinductivism, Aristotelianism, Platonism, empiricism, and apriorism; it is a somewhat formal feature, somewhat empty of content, and yet it is not uninformative. It is very difficult to make precise by way of definition; attempts to do so tend to destroy the great explanatory power it possesses. Instead it is better to give content to it by showing it to be inherent in various aspects of Galileo's work, which is what Clavelin does. At the same time, the concept has an openness which allows one to apply it in new more complex ways, or to new unexamined aspects of Galileo's work; again, such an application is what Clavelin has superbly carried out. I am convinced that it is this type of interpretation of Galileo's scientific practice or method that will stand the test of time.

Let us now look at Clavelin's account of Galileo's philosophy of science, or theory of scientific practice. It occupies the first part of the chapter on 'Reason and Reality', and, as the second quotation above suggests, four elements are involved. The first is a concept of explanation as rational reconstruction of the phenomena;[43] the second is an identification of physical with mathematical intelligibility;[44] the third is the principle of correspondence between rational and physical necessity;[45] and the fourth is the principle of simplicity.[46] It is not clear how, for Clavelin, these four elements are related to each other and as a whole to Galileo's scientific practice. Regarding the latter question, the only answer I can find is that Clavelin seems to think that Galileo's philosophy of science represents an 'ideal' for his actual procedure. Regarding the former question, I think the following would be a plausible reconstruction of Clavelin's view of the inter-relationship of those four elements: intellectual reconstruction of the pheno-mena and mathematical analysis (in the sense of formulation of physical problems in mathematical terms so as to make possible mathematical mani-pulations and inferences) would be part of what Galileo *means* by explana-tion, whereas the principles of simplicity and of correspondence between rational and physical necessity would be a priori necessary conditions for the possibility of explanation in the sense specified. However, I shall not pursue such questions any further because, however interesting and complex Clavelin's view of Galileo's philosophy of science is, I don't think it goes far enough. There are three main problems with it.[47] First it does not do justice to the wealth of Galileo's remarks and reflections on scientific inquiry. In other words, though Clavelin's account is the best available, it is still an over-simplification of Galileo's philosophy of science as shown by my earlier analysis of the methodological content of the *Dialogue Concerning the Two Chief World Systems*. Second, the account is still a premature systematization, that is a systematization of Galileo's philosophical remarks before making a systematic attempt to identify and to state clearly all such remarks; this means that such a philosophy of science is more of a *theoretical elaboration* of Galileo's own philosophy, than an account of it; in other words it is more of a *Galilean* philosophy of science, rather than Galileo's philosophy of science. Third, in most of his analysis Clavelin takes Galileo's philosophical remarks out of context, being interested in relating those remarks to each other rather than to the live, concrete scientific practice out of which they arose.

What I have argued is that the philosophical content of Galileo's work is much richer and much more complex than anyone has suggested so far. It is, however, possible to define the essential characteristics of Galileo's philo-sophy of science: Galileo did not here have a philosophy of science, in the sense of "philosophy" common among some philosophers of science. In the theory of scientific method and knowledge, Galileo is not a systematic thinker in the sense of having a relatively small number of basic principles, systematically unified, and in terms of which all the facts to which they refer can be understood. This does not mean that he is not a philosopher, but rather that he is a philosopher of a particular kind, namely an unsystematic philosopher, very much like Socrates was in moral philosophy. What characterizes such unsystematic (or, to use a more positive term, Crocean) philosophy is a constant and sustained willingness to reflect on one's activities in order to understand and justify them, as the need arises in a concrete situation, and a willingness to avoid inconsistencies, when these become apparent. When such reflections center around concepts such as truth, knowledge, method, nature, physical reality, mathematics, sense-experience, etc., we have Galileo's case of Crocean philosophy; when the concepts are wisdom, goodness, justice, piety, human life, etc., we have Socrates's case. In other words, what characterizes Galileo's philosophy of science is his skillful combination of scientific inquiry and reflection upon this inquiry, reflection in the sense of remarks, judgments, explanations, justifications, etc., in short his skillful synthesis of scientific practice and philosophical theorizing (about it). Or equivalently, in order not to bias the description in favor of the primacy of practice over reflection, we may speak of Galileo's skillful application of philosophy of science to science, that is concrete putting into practice of epistemological and methodological concepts. This feature of Galileo's philosophy of science is supported by all the evidence which does support other interpretations and by all the interpretations I provided in my methodological analysis of the *Dialogue*. It has all the advantages that the feature described as "proper combination of reason and sense-experience" has at the level of scientific practice or method. It is both informative and yet non-specific and able to take into account the wealth of philosophical content that we find in Galileo. I believe it is the only feature rich enough to do justice to the riches under consideration.

NOTES

[1] These pages are those of G. Galilei, *Opere,* edited by Favaro, Vol. 7. In the rest of this section, including Tables I–IV, all pages references are to this book.

[2] R. H. Thomason, ed., *Formal Philosophy, Selected Papers of Richard Montague.* The claim can be read in a Yale University Press advertisement on p. xiii of the APA Eastern Division Program for 1972.

[3] B. Croce, *Teoria e storia della storiografia,* published in Britain under the title *Theory and History of Historiography*, and in the United States under the title *History: Its Theory and Practice,* now reprinted by Russell & Russell.

[4] Croce, *History: Its Theory and Practice,* p. 164.

[5] B. Croce, *Storia dell'età barocca in Italia,* pp. 60–64, especially p. 62.

[6] F. Ueberweg, *History of Philosophy,* Vol. ii, *Modern Philosophy,* p. 28. In addition to Ueberweg's four lines, one finds in this book two pages on Galileo in a so-called

'Appendix II. Historical Sketch of Modern Philosophy in Italy', written by Vincenzo Botta of the University of Turin.

[7] E. Bréhier, *The History of Philosophy, The Seventeenth Century*, pp. 10–11.

[8] F. Copleston, *History of Philosophy*, Vol. iii, *Late Medieval and Renaissance Philosophy*, pp. 280–81, and 284–86.

[9] W. T. Jones, *A History of Western Philosophy*, pp. 620–31.

[10] D. Hume, *Dialogues Concerning Natural Religion*, pp. 136, 138, 150–51, 162–63, 214. Some of the connections are discussed by R. H. Hurlbutt III, *Hume, Newton, and the Design Argument*, especially pp. 125–26, 140–46, 153–56. See also C. de Remusat, *Bacon, sa Vie, son Temps, et son Influence jusq'à nos Jours* (2nd ed.; Paris, 1858), p. 396, who is reported to give Hume credit for being the first English writer to appreciate Galileo as superior to Bacon; this is reported by C. W. Hendel, *Studies in the Philosophy of David Hume*, p. 287, footnote 35. See also, the Introduction, above.

[11] I. Kant, *Critique of Pure Reason*, 'Preface to the Second Edition', p. 20 (Bxii–xiii). See also the Introduction above.

[12] E. Husserl, *The Crisis of European Sciences and Transcendental Phenomenology*, pp. 23–59.

[13] J. Ortega y Gasset, *En torno a Galileo*. See also the Introduction above.

[14] Quoted by V. Botta in his Appendix to Ueberweg's *History*, p. 472, presumably from Sir David Brewster, *Lives of Copernicus and Galileo* (*Edinburgh Review*, 1830) or his *Martyrs of Science* (London, 1841); cf. Ueberweg, p. 473.

[15] E. A. Burtt, *Metaphysical Foundations of Modern Physical Science*, pp. 103–104.

[16] Ueberweg, *History*, p. 28. The date of Galileo's death is printed by Ueberweg as 1641, whereas the correct date is 1642. I suppose this is another indication of his slight interest in Galileo.

[17] One history-of-philosophy book which, while not Crocean, comes close to doing this sort of thing is K. Jaspers's *The Great Philosophers*.

[18] G. Vlastos, 'Introduction: The Paradox of Socrates', pp. 1–21.

[19] *Ibid.*, p. 20.

[20] *Ibid.*, pp. 19, 20.

[21] *Ibid.*, p. 20.

[22] *Ibid.*, p. 21.

[23] *Ibid.*, p. 20.

[24] A. C. Crombie calls him a philosophical symbol in 'Galileo Galilei, A Philosophical Symbol'.

[25] Paul K. Feyerabend, 'Machamer on Galileo', p. 297. Feyerabend's counterinduction should be distinguished from his anarchism. I believe the former is regarded by him as merely a rule among other rules, which applies to the case of Galileo but may not apply to others, whereas anarchism would be the formal super-rule (somewhat empty of content). Thus Feyerabend's anarchism may be regarded as an extreme formulation of the concrete, Socratic-Crocean methodology of science which I am supporting here and which emphasizes a move away from method and in the direction of judgment. The present critique applies only to the conterinductivist methodology.

[26] *Ibid.*, p. 298.

[27] W. A. Wallace, 'Galileo and Reasoning *Ex Suppositione*: The Methodology of the *Two New New Sciences*', pp. 82–88. Further qualified support for Galileo's Aristotelianism is given in H. I. Brown, 'Galileo, the Elements, and the Tides'.

[28] Wallace, 'Galileo and Reasoning *Ex Suppositione*', p. 95.

[29] Howard Stein, 'Maurice Clavelin on Galileo's Natural Philosophy', p. 397; see also William Shea, *Galileo's Intellectual Revolution,* pp. 150–55.

[30] Thomas B. Settle, 'An Experiment in the History of Science'; Stillman Drake, 'Galileo's Experimental Confirmation of Horizontal Inertia: Unpublished Manuscripts'; and James MacLachlan, 'The Test of an "Imaginary" Experiment of Galileo's'. In labeling these three historians 'empiricists' I am referring more to how they are perceived in the philosophical community, rather than to an explicit and articulated empiricist philosophy in their work, which could be described more fairly as merely aphilosophical.

[31] R. Naylor, 'Galileo: Real Experiment and Didactic Demonstration', pp. 398–99; idem, 'Galileo and the Problem of Free Fall', pp. 133–34; idem, 'Galileo's Simple Pendulum'.

[32] See Chapter 8, below, for a critical examination of this claim.

[33] D. Shapere, *Galileo.*

[34] See Michael S. Mahoney, 'Galileo's Thought'; Charles B. Schmitt, 'Review of Shapere's *Galileo*'; and my 'Philosophizing About Galileo'.

[35] See my 'Philosophizing About Galileo'.

[36] C. B. Schmitt labels 'scholastic in the bad sense' certain features of the book.

[37] Another good example is E. Agazzi's 'Fisica galileiana e fisica contemporanea', expecially pp. 46–47.

[38] Maurice Clavelin, *The Natural Philosophy of Galileo;* p. 383.

[39] *Ibid.,* pp. 453–4.

[40] *Ibid.,* pp. 424–53.

[41] *Ibid.,* pp. 424–32.

[42] *Ibid.,* pp. 432–48.

[43] *Ibid.,* pp. 404–9.

[44] *Ibid.,* pp. 409–17.

[45] *Ibid.,* pp. 417–20.

[46] *Ibid.,* pp. 421–4.

[47] For a detailed criticism, see below, Chapter 10.

CHAPTER 7

THE PRIMACY OF REASONING:
THE LOGICAL CHARACTER OF
GALILEO'S METHODOLOGY

It has turned out (Chapter 2) that it is possible to reconstruct the whole *Dialogue* from the point of view of the internal relationships among the propositions it contains. From this point of view the proposition that the earth moves turns out to be the crucial one in the sense that the whole book is a complex argument of which it is the conclusion. This is the logical point of view, namely the point of view of reasoning or logical practice. It should also be noticed that the methodological analysis of the *Dialogue* showed that the topic of logic and reasoning was one of the most frequent ones; in fact, the number of its discussions was exceeded only by the topic of experience, the excess being merely one. In other words logic and reasoning are one of the central methodological topics in the *Dialogue*. Because of these two 'logical' features of the *Dialogue,* it will be useful to inquire whether or to what extent it is possible to interpret the other methodological topics in terms of logic and reasoning, namely to what extent one can reduce the book's methodological discussions to discussions concerning logic and reasoning.

Let us first look at the specific topics[1] discussed in those passages where logic and reasoning are explicitly mentioned. In 59–71[2] we find a discussion of the distinction between practice and theory in logic, or logical practice and logical theory, or logic-in-use and reconstructed logic, or between reasoning and logic in a narrower sense. In 290–9 we find a discussion of the distinctions between an argument and a rhetorical appeal, and between the form and content of an argument. In 57–9 and 169–80 we find a discussion of some relationships between reasoning and sense-experience; in 169–80 the relationship between reasoning and the Socratic method is discussed; the passage on 150–9 is a discussion of the value of reasoning; the one on 159–69 is an illustration of the analysis and evaluation of arguments; and the passages on 139–50, 196–206, 399–406, and 462–70 are discussions of special types of reasoning, namely probable arguments, hypothetical reasoning, *ad hominem* arguments, and reasoning *ex suppositione,* respectively. Such explicit discussions of logical topics are clear evidence of Galileo's methodological self-awareness about logic and reasoning. If one wanted to 'systematize' such remarks, the following would be a possibility.

Reasoning is the subject matter of logic (i.e., of logical theory) in the sense that it is something whose structure and validity logic studies (59–71). The basic structure of reasoning consists of the inferential interrelationships among sentences or propositions, and the validity of reasoning (or logical validity) refers to the propriety of such interrelationships (159–69 and 290–3). The construction of arguments is distinct but not separable from the analysis of arguments, in the sense that arguments are never constructed in a vacuum but rather in the context of counterarguments, and in the sense that the analysis of counterarguments is the best practice and preparation for the construction of one's own positive arguments (150–9). The examination of the consequences of more or less arbitrary assumptions is another valuable practice, though one must be careful and realize that genuine hypothetical reasoning is very difficult (196–206). One must also realize that reasoning that proceeds from the unexamined assumptions of opponents has limited methodological value, though great rhetorical effectiveness (399–406). Moreover, so-called reasoning *exsuppositione* is rather tricky in that its validity depends on whether the supposition in question is contextually free of objections (462–70). Finally one of the things to be contrasted to reasoning is sense experience; the following are two principles that should be kept in mind: (1) whether or not sensory appearances correspond to reality is to be decided by reasoning (57–9), and (2) whether or not there is a need to use one's senses is to be decided also by reasoning (180–8).

Experience. Let us now examine the discussions of other topics to see how these other topics relate to reasoning. Let us begin with experience. The discussion of conceptual frameworks in 38–57 relates to experience only indirectly since it pertains primarily to conceptualization. Hence this is a discussion of reasoning insofar as conceptualization is a type of reasoning, which will be examined later. Here it should be noted that the discussion emphasizes the harm that sense experience can cause.

In the next passage (57–9) the emphasis is on being critical of experience, in the sense that we should ask ourselves whether what the senses reveal to us corresponds to reality. An example referred to here, but not explicitly discussed till later (159–69), is the vertical fall of bodies. Experience shows us that bodies appear to fall vertically. Only reasoning can show whether or not this implies that they actually do. The implication does not hold because it can be easily shown that on a rotating earth, apparent vertical fall would imply actually slanted fall.

The next discussion of experience (95–112) does show that for certain purposes there is no substitute for it. It is true that the same discussion shows that what conclusions one draws from experience is a matter of reasoning.

However, this does not change the fact that the use of the senses is some-times essential. The proper use of this discussion is not to try to do away with the observational element, but rather to use it by way of contrast in order to get a better idea of what reasoning is and what it is not. In other words, though Galileo regards reasoning as very important, he does not want to say that it is the only activity in natural philosophy. To realize that other and different activities exist, and that they are sometimes necessary, gives more content to the emphasis on reasoning that Galileo practices and preaches. We may say that this discussion of experience is part of an attempt at a definition of reasoning, a definition by contrast.

The next discussion of experience (169–80) shows how only reasoning can decide whether or not an experiment is called for. Again this presupposes a distinction between the two, but it is evidence of Galileo's emphasis on reasoning rather than experience.

The passage on pages 206–9 discusses some of the ways in which one should be critical of experimental results. The main connection with reason-ing is that only by reasoning can one decide whether a certain natural effect is large enough to be detectable. This is another limitation of experience vis-à-vis reasoning.

In the next passage (209–14) the over-all trend is toward emphasizing the importance of experiments, especially in order to observe phenomena that presumably couldn't possibly happen according to certain lines of reasoning. Here experience does perform an essential function, but even here it is reasoning that has to choose the proper time or occasion for making the experiment.

Concerning the emphasis on experience in the passage on physico-mathe-matical synthesis (244–60), the following points should be noted. First, some of the experience referred to (e.g., the times-squared law of fall) involves observations which are mathematically described and analyzed, and hence is experience filtered through a type of reasoning, namely mathematical reason-ing. Second, some of the experiences referred to (e.g., the body oscillating in a tunneled earth) are thought-experiments rather than actual ones; but a thought experiment is really a piece of hypothetical reasoning. Third, in the synthesis of physical considerations and mathematical analysis, the element of experience obviously enters in the physical part; but equally obviously it cannot be experience itself, but rather reasoning, that determines the proper combination – how, when, and exactly to what extent the two elements are to be combined.

The discussion of the deception of the senses (272–81) is really a discussion of reasoning since Galileo is claiming that the deception comes from our being uncritical about our sense experience and from incorrectly reaching conclusions about reality on the basis of sensory appearances; that is, the so-called deception of the senses is deception of reasoning. Insofar as experience is being mentioned, what we have here is again a contrast between it and reasoning.

The discussion of simplicity, insight, and sense experience (349–68) is one of the passages with the most definite emphasis on the importance of experience, even though it is often misinterpreted as emphasizing 'reason'. Galileo is saying that he is enough of an empiricist that if it had not been for the experiences possible with the telescope, he could never have accepted the Copernican system. He is indeed *awed* by the *insight* of those who believed it in the light of clear empirical counterevidence; he is *not* saying that it was their superior *reasoning* that led them to what he now believes to be the truth. Galileo's point is that reasoning ought not to do that kind of violence to sense experience, that is the kind of violence exemplified by Copernicus' procedure. Hence it is clear that the discussion in this passage is about reasoning, as much as it is about experience, which is all that I wish to claim. For my examination of the extent to which the topic of logic and reasoning prevails in the book is not predicated on any assumption that would equate reasoning with "reason" in the sense of a priori reason. Though different from sense experience, reasoning uses sense experience as much as needed to avoid errors and arrive at the truth. In other words there is no reason to equate reasoning with a priori reasoning; the latter does exist, but it is merely a special case of the former; hence the present interpretation of Galileo, as a logician in the sense of a conscious and skillful practitioner of reasoning and of logical analysis, is not a 'rationalist' interpretation in the sense of portraying him as an apriorist.[3]

The discussion of the logic and methodology of theoretical explanation (372–83) relates to experience insofar as it suggests that, in certain contexts, sense experience has much more of a methodological than logical function. That is, in the theoretical explanation of sense experiences, these cannot properly provide the basis of reasoning to theoretical conclusions about reality, but rather the basis of reasoning to conclusions about how to proceed in natural investigation. In a sense, what we have here is another fact about reasoning, a fact that can appear as a 'limitation' to those who may have been inclined to overestimate its power. This contrast between logic and methodology is a contrast between reasoning and *acting* in natural investigation, and hence it is in a sense part of a contextual clarification of the concept of reasoning.

The last passage dealing with experience is a discussion of theories and observations (409—14). Here reasoning is indirectly discussed in a way similar to the passage just analyzed; acceptance and rejection of a hypothesis are decisions to act in a certain way, namely to do and to decline from doing further work on it, respectively.

Cause, Explanation, and Understanding. The topic of causes, explanation, and understanding is one of the most frequently discussed ones in the book (eleven times). One of these passages (372–83) is identical with one already examined under experience and hence will not be analyzed here. Let us look at the other ten.

The discussion of the contrast between understanding and information and of the Socratic method (180—8) is easily reduced to a discussion about reasoning. For the understanding conveyed through the Socratic method involves deriving what was previously not accepted from propositions that are accepted, and this is precisely what reasoning is.

The discussion of criticizing and understanding (214—7) claims that criticism is more effective when grounded on understanding. The connection with reasoning is that Galileo seems to be claiming that if one wants to be rational in one's criticism, if one wants to reason about what he criticizes, then he should make sure that he understands it. Hence we have here a plea for reasoning.

The first discussion of causal investigation (237—44) may surprise those who believe that Galileo did away with such investigations; actually there are several such discussions. From the point of view of reasoning, causal investigation is simply reasoning about causes, which makes it easy to see that there is nothing in principle illegitimate about it. Hence the several discussions of causal investigations can be easily understood from such a logical point of view. The present passage also offers the occasion to note the ambiguity of the phrase 'reason for', which can mean either 'causal explanation for' or 'logical justification for'. Hence, reasoning, that is the search for logical justifications, is to be distinguished from the search for causal explanations. Here we then have another partial definition of reasoning by contrast.

The discussion of comprehensibility and truth (260—72) relates to reasoning as follows. First, it is the discussion of a number of principles of reasoning about comprehensibility and truth: that comprehensibility does not imply truth, that incomprehensibility does not imply falsehood, that naming does not imply comprehension, and that causal inexplicability does not imply incomprehensibility. Second, though reasonable truth, i.e., truth to which one can arrive by a process of reasoning, is a type of comprehensible truth,

one cannot equate the two because of the possibility of intrinsic comprehensibility. Hence, understanding is one of the things which is to be distinguished from reasoning, though they are also related.

The discussion of inconceivability-claims and facts (281–8) is a discussion
of the following principle of reasoning: that it is wrong to argue that something is not the case because there is no way to *conceive how* it *could* be
the case.

The discussion of the role, discovery, and justification of causes (431–6)
is, like the previous discussion of causal investigation, partly a discussion of
reasoning about causes and partly a discussion of the difference between
reasoning and (causal) explaining.

The passage on the choosing of the effect in a causal investigation (442–5)
is a discussion of a preliminary step to take in reasoning about causes. The
next passage (445–62) is a discussion of the following principle of reasoning:
that a proposed cause should be rejected if it does not allow an artificial
reproduction of the effect, and it should be accepted if such an artificial
reproduction is possible.

The discussion of confessed ignorance and occult qualities (470–1) can
be interpreted as involving the principle that it is sometimes better to refrain
from reasoning rather than arrive at a conclusion with little foundation.

The passage on the method of concomitant variations (471–84) is a discussion of a principle of what some would call 'inductive reasoning'. Or we
may say that it is a principle for the discovery (or conception) of causes.

Socratic Method and Unconscious Knowledge. The topic of Socratic
method and unconscious knowledge is discussed five times, and it is easily
reducible to logic and reasoning. The passage on cognitive awareness (112–7)
illustrates how the intellectual contents of the human mind are not always
open to introspection or known best to the person involved. When a second
person undertakes a Socratic cross-examination, the person involved is being
forced to reason about the rationale underlying his beliefs. In general,
Socratic cross-examination is one of the best instances of reasoning, of logic
at work. The second passage on this topic (169–80) has already been
examined in the context of the analysis of the discussions of experience; it
was found to be a paradigm example of a discussion of reasoning. The same
has been done for the third passage (180–8) in the context of our examination of the passages dealing with understanding. The discussion of unconscious
knowledge (193–6) is almost a definition of a type of unconscious knowledge
as that to which one can arrive by reasoning. The last discussion of the topic
of Socratic method and unconscious knowledge (217–23) is also the most

explicit. Equally explicit is the emphasis on reasoning, for Galileo is saying that knowledge is a kind of recollection in the sense that it is sometimes acquired by reasoning based on facts ascertainable by reflection upon certain chosen aspects of one's experience.

Mathematics. Let us now examine the discussions about mathematics. The first thing to notice about them is that mathematics is not at all the book's main methodological topic; it is discussed only five times. The first discussion (33–8) is pretty clearly a plea for mathematics in the sense of mathematical reasoning as opposed to number mysticism. In other words, mathematics is simply a special case of what I mean by logic in this book, the case where the reasoning deals with numbers, quantities, and geometrical figures. The discussion of geometry and philosophy (223–9) is a discussion of the importance of geometrical reasoning in natural philosophy. The passage on mathematics and physical reality (229–37) discusses the applicability of mathematics; in so doing Galileo mentions another special feature of mathematical reasoning, namely its necessity; however, he is clear that this necessity does not extend to physics, even mathematical physics, since the latter is the application of mathematics to nature, and every such application lacks necessity. In other words, Galileo here seems to be discussing the nature and status of mathematical reasoning: he is saying about its nature that it is apodictic; and about its status he is saying that it is relevant to physics, exactly how and why the application is possible, but that this application is nonapodictic. Or we may say that Galileo is discussing apodictic reasoning, claiming that it exists in mathematics, but that it has limitations since the *mere application* of it involves nonapodictic reasoning.

The discussion of physico-mathematical synthesis (244–60) has been examined before, as a discussion about experience. Here we may add that such synthesis may be regarded as the application of mathematical reasoning to physical reality.

The passage on the role, discovery, and justification of causes (431–6) has been examined before from the point of view of the topic of cause, explanation, and understanding. The references to mathematics in this passage are not central, but they do deserve some discussion here. One of Galileo's claims is that practice with mathematical demonstrations makes one less willing to accept a causal explanation as conclusively proved; this is easy to understand now, since we have just seen that for Galileo mathematical demonstrations are apodictic. Another claim that he makes here is that the direct experimental confirmation of a cause can give a causal explanation a certainty comparable to that of mathematics; here he seems to be saying that if in the domain of reasoning it is sometimes possible to have necessity, namely with mathematical reasoning, in the domain of sense experience it is possible to have a comparable certainty, namely with the clear evidence of the senses.

Authority. The topic of authority is discussed five times, four as it relates to Aristotle and once briefly as it relates to the Bible. Generally speaking, Galileo's plea against authority can be related to reasoning as follows: appealing to authority is a good example of nonreasoning, but such nonreasoning, unlike for example sense experience, is undesirable. It is interesting to note that three of Galileo's critiques proceed by focusing attention on Aristotle's practice, as opposed to his mere words, and then arguing that if we take this practice seriously, then we need not take seriously certain of his claims.

The discussion of Aristotle's authority as a logician (59–71) has already been examined, since it is a direct and explicit discussion of logic and reasoning. From the point of view of authority, Galileo is saying that insofar as one is using authority in one's reasoning, one has to make sure that the authority is relevant, hence appealing to authority is no substitute for reasoning.

The next passage (71–82) discusses the role of authority in natural philosophy. Here Galileo gives another reason why appealing to authority is no substitute for reasoning. If one is appealing to an authority, it is more adequate to look not only at the authority's conclusions but also at his reasons and arguments. In so doing one will be examining the authority's manner of reasoning. Now, in some situations the facts may be such that, by applying this manner of reasoning to them, one is led to a different conclusion. In such a case there is a conflict between appealing to the authority's conclusion and appealing to his reasoning. Then one has to use his own judgment to decide which part of the authority to follow. Normally, the manner of reasoning is a much more important aspect, and it should be followed. In the present passage Galileo discusses this point by reference to Aristotle's claim that the heavens are inalterable and to his argument that this is so because no changes have ever been observed in the heavens. In another passage (346–9) Galileo discusses the same point by reference to the example of Aristotle's claim that the earth is the center of the world and his argument that this is so because the earth is the center of celestial revolutions.

The passage on independent-mindedness (132–9) also shows that what Galileo is really objecting to is the mindless appeal to authority, that is appealing merely to the conclusions and not to the reasons or evidence. Finally the passage on the Bible (383–5) merely dismisses it as a possible authority in natural phenomena, since it does not contain reasoning about natural phenomena and the claims it makes about them are unreasoned.

Method. There are five passages, interspersed throughout the book, which in one way or another discuss the idea of method. They are the passages on the role of authority in natural philosophy (71–82); physico-mathematical synthesis (244–60); procedure vs results, and Aristotle's authority (346–9); logic vs methodology in theoretical explanation (372–83); and searching vs results: the supremacy of method (484–9). All of these passages except the last one have already been examined insofar as they contain discussion of various other philosophical topics; when grouped together, they suggest that Galileo is practicing and preaching the idea that the method of arriving at a result is much more important than the result itself. This is something that could be called methodologism; it is also a species of formalism, insofar as the method/result distinction reminds me of the form/content distinction. It is also a species of Socraticism, insofar as the maieutic character of the Socratic method represents also an emphasis on method as opposed to end-result.

How does such methodologism relate to the logicism that seems to be the essential feature of the *Dialogue*? It would be wrong to equate the two in general since methodologism could take forms different or incompatible with logicism. However, it is clear that to emphasize reasoning is a type of method-ologism, since reasoning involves the interrelationships among facts, ideas, or beliefs, and the proper relationships can exist even when the specific ideas are changed. It is also clear that the content of Galileo's methodologism, i.e., that the specific type of his methodologism is logicism, since in three out of the five passages (71–82, 346–9, and 484–9) he is explicit that the method he has in mind is reasoning; and one other passage (244–60) has already been examined and reduced to reasoning. Admittedly, the remaining passage (372–83), discussing the logic vs the methodology of the theoretical explan-ation, does use a different concept of method, relating to acting rather than reasoning. But this represents no real problem since the discrepancy could properly be used to take this passage out of this group, because after all it was so grouped only from an initial impression that it had something in common with the other four. Moreover, the passage has already been examined, for its relation to reasoning, in the context of the topic of sense experience.

Anthropocentrism. Teleological anthropocentrism is my label for the thesis criticized by Galileo that the universe exists for the sake and benefit of man. The topic is discussed on four different occasions. The first passage (82–95) is easily reduced to a discussion of logic and reasoning. For what he does in this passage is to begin by asking what is the rationale on which the belief is based and what are the conclusions that might be drawn from it; he criticizes the rationale insofar as it consists of the earth/heaven distinction, and he criticizes the alleged implication that celestial bodies are perfectly round. In a sense Galileo is here making a plea for objectivity by saying that, no matter how much we like anthropocentrism, we cannot let our emotions prevail over reason in this case. In other words, one activity that can be contrasted to reasoning is wishful thinking, even where the wish would extend to all mankind. So one of the things that reasoning is *not,* is feeling or emotion; and with respect to some issues, such as the present one, reasoning must prevail; and finally, even when feeling may be allowed to prevail, only reasoning can decide this.

In the discussion of the powers of human understanding (124–31), we see that another version of anthropocentrism is the thesis that the human mind is a measure of what can occur in nature. Galileo's criticism of this shows that his emphasis on logic and reasoning, which may be called logicism, is not a species of "rationalism" as ordinarily understood. For one type of rationalism would make optimistic, extravagant, and unrealistic claims about the ability of man to know about nature merely by the exercise of his mind; and such rationalism impresses Galileo as arrogant anthropocentrism, to be rejected. He makes a similar point in the discussion of the primacy of nature over man (289–90): since nature is prior to man, a priori reasoning has a limited function.

The passage on objectivity and the concept of size (385–99) is a discussion of some of the ways in which reasoning is dependent on sense experience and on the faculty of imagination; for example, our reasoning about sizes depends on our perception of sizes and on our ability to imagine various sizes. However, such dependence is not equivalent to a dependence on human interest; reasoning may have to take into account human interests, but it cannot be subservient to them; instead it must rule over them.

Conceptualization. The three passages on conceptualization deal with conceptual frameworks (38–57), inconceivability-claims and facts (281–8), and the superficiality of abstract answers (293–8). Conceptualization was the label under which these three passages were grouped together in the context of the analysis of the book's main methodological themes; the label was meant to refer to conceptual thinking. From the point of view of reasoning, we can say that the topic in these three passages is a priori reasoning. In his discussion of conceptual frameworks Galileo is suggesting that on certain occasions a priori reasoning is proper; in fact, on such occasions what is wrong is to infect it with empirical considerations. The occasions are situations

when what is at issue is the basic concepts in terms of which one is to interpret empirical material. The question of what is regarded as 'natural motion' is
an example. In the other two passages Galileo criticizes certain types of a priori
reasoning: in one (281–8) we have an empirical argument designed to throw
some doubt on whether one can ground factual falsehood on conceptual
difficulty; the last passage (293–8) suggests that often a priori reasoning is
misapplied, in the sense that relevant factual considerations are neglected.

Open-mindedness. The first passage (150–9) is an explicit discussion of
reasoning, and of its essential connection with open-mindedness, and hence it
has been examined above. Here it should be pointed out that in that passage
Galileo stresses how open-mindedness leads to reasoning. In the passage on
open-mindedness, intellectual cowardice, and curiosity (425–31) Galileo emphasizes the converse connection, namely how closed-mindedness leads to a
failure to reason by summarily dismissing other views; this is also discussed
by Galileo by suggesting how reasoning can lead to open-mindedness, namely
that by examinating the argument in support of a view one may become interested or intrigued by the argument, and as a consequence by its conclusion.

Ignorance. The four passages grouped under the label of ignorance contain
discussions of the desirability of modesty in one's claims to knowledge. In
none of these passages except the last one (470–1) is such modesty the main
topic of discussion; consequently, three out of these four discussions have
already been examined from other points of view. Nevertheless, this theme of
epistemological modesty, mentioned on four occasions, is further evidence
that Galileo's emphasis on reasoning, which we may call his logicism, is not
equivalent to 'rationalism', as ordinarily understood. It is interesting to
extract Galileo's reasons for such epistemological modesty: from the passage
on the powers of human understanding (124–31) we get that rationalistic
arrogance is a sign that one has never understood or tried to understand any
one topic fully; from the discussion of independent-mindedness and the
misuse of authority (132–9) we get that if one doesn't claim to know everything, then he will be taken more seriously when he does claim to know
something; from the passage on open-mindedness, intellectual cowardice, and
curiosity we get that modesty makes one more open-minded and arrogance
more dogmatic and rigid; finally, from the explicit discussion on confessed
ignorance and occult qualities (470–1), we get that epistemological arrogance
is likely to be caused by a failure to use reasoning and by the exercise of idle
imagination, vanity, and verbosity.

Simplicity. Simplicity appears as one of the main topics of discussion in
two passages (139–50 and 349–68) and as the single central topic in another

(416—25). The passage on simplicity and probability (139—50) has already been examined since it contains an explicit discussion of reasoning, specifically of probable arguments; it is worth repeating here that Galileo claims that reasoning depending on the principle of simplicity is merely probable and not apodictic. Both in this passage and in the one on simplicity, insight, and sense experience (349—68) Galileo says that in the light of empirical counterevidence this probability reduces to zero, in the sense that it is not a sound principle of reasoning to infer even probable truth from simplicity when there are clear and unanswerable empirical difficulties. Such reasoning is unsound in the sense that it is a bad policy to follow regularly, though of course it has unpredictable exceptions, like Copernicus. That is, one cannot but marvel at their insight, but one can speak of *insight* only after sound reasoning can justify their conclusion, for example after one has been able somehow to remove the difficulties. So reasoning based on simplicity must be handled with care for two reasons: at best it is merely probable, since nature *tends to act,* but we can't be certain that it always does act, by means of the simplest operations (139—50); and one must watch out for unanswerable empirical counterevidence, which can easily offset these degrees of probability. A third problem with reasoning based on simplicity is due to the complexity of the concept, on which we have an explicit discussion (413—25): it is not always possible to agree on which idea is simpler, since different criteria of simplicity are often used, two frequently differing criteria being an abstract-mathematical one and an ontological-physical one.

Astronomy. The four passages on the nature of astronomical investigations are, from the point of view of reasoning, discussions of some problems and limitations with quantitative reasoning and of the role of precision and exactness in reasoning. The passage on quantitative data and qualitative conclusions in astronomy (299—337) discusses the problem that sometimes quantitative astronomical data are so inconsistent that no quantitative conclusion can be drawn from them. The passage on the reliability of astronomical measurement (337—46) discusses the problem that sometimes the data are based on unreliable measurements. The discussion of computational vs philosophical astronomy (368—72) is a critique of quantitative-mathematical reasoning in astronomy, insofar as such reasoning can lead to the neglect of qualitative-physical reasoning. The passage on the inexactness of astronomical instruments (414—6) discusses how sometimes the effect to be measured is so small that great precision of instruments is essential; but even here the great precision often derives from the great sophistication, and this sophistication is often the result of a new qualitative idea.

Criticism. Out of the four passages under the heading of criticism, one (206–9) has already been examined, in the discussion of sense experience, and also it raises no new point in the present context. The remaining three relate to reasoning as follows. The discussion of different categories of error (117–24) suggests that in criticism sometimes there are things other than reasoning to be evaluated, one of these external considerations being whether or not an erroneous idea is being defended at a time when a better one has been published. The discussion of criticizing and understanding (214–7) is saying that, in criticizing reasoning, sometimes one should strengthen it before finding fault with it. The discussion of the value of criticism (406–9) suggests that criticism can benefit reasoning, in the sense that reasoning can be improved as a result of answering objections made to it.

In conclusion, we may say that all the methodological discussions in the book can be interpreted as centering around the topic of logic and reasoning; what is being discussed directly or indirectly are such things as the relation between logic and reasoning; the nature of reasoning; types of reasoning; the relation between reasoning and other things such as sense experience, appealing to authority, explaining, and feeling; and a number of principles of reasoning.

NOTES

[1] In this chapter, the discussion is centered around the topics as systematized in Tables III and IV of the last chapter.

[2] The numbers are page references to G. Galilei, *Opere,* edited by Favaro, Vol. 7. Similarly for the rest of the discussion in this chapter. See Tables III and IV of the last chapter.

[3] For a further discussion of the distinction between reasoning and a priori rationalism, see my criticism of Koyré in Chapter 9.

CHAPTER 8

THE RATIONALITY OF SCIENCE AND
THE SCIENCE OF RATIONALITY:
CRITIQUE OF SUBJECTIVISM

CHANGE, PROGRESS, AND RATIONALITY IN SCIENCE

Is science rational? What does its rationality consist of? Does it consist, for example, of the effectiveness of its method? Does it consist of something peculiar to science and demarcating it from other cognitive activities, or does it consist essentially of a special case of general cognitive rationality? Three types of answers are possible to such questions. One may say that science is rational insofar as it evolves and develops in a way that can be understood and explained, or insofar as one can show that the changes that it undergoes are changes for the better, or insofar as it is the result of individuals thinking reasonably and acting with good reasons. The rationality of science is reduced to the *explicability* of scientific *change* in the first approach, to the *provability* of scientific *progress* in the second, and to the rationality of *scientists* in the third.

These three approaches derive from the types of questions that it is possible to ask about a given scientific episode. One may ask, how and why the episode took place, whether and how its occurrence was a good thing (i.e., a change for the better), and whether the episode was rational in the sense that persons involved in it acted rationally. Hence we may also say that the three approaches derive from three distinct aspects of science, namely, change, progress, and rationality (in a restricted sense of 'rationality', according to which only conscious, intelligent beings can act rationally or irrationally). Since this restricted sense of 'rationality' is the primary one, and since the concepts of change and of progress apply rather well and specifically to the two other aspects of a scientific episode, it is best to restrict the term 'rationality' to its primary sense, as we shall do below. We may also say that the three approaches constitute three different types of inquiry, which are respectively explanation, evaluation, and rational reconstruction (in the sense of reconstruction of reasons).

These three approaches are related but distinct. The relations are as follows. Sometimes it is possible to explain the occurrence of an episode as being the result of scientists acting rationally; and moreover to show that a given scientific change was one made by rational agents would define a sense in which it was a change for the better. In other words, the rationality of a scientific change can *sometimes* explain its occurrence and *sometimes* justify it as being good. The three approaches are however different, for obviously not every change need be a change for the better; moreover, not all progress derives exclusively from the rationality of the agents involved, since it could be defined as approach to a certain desirable goal, and this may occur even if the agents happened to have acted somewhat irrationally. Hence there is nothing wrong with a certain amount of division of labor and with restricting oneself to just one type of inquiry. However, just as the pursuit of a single one of these topics does not guarantee success, so the simultaneous investigation of all three does not necessarily involve confusion. For what one must do is the following: if one's claims and conclusions are about change then one's evidence and reasons must pertain to change and not progress or rationality; if one's conclusions are about progress then one's evidence must pertain to progress and not to change or rationality; and similarly for rationality.

This tripartite distinction helps to make sense of the work of recent philosophers of science who are historically oriented.[1] It is also the one that seems to be adumbrated by Stephen Toulmin's project on human understanding. His first volume[2] dealt with what he calls "the collective use and evolution of concepts" and corresponds basically to a study of scientific change. The second projected volume was supposed to deal with "the individual grasp and development of concepts"[3] and would correspond to a study of scientific rationality (in the primary sense). The third projected volume will presumably deal with "the rational adequacy and appraisal of concepts"[4] and promises to be a study of scientific progress. The distinction also corresponds to Karl Popper's talk of the three worlds.[5] His first world corresponds to that of scientific change, his second to that of rationality, and his third to that of progress.

The distinction could also be interpreted as follows: the study of scientific change corresponds to history of science, the study of scientific rationality to the psychology of science, and the study of scientific progress to the epistemology of science. I would have no objection to this terminology if it were not used in an invidious and prejudicial sense. However, usually talk of the history, psychology, and epistemology of science goes together with excluding the former two from the domain of philosophy of science, which is thereby equated with the third. However, there is no reason why the study of scientists' reasoning cannot be carried out in a philosophical manner, nor is there any guarantee that the epistemology of science will not degenerate into a nonphilosophical linguistics of science, for example. And as regards the history of science, I have argued elsewhere[6] that the explanation of scientific change involves special methodological problems, of a kind that requires philosophical sensibility, though the philosophy in question is not a philosophy of the historical development of science, but rather an ability to perceive and use conceptual distinctions concerning the given subject matter, so that such concrete philosophizing may be no more common among scholars who call themselves 'philosophers' than among those who call themselves 'historians'.

However, this raises the question of the connection between my tripartite distinction and the one usually made between historically oriented and logically oriented philosophy of science. My distinction is *not* a simple expansion of the usual dichotomy. In fact, I would analyze most logically oriented philosophy of science as being a mixture of mostly studies of rationality, of some studies of progress, and a little history (mostly of contemporary science); whereas the historically oriented philosophy of science consists mostly of studies of change, with some studies of progress, and a few studies of rationality; so that the difference is one of emphasis. However, a few historical philosophers have emphasized the problem of rationality, notably Mary Hesse[7] and Paul Feyerabend;[8] while a few philosophers who began with a logical orientation, e.g., Karl Popper[9] and Toulmin,[10] ended up dealing mostly with scientific change and progress.

SCIENTIFIC RATIONALITY

The main concern of the present investigation is scientific rationality. Few recent philosophers have been as involved in solving, or at least dissolving, this problem as Feyerabend; thus, since his views are relatively well-known and accessible in several languages,[11] I shall formulate mine in the course of a critical examination of his recent book *Against Method.*

One of the most attractive features of Feyerabend's approach is what I shall call its *concreteness.* This is not his word, but it is obviously what he has in mind when, for example, he criticizes Carnap's excessively abstract approach (p. 183, n. 7).[12] Moreover, Feyerabend's use of historical evidence is an expression of this concreteness, for he is not interested in history as such, but insofar as it puts the philosopher in better contact with scientific practice.

Second, I agree with the *anthropological* aspect of Feyerabend's approach (Ch. 17, esp. pp. 249–60, and p. 252). This partly corresponds to what some would call the phenomenological method (in the sense of existential phenomenology), and partly to the historical method, in one sense of 'historical', according to which one suspends one's own point of view and takes the point of view of the historical agents. To be sure, Feyerabend does not always practice this anthropological method that he preaches. For example, he portrays Galileo as a counterinductivist, which is behavior in accordance with the rule that "advises us to introduce and elaborate hypotheses which are inconsistent with well-established theories and/or well-established facts" (p. 29). However, none of this evidence, even if it were otherwise acceptable, tends to show that Galileo introduces hypotheses which *he believes* are inconsistent with well-established theories and/or facts; in other words, Feyerabend gives no evidence that Galileo agrees with Feyerabend and with the Aristotelians in thinking that the Ptolemaic system was well established. For someone following the anthropological method it is Galileo's actual thought and beliefs that should be reconstructed. However, in spite of such lapses, Feyerabend's 'anthropological' orientation is to be welcomed.

A third acceptable feature of Feyerabend's approach is his normative, critical orientation. He wants to find out not only what scientific rationality is but whether it is good or bad, and he wants to reform it insofar as it is bad. This may seem to conflict with the anthropological approach insofar as the latter could be interpreted as emphasizing description, rather than prescription. There is no conflict, however, because Feyerabend is critical primarily of contemporary science anthropologically understood, and he grounds this criticism on past science also anthropologically understood. In other words, he uses certain parts (historical stages) of science against other parts. To be sure, Feyerabend also tends to exaggerate his criticism. For example, noting that a streak of intolerance has developed within modern science, he bemoans the fact "while an American can now choose the religion he likes, he is still not permitted to demand that his children learn magic rather than science at school" (p. 299). However, there is no good reason to follow him in going that far. Moreover, some of his criticism is grounded on extrascientific factors, for example that parts of modern scientific education "cannot be reconciled with a humanitarian attitude" (p. 20); but such criticism, however intrinsically plausible, derives its relevance from the other grounded on the history of science. Thus, Feyerabend's destructive, antiscientific pronouncements should not be confused with the healthy, reformist, critical stance.

Fourth, I agree with Feyerabend's argument that scientific rationality does not consist of normative principles stated in terms of the theory/observation distinction. His argument would be that as long as we use such concepts, then we are bound to value theories which are well-grounded on facts, and hence to be intolerant toward theories that conflict with facts or other well-established theories. But such intolerance will make it difficult or impossible for better theories to be formulated or new facts discovered because a principle which Feyerabend labels 'counterinduction' can be shown to be a very effective method. Feyerabend supports the counterinductive rule with an historical and a philosophical argument. The historical argument consists of evidence that most of the great discoveries in the history of science have had this feature of conflicting with previously well-established theories and/or facts. His philosophical argument has two parts, one relating to theoretical counterinduction, the other to factual counterinduction. First, given a very general theory, some facts cannot be discovered except with the help of a conflicting alternative theory; hence if we value new facts, and if we want to be critical toward a well-established theory, we must consider alternatives; in short, the improvement of a general theory, no matter how well established, is best done by the external criticism of contrast to other theories, rather than by internal criticism of comparison with experience. As for factual counterinduction, historical investigation shows that there is not a single interesting theory that agrees with all known facts in its domain; and epistemological analysis shows that every observational report presupposes some theoretical assumption; therefore it is a mistake to take 'facts' too seriously, so seriously as to prevent the consideration of theories that conflict with them. From such arguments Feyerabend does *not* conclude that counterinduction ought to be the new rule for the conduct of scientific research, but rather that even such an obvious rule as the inductive principle which counterinduction opposes is no guarantee of success (pp. 32–33). However, he does conclude that there can be no rules of any kind, whereas the only thing that follows is that there are no rules of the kind considered by him, namely formulated in terms of the theoretical/observational distinction. In other words, Feyerabend's arguments 'against method', as he puts it, are really arguments that we ought to do away with the theory/observation distinction (cf., e.g., p. 168). However, as he himself argues in his book, theories are not rejected until there is an alternative to replace them; applying this to the theory of the 'theory/observation' distinction, we note that he has not provided an alternative way of studying scientific rationality, which would replace the philosophy of science based on the theory/observation distinction. The present investigation is in part an attempt to provide such an alternative by using the premise/conclusion distinction of logicians, that is by using the concepts of argument and reasoning of elementary logic.

This brings us to a consideration of Feyerabend's discussion of the role of reason in science, for after all his book is full of the rhetoric of irrationalism. The first point to be made here is that even if he were otherwise right, his irrationalism would be wrong. That is, even if it were true that *"essential ingredients of modern science survived only because reason was frequently overruled in their past"* (p. 145, ital. in original; cf. p. 155), this would only show that it is sometimes reasonable to overrule 'reason', which can therefore be only what was thought to be, but is not really, reason. Thus, in this type of argument, far from it being the case that Feyerabend can't lose (as some critics have pointed out in other contexts),[13] he can't win.

Moreover, Feyerabend frequently (e.g., pp. 153–4) fails to distinguish the nonrational from the irrational, and just because a certain procedure is not in accordance with reason, he concludes that it is in violation of reason. Just as frequently, Feyerabend concludes that something is irrational just because it goes against what his opponents (be they empiricists, inductivists, or Popperians) would call 'rationality' (cf., e.g., pp. 179–80). The best example of this is his analysis of Galileo, which will be criticized in detail below. Here I merely wish to point out that Feyerabend is at pains to point out that *he* thinks that Galileo's procedure was highly desirable, though it supposedly goes against widely accepted methodological principles, so that *those who accept them* would have to regard it as irrational.

Let us look at some of Feyerabend's specific points against 'reason'. His considerations can be divided into three arguments, which we may call the insufficiency of reason argument, the argument from incommensurability, and the historical argument. In the first Feyerabend argues as follows. Since the teaching of small children is not exclusively a matter of argument, and since there are processes like the mastery of a language which look like the result of reason but are due partly to indoctrination and partly to *natural* processes of growth, it follows that nonrational growth is possible both in adults and in institutions. Therefore "even the most puritanical rationalist will then be forced to stop reasoning and use *propaganda* and *coercion*" (p. 25) whenever conditions are such that "forms of argumentation turn out to be too weak" (p. 25) to accomplish his ends. Feyerabend's argument may be accepted with the following qualifications. There is nothing irrational in sometimes stopping reasoning; there would be only if one assumed that reasoning is the only proper activity human beings can engage in, an assumption no rationalist needs to make. Moreover, there is no need to resort to propaganda and coercion, for propaganda is a perversion of rhetoric, while coercion is a misuse of noncognitive causes, and such perversions and misuses may or may not be used by the rationalist; the only thing that follows is that rhetoric and noncognitive causes may have to be used; but such things are nonrational rather than irrational, and they are susceptible of being handled properly as well as improperly. So the conclusion to be reached is that reasons and arguments are not the only causes that affect human actions and thought; there are also rhetoric and noncognitive causes that operate.

Now at this point, Feyerabend would like to add that therefore it would be very strange if arguments were the only thing operating in scientific rationality, so strange that even if that *were* the case, then one should rebel against such unnatural restrictions and help to create a more human science in which the other factors are not excised. To this we may reply by asking why should every human activity be a microcosm of the entire human life? Why can't there be some activity, e.g., science, where the only rules of the game involve argumentation? As long as we remember that there are other things in life, why should this activity of reasoning which we have labeled "science" be mixed with the others? I believe Feyerabend's only plausible answer could be, because as a matter of (historical) fact what we label 'science' does contain a mixture of elements. Here we would leave Feyerabend's general considerations and go to his historical argument, which therefore is more crucial to his position than he makes it seem.

Before examining Feyerabend's historical argument, however, let us look at the one from incommensurability, which he thinks "creates problems for all theories of rationality" (p. 214). Feyerabend calls two theories incommensurable (pp. 223, 228–9, 269) when the subject matter to which they refer differs so radically that (1) the existence of the entities and processes presupposed by one theory implies the nonexistence of the entities and processes presupposed by the other theory, and (2) it makes no sense to say that one theory is a generalization of the other, or that they overlap. The main step in the argument consists in trying to establish that there are scientific changes where the theories involved are incommensurable, e.g., Aristotelian and Galilean physics, classical and quantum physics, and Newtonian and relativistic mechanics (pp. 224–5, 271, 276–7); given this incommensurability, Feyerabend concludes that one cannot say that the new theory which replaces the old is epistemologically better. However, if we distinguish progress from rationality, then we realize that Feyerabend's conclusion does not imply that the agents involved in the transition did not act rationally. In fact, while answering a number of philosophical objections against his incommensurability claims (pp. 277–85), Feyerabend himself argues cogently that, in spite of incommensurability, it is still possible to behave rationally;

for example, a self-inconsistent observation report would disconfirm the corresponding theory (p. 278); one could learn the meaning of the new theory the way anthropologists learn the language of newly-discovered tribes (pp. 278–82); and crucial experiments are still possible (pp. 282–3). So Feyerabend has not shown that incommensurability creates problems for theories of rationality, but rather that it creates problems for theories of progress. Moreover, since, besides being distinct, rationality and progress are related, the possibility of rationality opens up a minimal possibility of progress, namely that changes involving incommensurable theories are changes for the better in the sense that they result from the behavior of rational agents.

But how shall we test for the actuality of these possibilities still left open after Feyerabend's arguments? The answer, whose suggestion can be found in Feyerabend himself, is that the test is to be conducted by anthropological field work: "Let us commence field work in this domain also, and let us study the language of new theories not in the definition-factories of the double language model, but in the company of those metaphysicians, physicists, playwrights, courtesans, who have constructed new world views!" (p. 282). When the event under consideration is relatively far into the past, like the emergence of modern science in the 17th century, the only kind of field work possible is the analysis of historical records. However, these historical records must be sufficiently rich, and the analysis must be conducted with an appropriate 'anthropological' spirit. For the emergence of modern science a sufficiently rich record can be found in Galileo's *Dialogue,* parts of which have therefore been examined by Feyerabend, though, alas, without a sufficiently 'anthropological' attitude, as I shall soon show.

This brings us to his historical argument against reason. Here he tries to show that *"essential ingredients of modern science survived only because reason was frequently overruled in the past"* (p.145). When we look at the details of the argument (e.g., pp. 145–61, 179–80, 196–200) we discover, as mentioned earlier, that the reason which was overruled was usually what empiricists, inductivists, Popperian critical rationalists, or Imre Lakatos *think* is reason. In other cases, there was no *overriding* of reason but rather a combination of it with nonrational factors; nor does Feyerabend show that such nonrational factors were improper from their own point of view. All of the strengths and all of the limitations of Feyerabend's position are present in his analysis of Galileo, and so it will be valuable to examine this in detail.

GALILEO

The longest case study on which Feyerabend grounds many of his philosophical conclusions concerns Galileo. The account consists of four elements: an interpretation of Galileo's work in terms of a definite procedure which he allegedly follows; a description of this procedure in terms of such concepts as theory, observation, propaganda, appeals to emotion and prejudices, etc; a historical explanation of the *fact* of Galileo's success as resulting from the procedures he used; and a favorable evaluation of Galileo insofar as the procedure he follows is *'perfectly reasonable'* (p. 143).

One need not have mentioned the first item in this list were it not for the fact that all of Feyerabend's talk against method may easily be interpreted as showing that the concept of method is a useless one in the philosophy of science. It is clear, however, from Feyerabend's 'actual procedure' (*his* method, as it were) that he is only arguing against "the idea of a method that contains firm, unchanging, and absolutely binding principles for conducting the business of science" (p. 23), i.e., "the idea that science can, and should be run according to fixed and universal rules" (p. 295). In a few places (e.g., pp. 145, and 163) he even uses the forbidden word 'method' to refer to Galileo's procedure. So Feyerabend's point is really that no general theory of method is defensible, which says that all scientists in all situations use a certain definite method; instead different scientists use different methods on different occasions. It is important to note, however, that Feyerabend's account presupposes that the concept of method is a very useful one in understanding what a scientist does; a method becomes the effective cause (in a nondeterministic sense) in the historical explanation of a given scientific success. I do not think that such method-explanations of success are objectionable on general grounds; I think that such accounts provide historical understanding and can even be used to ground philosophical theses.[14] However, in this case the following questions must be raised.

First, it may be that Feyerabend is not being radical enough in ascribing a certain definite method to Galileo; that is, it may be that not even the same scientist follows the same method to any significant extent in his activities. Perhaps there is 'anarchy' within the work of a given scientist. What I am saying is that, for Feyerabend, science is an 'anarchical' enterprise, but individual scientists are not.[15] There is no logical inconsistency here; but there is a tension within Feyerabend's account since *he* tends to transfer the characteristics from science-the-institution to the individual scientist. I believe there is textual evidence for this tendency in Feyerabend, but let me illustrate the problem as follows. Feyerabend presupposes that one could not understand Galileo's work unless one found *some* order in it, enough to speak of a method or procedure. But if one does this with an individual scientist, why shouldn't he do the same with science as a whole? Of course, one is *not* logically bound to conclude that there is at least as much order in science as in a scientist, but if he chooses to order a given scientist's activities by reference to a definite method, why should he not do enough work so as to order science as a whole into a method? Feyerabend might answer, because this *can* be done in the case of a given scientist, e.g., Galileo, but not for the whole of science. What follows from this is that one has to be very careful about whether or not there is method within a scientist's activities. To some extent this problem reduces to that of describing the method in a sufficiently complex and sophisticated way. However, if this complexity was too great then we wouldn't have a method. In these investigations I am partly testing this assumption that Feyerabend seems to make.

Second, once we realize that Feyerabend depicts an anarchical science consisting of methodical scientists, the possibility arises for a methodical science consisting of anarchical scientists. Given these two abstract alternatives, it is not clear which Feyerabend would choose. Perhaps he would opt for a third possibility of an anarchical science of anarchical scientists; though this may be his wish, it would go against his actual account in *Against Method*.

Third, Feyerabend's method-explanation of Galileo's success never takes seriously the question of the *connection* between the procedures and the success. That is, even if Feyerabend is right about the methods he attributes to Galileo, there is no argument that it was *because* of these methods that he succeeded. There is no question that Feyerabend is committed to such a causal claim (e.g., p. 112), which therefore requires careful examination. One way to conduct this examination is to ask whether the situations in which Galileo proceeded à la Feyerabend involved activities which would be regarded as successful. For example, he examines mostly Galileo's attempts to prove the motion of the earth; however, few people, be they 17th century contemporaries of Galileo, or present-day scholars, would agree that Galileo *succeeded* in his attempts. Feyerabend would be the first to claim that Galileo failed here. But if he failed when he was proceeding as Feyerabend claims, then this is, if anything, evidence that such Galilean procedures led to failure, not that they led to success.

Prescinding now from such problems, let us see how Galileo's method is described. Feyerabend deserves great credit for enlarging Galileo's method beyond the epistemological factors usually considered and for including rhetorical and artistic components. The epistemological factors considered by Feyerabend are the following. (1) Galileo introduces and elaborates hypotheses (e.g., motion of the earth, relativity of motion, celestial reliability of telescopes) which are inconsistent with well-established theories (e.g., geocentrism, operative character of motion, optics) and with well-established facts (e.g., vertical fall, apparent size of Venus and Mars); this procedure is labeled 'counterinduction' (pp. 29, 77, 99–101). (2) Galileo lets inadequate views mutually support each other, e.g., (a) "an inadequate view, the Copernican theory, is supported by another inadequate view, the idea of the nonoperative character of shared motion" (p. 89), (b) he lets one refuted wiew — Copernicanism — support another refuted view — the idea that telescopic phenomena are faithful images of the sky (pp. 141–2), and (c) "Galileo changed his view about the 'neutral' motions — he made them permanent and 'natural' — in order to make them compatible with the rotation of the earth" (p. 96); Feyerabend gives no special name to this aspect of Galileo's procedure, which I shall label 'dialectical', to follow some terminology suggested elsewhere.[16] (3) Ad hocness: Galileo uncritically accepts any and all ideas and observations which support Copernicanism (e.g., p. 90, 93–8). (4) Galileo drastically reduced the *content* of dynamics by replacing the Aristotelian comprehensive theory of change, including locomotion, qualitative change, and generation and corruption by a theory dealing merely with the locomotion of matter (pp. 90–100, 160–1); this is what Feyerabend labels the 'backward step' (pp. 176, 153, 113).

The rhetorical aspects of Galileo's method are allegedly the following: deceptive tactics (pp. 70, 81, 87, 160); utterances which are arguments in appearances only (p. 81); propaganda (pp. 81, 90, 157, 160); psychological tricks (pp. 81, 88, 154); persuasion by confusion (p. 84); clever techniques of persuasion (pp. 141, 143); misleading insinuations (p. 160); distortion (p. 160); appeal to emotion and prejudice (p. 154); jokes (p. 154); non sequiturs (p. 154).

The artistic, aesthetic, literary factors include "style, elegance of expression, simplicity of expression, tension of plot and narrative" (p. 157), "a sense of humor, an elasticity . . . and an awareness of the valuable weaknesses of human thinking" (p. 161).

The epistemological factors emphasized by Feyerabend are ones that go against almost every principle held by orthodox philosophers of science; the rhetorical factors seem to violate the ideas of the basic honesty and decency of science, widely held by scientists and laymen alike; the aesthetic factors contradict the alleged gap between the arts and the sciences. Indeed Feyerabend delights in being contrary. His contrariness reaches its highest pitch when he argues that such Galilean procedure is 'perfectly reasonable' in itself (pp. 143, 145–61) and fruitful in other fields (pp. 163–4). In other words, Feyerabend's *evaluation* of Galileo's method is the orthodox one, it is his descriptive interpretation of the features of that method that is unorthodox; and since his evaluation is partly grounded on his interpretation, that means that his *reasons* for his evaluation are also unorthodox. Feyerabend does not always *effectively* distinguish his description and his evaluation, though he often seems to or tries (pp. 143, 156). Thus we often find him arguing that because it can be shown (by Feyerabend himself) that a certain view was refuted or inadequate, and because therefore Galileo *should* have seen that it was inadequate, therefore it follows that Galileo regarded it as inadequate, and hence that he was acting counterinductively; or that because it can be shown (by Feyerabend himself) that a certain argument is incorrect, and because therefore Galileo *should* have known that it was incorrect, therefore it follows that he knew it to be incorrect, and hence that his using it amounted to deceptive trickery. We might say that Feyerabend's alleged descriptive interpretations are really evaluations, and hence his whole case is historically unfounded; and since, as we argued earlier, his two other arguments (from the insufficency of reason, and from incommensurability) depended on the historical argument, we might conclude that Feyerabend's views do not merit any further consideration.

This, however, would be a mistake. We need not deny our inclination to think that Feyerabend is doubly in the wrong, namely that his historical interpretations are descriptively wrong, and that *if* they were right, then that would show that *Galileo* was wrong (irrational). The fact is, or rather my suspicion is, that Feyerabend turns out to be *right in spite of himself*. That is, his account of Galileo is not really irrationalistic, but pseudo-irrationalistic, and in reality Galileo proceeds rationally for Feyerabend; however, our idea of scientific rationality must be expanded first to allow aesthetic and rhetorical factors, and second epistemological practices proscribed by orthodox philosophies of science; moreover the aesthetic and rhetorical factors are by themselves merely *alogical*, and they must be judged by their own criteria; finally, whatever unorthodox epistemological practices Galileo may engage in, the real test of their rationality or propriety is their correspondence to basic and elementary forms of reasoning and argumentation, rather than to philosophically articulated *theories* of scientific rationality, most of which presuppose the theory/observation distinction, for example.

Keeping this in mind, I now proceed to test my suspicion. I shall do so by examining Feyerabend's account of the tower argument.

THE TOWER ARGUMENT

The tower argument was one of the classical objections to the earth's rotation. Galileo states and criticizes it in the following passage:

SALV. . . . Aristotle says, then, that a most certain proof of the earth's being motionless is that things projected perpendicularly upward are seen to return by the same line to the same place from which they were thrown, even though the movement is extremely high. This, he argues, could not happen if the earth moved, since in the time during which the projectile is moving upward and then downward it is separated from the earth, and the place from which the projectile began its motion would go a long way toward the east, thanks to the revolving of the earth, and the falling projectile would strike the earth that distance away from the place in question. Thus we can accommodate here the argument of the cannon ball as well as the other argument, used by Aristotle and Ptolemy, of seeing heavy bodies falling from great heights along a straight line perpendicular to the surface of the earth. Now, in order to begin to untie these knots, I ask Simplicio by what means he would prove that freely falling bodies go along straight and perpendicular lines directed toward the center, should anyone refuse to grant this to Aristotle and Ptolemy.

SIMP. By means of the senses, which assure us that the tower is straight and perpendicular, and which show us that a falling stone goes along grazing it, without deviating a hairsbreadth to one side or the other, and strikes at the foot of the tower exactly under the place from which it was dropped.

SALV. But if it happened that the earth rotated, and consequently carried along the tower, and if the falling stone were seen to graze the side of the tower just the same, what would its motion then have to be?

SIMP. In that case one would have to say 'its motions', for there would be one with which it went from top to bottom, and another one needed for following the path of the tower.

SALV. The motion would then be a compound of two motions; the one with which it measures the tower, and the other with which it follows it. From this compounding it would follow that the rock would no longer describe that simple straight perpendicular line, but a slanting one, and perhaps not straight.

SIMP. I don't know about its not being straight, but I understand well enough that it would have to be slanting, and different from the straight perpendicular line it would describe with the earth motionless.

SALV. Hence just from seeing the falling stone graze the tower, you could not say for sure that it described a straight and perpendicular line, unless you first assumed the earth to stand still.

SIMP. Exactly so; for if the earth were moving, the motion of the stone would be slanting and not perpendicular.

SALV. Then here, clear and evident, is the paralogism of Aristotle and of Ptolemy, discovered by you yourself. They take as known that which is intended to be proved.

SIMP. In what way? It looks to me like a syllogism in proper form, and not a *petitio principii.*

SALV. In this way: Does he not, in his proof, take the conclusion as unknown?

SIMP. Unknown, for otherwise it would be superfluous to prove it.

SALV. And the middle term; does he not require that to be known?

SIMP. Of course; otherwise it would be an attempt to prove *ignotum per aeque ignotum.*

SALV. Our conclusion, which is unknown and is to be proved; is this not the motionlessness of the earth?

SIMP. That is what it is.

SALV. Is not the middle term, which must be known, the straight and perpendicular fall of the stone?

SIMP. That is the middle term.

SALV. But wasn't it concluded a little while ago that we could not have any knowledge of this fall being straight and perpendicular unless it was first known that the earth stood still? Therefore in your syllogism, the certainty of the middle term is drawn from the uncertainty of the conclusion. Thus you see how, and how badly, it is a paralogism.[17]

The argument which according to Galileo's spokesman, Salviati, begs the question is the one contained in Salviati's first speech in this passage; it may be reconstructed as follows:

(1)   Bodies fall vertically.

(2)   If the earth rotated, bodies could not fall vertically,

because

(3)   If the earth rotated, then while a body was falling the place on the earth directly below it would be carried along toward the east and the body would land to the west of where it was originally dropped.

Therefore,

(4)   The earth does not rotate.

Galileo has no objection to the (formal) validity of the last step in this argument (the one from (1) and (2) to (4)); however, he questions its soundness in terms of whether premise (1) is justified. The Aristotelian spokesman, Simplicio, gives the following justification:

(1)   Bodies fall vertically

because

(5)   Bodies *are seen* to fall vertically.

Now, with respect to this argument Galileo accepts the premise but questions its (formal) validity. This validity depends on whether or not apparent vertical fall implies actual vertical fall. This implication needs justification because it would *not* hold on a rotating earth. In fact, if the earth *were* rotating and bodies were *seen* to fall vertically, then in actuality they would be following a path slanted to the earth's surface; that is, on a rotating earth, apparent vertical fall would *not* imply actual vertical fall. Now, how could one possibly justify the implication? The only relevant argument would seem to be the following one:

(6)   If the earth does not rotate, then apparent vertical fall implies actual vertical fall.

(7)   The earth does not rotate.

Therefore,

(8)   Apparent vertical fall implies actual vertical fall.

Premise (6) is indeed true, and the argument is (formally) valid. Unfortunately premise (7) is identical with conclusion (4) of the original argument, which is being examined. In short, the vertical fall argument, from (1) and (2) to (4), is such that one of its premises (1) is being justified on the basis of the same proposition (7) [=(4)] it has for conclusion.

The important issues can be put into focus in terms of the following argument, which combines the three separate segments just discussed, and where some obvious symbolic abbreviations have been made and the proposition numbers correspond:

(6)   If not-R, then S implies V.
(7)   Not-R.
(8)   ∴ S implies V.
(5)   S.
(1)   ∴ V.
(2)   If R, then not-V.
(4)   ∴ Not-R.

Logically speaking, this is a very interesting deduction; in particular, its three steps are valid. Rhetorically speaking, however, the identity of propositions (7) and (4) makes it worthless: someone who does not adhere to (4), will not be any more favorably inclined toward it after this argument, which requires such adherence already at the very beginning.

In Feyerabend's account four main theses are involved. First (pp. 70–75), the Aristotelian tower argument presupposes naive realism with respect to motion (p. 75); and it is important to notice that this Aristotelian presupposition takes the form of what Feyerabend calls a 'natural interpretation'; that is, "one does not first distinguish the apparent motion from the real motion and then connect the two by a correspondence rule. One rather describes, perceives, acts towards motion as if it were already the real thing" (p. 75). Second (pp. 75–8), Galileo discovered this Aristotelian natural interpretation counterinductively. Third (pp. 78–81), Galileo tests and examines the Aristotelian identification of real and apparent motion by introducing a different interpretation which "restores the senses to their position as instruments of exploration, *but only with respect to the reality of relative motion.* Motion 'among things which share it in common' is 'non-operative', that is, 'it remains insensible, imperceptible, and without any effect whatever'" (p. 78). Fourth (pp. 81–92), Galileo uses, and needs to use, propaganda, psychological tricks, and deception in elaborating this new concept of the relativity of motion.

It is a great insight for Feyerabend to have arrived at the first of these conclusions, though it must be qualified to restrict it to vertical fall, so as to say that the tower argument presupposes, as a 'natural interpretation', the identification of real and apparent *vertical fall.* Feyerabend himself goes through the motions of making the qualification (p. 75), but he then forgets its importance in the rest of his discussion. At any rate, it is possible to give a textual and logical proof of the qualified conclusion, which will be done later in another context.[18] However, that same analysis also shows that Feyerabend is completely wrong in his second thesis; Galileo is simply not aware of this Aristotelian presupposition of the tower argument, so he neither discovers it counterinductively, nor does he discover it at all. *We* can discover it by logical manipulations, but these are merely suggested by the text, nowhere contained in it.[19] Feyerabend formulates his second thesis because he approaches the philosophy of science with the spectacles of the theory/observation distinction; so he wants to reconstruct some theory which the Aristotelians held, and the one that Galileo introduced. If Feyerabend had approached the tower argument as an *argument,* which is the way Galileo does, then he would have been inclined to interpret Galileo's discovery as one about an argument, namely that the tower argument is a *petitio principii.* Though Galileo could have proceeded to inquire about general theories and natural interpretations, the fact is that he does not do so; and he does not do so because he seems to be interested in *arguments.* Thirdly, Galileo introduces his concept of the relativity of motion *not* to test a presupposition of the tower argument, but to answer the objection to the earth's motion from the deception of the senses. Feyerabend thinks as he does because he fails to distinguish two different arguments against the earth's motion, namely the one from vertical fall and the one from the deception of the senses. The logic of the two arguments is different. Galileo correctly criticizes the first as being circular, the second as being groundless,[20] in the sense that the earth's motion would *not* involve any deception of the senses; and there would be no deception of the senses because of the relativity of motion; in other words, it's not a case where one does not see something he ought to see (the real slanted path on a rotating earth); there is no reason why one ought to see such a slanted path because our eyes can only detect motion relative to them; that our eyes are so built can be proved by experiences on moving systems such as ships. Such being the logic of the argument from the deception of the senses, and Galileo's use therein of the relativity of motion, Galileo needs neither propaganda, nor tricks, nor deceptions, which are figments of Feyerabend's imagination, as we will see below. He so imagines because he thinks wrongly that Galileo is using the relativity of motion in discussing the tower argument. Feyerabend does not realize that Galileo's analysis of this argument is not designed to produce a positive, substantive conclusion about phenomena, but rather a negative destructive criticism of the Aristotelian argument. For this purpose Galileo does not *need* the relativity of motion. *If* Galileo *were* using the relativity of motion in connection with the tower argument, and *if* he *were* thereby trying to prove the earth's motion rather than merely refuting the argument, *then* Galileo would have to resort to propaganda, tricks, and deception, since that task cannot be accomplished by legitimate means. However, such a conditional is counterfactual.

At this point the question arises, whether Galileo is doing anything wrong in the context of his discussion of the deception of the senses argument, where he admittedly uses the relativity principle. In that context, Galileo refutes the objection by arguing that there would be no deception of the senses if we experienced apparent vertical fall on a rotating earth, because only motion relative to the observer is perceivable. In this critique Galileo is certainly subsuming vertical fall under what Feyerabend calls 'Paradigm II' (p. 87), namely the paradigm of motion in a moving system, like a boat. Is he therefore being deceptive, etc. in so doing? Not at all, because this is the same paradigm used in the statement of the objection, which is that *on a rotating earth* our senses would be deceived in experiencing the appearance of vertical fall while in actuality the fall was over a slanted path. In other words, this *different* argument against the motion of the earth cannot even be made without using Paradigm II. Hence, in answering it, it is quite proper to use the same paradigm.

And this brings us to Feyerabend's fourth thesis. He is certainly right that Galileo uses rhetorical alogical considerations which are interesting to examine and which play an important function. However, their propriety is a very complicated affair. In order to begin to understand this matter, it is essential to have in front of us the text of the dialogue immediately following the above quotation:

SAGR. On behalf of Simplicio I should like, if possible, to defend Aristotle, or at least to be better persuaded as to the force of your deduction. You say that seeing the stone graze the tower is not enough to assure us that the motion of the rock is perpendicular (and this is the middle term of the syllogism) unless one assumes the earth to stand still (which is the conclusion to be proved). For if the tower moved along with the earth and the rock grazed it, the motion of the rock would be slanting, and not perpendicular. But I reply that if the tower were moving, it would be impossible for the rock to fall grazing it; therefore, from the scraping fall is inferred the stability of the earth.
SIMP. So it is. For to expect the rock to go grazing the tower if that were carried along by the earth would be requiring the rock to have two natural motions; that is, a straight one toward the center, and a circular one about the center, which is impossible.
SALV. So Aristotle's defense consists in its being impossible, or at least in his having considered it impossible, that the rock might move with a motion mixed of straight and circular. For if he had not held it to be impossible that the stone might move both toward and around the center at the same time, he would have understood how it could happen that the falling rock might go grazing the tower whether that was moving or was standing still, and consequently he would have been able to perceive that this grazing could imply nothing as to the motion or rest of the earth.
Nevertheless this does not excuse Aristotle, not only because if he did have this idea he ought to have said so, it being such an important point in the argument, but also, and more so, because it cannot be said either that such an effect is impossible or that Aristotle considered it impossible. The former cannot be said because, as I shall shortly prove to you, this is not only possible, but necessary; and the latter cannot be said either, because Aristotle himself admits that fire moves naturally upward in a straight line and also turns in the diurnal motion which is imparted by the sky to all the elements of fire and to the greater part of the air. Therefore if he saw no impossibility in the mixing of straight-upward with circular motion, as communicated to fire and to the air up as far as the moon's orbit, no more should he deem this impossible with regard to the rock's straight-downward motion and the circular motion natural to the entire globe of the earth, of which the rock is a part.
SIMP. It does not look that way to me at all. If the element of fire goes around together with the air, this is a very easy and even a necessary thing for a particle of fire, which, rising high from the earth, receives that very motion in passing through the moving air, being so tenuous and light a body and so easily moved. But it is quite incredible that a very heavy rock or a cannon ball which is dropping without restraint should let itself be budged by the air or by anything else. Besides which, there is the very appropriate experiment of the stone dropped from the top of the mast of a ship, which falls to the foot of the mast when the ship is standing still, but falls as far from the same point when

the ship is sailing as the ship is perceived to have advanced during the time of the fall, this being several yards when the ship's course is rapid.

SALV. There is a considerable difference between the matter of the ship and that of the earth under the assumption that the diurnal motion belongs to the terrestrial globe. For . . . [21]

It was certainly a clever move for Galileo to begin his critique of the tower argument (in the earlier quotation) by interpreting it to refer to *actual* vertical fall. Feyerabend is right in saying that the Aristotelian argument simply interchanges apparent and actual fall without distinguishing them; so Galileo could have begun with either version of the argument. The fault of the argument from *apparent* vertical fall (Sagredo's first speech in the passage just quoted) is that it depends on a premise which is as much in need of proof as the conclusion at issue. This fault is less serious than the circularity of the argument from actual vertical fall. It is indeed more effective to start with the more serious criticism, and so Galileo's first rhetorical move is a judicious one.

Having focused on the actual vertical fall version, Galileo had in front of him a formally valid instance of denying the consequent (*modus tollens*). Of the two premises of this argument, propositions (1) and (2) above, Galileo could have questioned either one. If he had questioned the conditional proposition, the circularity could not have been exhibited as easily since the justification of this conditional premise is circular only insofar as it depends ultimately on the nonconditional premise of the original *modus tollens*, namely the proposition (1) that bodies really fall vertically.[22] It is certainly rhetorically effective, but certainly not improper, to exhibit the failure of an argument *in the easiest possible way*.

Third, when Galileo comes around to discuss the apparent vertical fall version of the argument, this is made to look as a *revised* version of the original argument, and hence to some extent as *ad hoc*. This seems to be a merely rhetorical fault, rather than a logical one. However, the argument is also shown to be dependent on a premise which is as much in need of proof as its conclusion. This feature, though not purely logical, is not purely rhetorical either. For though questions of the comparative knowability of propositions are context-dependent, the context is an epistemological one, rather than one dependent merely on who is speaking to whom.

Fourth, the way that the discussion changes from apparent vertical fall to the ship experiment involves rhetorical considerations. Someone might say, following Feyerabend's evaluation of Galileo's rhetoric, that besides introducing the argument from apparent vertical fall in a prejudicial manner, he drops the topic in such a way as to make it look as if Simplicio was giving up on the argument and relying on a new one.[23] This rhetorical evaluation would depend on interpreting Galileo's criticism of the ship/earth analogy,[24] which follows the passage just quoted, as part of his criticism of the ship experiment argument against the earth's motion.[25] This interpretation in turn would involve attributing to Galileo another deception, namely that he would be trying to show that a rock dropped from the top of the mast on a moving ship would land at the foot of the mast, and similarly on a rotating earth, while he does not believe in the analogy. However, it is itself a prejudicial interpretation to regard the passage where Galileo criticizes the ship/earth analogy as a logical part of his discussion of the ship experiment argument. No sound reason can be given for this interpretation, whereas evidence can be given to show that Galileo wants his criticism of the analogy to be part of his criticism of the apparent vertical fall argument. The evidence consists, first, of the logical fact that such a connection exists; for the alleged results of the ship experiment can be cited as evidence that bodies refuse to move simultaneously toward and around the center, and hence as support for the corresponding premise in the apparent vertical fall argument, namely the proposition that if the earth rotated, bodies could not be seen to fall vertically. Moreover, there is textual evidence that this is what Galileo has in mind. For example, the first time that the ship experiment is mentioned[26] is in the context of an argument where the impossibility of mixed motion is given as the explanation for the alleged deflection on the moving ship. Finally, the way the switch is made from the criticism of the analogy to the criticism of the alleged experimental results[27] suggests the same thing: Salviati agrees that 'up to this point' the ship experiment as an argument purely from analogy has not been considered. In other words, in this passage Galileo is saying that though the ship experiment has just been criticized insofar as it is the basis of an explanatory argument against mixed motion, it is still standing as a pure analogy argument against the earth's motion. I believe Galileo is right here, and that a pure analogy argument has *some* force even though the mechanism of the analogy is unknown, i.e., even though we do not know *why* the analogy holds, as long as we are prepared to claim *that* it holds. At any rate, the conclusion is inescapable that interpretations of the rhetorical situation must be grounded on interpretations of the logical situation; and since it emerged earlier that logical analysis was crucial for the understanding of scientific rationality per se, this means that a systematic logical analysis of a crucial work like Galileo's *Dialogue* is very important for for the philosophy of science.

In summary, scientific rationality is distinct from, but related to, scientific progress and change. Feyerabend's approach can be easily misunderstood as being merely negative but suggests two very important insights. First, the anarchism of "anything goes" (which ought to be distinguished from the specific counterinductive rule) may be understood as a manner of speaking about judgment, and when so interpreted it receives the support of the evidence of what I earlier called Galileo's synthetic or "dialectical" methodology. Second, the propagandist-manipulative interpretation of scientific rationality may be taken as being itself a rhetorical exaggeration of the truth that rhetorical persuasion has an important role. Underlying both elements stands the logical dimension of science (logical in the sense of the theory and practice of reasoning), as a deeper level to which judgment is reducible, and on which rhetoric needs to be grounded.

NOTES

[1] Cf. I. Lakatos and A. Musgrave (eds.), *Criticism and the Growth of Knowledge* and my Essay-review of this book.
[2] S. Toulmin, *Human Understanding*, Vol. I: *The Collective Use and Evolution of Comcepts.*
[3] *Ibid.*, p. ii.
[4] *Ibid.*
[5] K. Popper, *Objective Knowledge: An Evolutionary Approach*, pp. 106–90, esp. pp. 106 and 153.
[6] *History of Science as Explanation.*
[7] M. Hesse, *The Structure of Scientific Inference.*
[8] P. K. Feyerabend, 'Explanation, Reduction, and Empiricism'; idem, 'Problems of Empiricism, Part II'; idem, 'Consolations for the Specialist'; idem, *Against Method.*
[9] K. Popper, *The Logic of Scientific Discovery*. This book is a translation of his *Logik der Forschung.*
[10] S. Toulmin, *The Philosophy of Science.*
[11] P. K. Feyerabend, *I probemi dell' empirismo;* idem, *Contro il metodo;* idem, *Einfuhrüng in die Naturphilosophie;* idem, *Ausgewählte Aufsatze;* idem, *Wider den Methodenzwangtheorie.*
[12] Hereafter references to *Against Method* will be made in parenthesis in the text.
[13] E. Gellner, 'Beyond Truth and Falsehood'.
[14] Cf. my *History of Science as Explanation*, pp. 223–8.
[15] Feyerabend has objected in private correspondence that he is not ascribing a single definite method to Galileo since "he does different things in the case of dynamics and in the case of optics (in the first case he changes the grammar of dynamical terms, in the second case he changes *sensations* by substituting the telescope for the eye". Though it is true that Feyerabend examines Galileo's work in these two contexts, and that in the sense just described different procedures are operative, it is also true that both are subsumed under Feyerabend's concept of counterinduction, the counterinductive idea in

the former case being the earth's motion, in the latter, the celestial reliability of the telescope. Hence, Galileo's activities are being ordered in a nonanarchical fashion.

[16] A. Funkenstein, 'The Dialectical Preparation for Scientific Revolutions', and my 'Dialectical Aspects of the Copernican Revolution: Conceptual Elucidations and Historiographical Problems'.

[17] G. Galilei, *Dialogue Concerning the Two Chief World Systems*, tr. Drake, pp. 139–140.

[18] See below Chapter 12.

[19] In my reconstruction (Chapter 12), twenty five steps are needed to arrive at a conclusion of this form: R if and only if not-(S if and only if V). Feyerabend's insight is to realize that this proposition is implied by Galileo's critique, but his error also occurs at this point. Being under the wrong impression that Galileo is *here* not merely trying to refute the tower argument, but also trying to prove the earth's motion, and confusing the tower and the deception of senses arguments, Feyerabend has Galileo support the right side of this biconditional, namely not-(S if only if V), on the basis of the relativity principle. Feyerabend's reconstruction is the following: apparent and actual vertical fall are distinct on moving systems like a ship, since only relative motion is visible in such systems; hence apparent and actual fall *with respect to the earth* are distinct. Feyerabend correctly points out that this amounts to subsuming vertical motion under 'Paradigm II: Motion of objects in boats, coaches, and other moving systems' (p. 87) instead of under 'Paradigm I: Motion of compact objects in stable surroundings of great spatial extension – deer observed by the hunter' (p. 87). He is also right in saying that the gap between these two paradigms is unbridgeable, in the sense that there is no noncircular way of choosing the former for the purpose of applying it to vertical fall. He is wrong, however, in attributing such an argument to Galileo, who wisely avoided it.

[20] For statements and critiques of the argument on the deception of the senses see Galilei, *Dialogue*, tr. Drake, pp. 167–71 and 233–47, and Galilei, edited by Favaro, *Opere* 7, 197–200, and 272–81.

[21] Galilei, *Dialogue*, tr. Drake, pp. 140–42.

[22] See my 'Galileo as a Logician', pp. 144–45. See also Chapters 12 and 14 below.

[23] Galilei, *Dialogue*, tr. Drake, p. 141; idem, *Opere* 7, 167.

[24] *Dialogue*, tr. Drake, pp. 141–43; *Opere* 7, 167–69.

[25] *Dialogue*, pp. 143–49; *Opere* 7, 169–75.

[26] *Dialogue*, p. 126; *Opere* 7, 151–52.

[27] *Dialogue*, p. 143; *Opere* 7, 169.

CHAPTER 9

THE HISTORY OF SCIENCE AND
THE SCIENCE OF HISTORY:
CRITIQUE OF APRIORISM

In emphasizing the role of reasoning in scientific rationality in the last chapter, our discussion was carried out in a context where the reasoning was taking the form of reason as opposed to unreason and was being distinguished (but not opposed to or separated from) method and rhetoric, as other elements of scientific rationality. The first danger that such emphasis carries is that of being misinterpreted as advocacy of a priori rationalism, so the most urgent thing to do now is to distinguish reasoning from rationalism. In accordance with our concrete approach, this distinction will be elaborated not by an abstract general definition of these two concepts, but by reference to views which advocate apriorist rationalism while confusing the two concepts, and to evidence materials which illustrate reasoning but not apriorism. The materials are certain arguments in Galileo's *Dialogue,* the views are those of Alexandre Koyré.

Few scholars in this century have had as great an impact on our understanding of science as Alexandre Koyré. It is perhaps no accident that his pioneering work on the subject is one dealing with Galileo. Entitled *Études galiléennes* the book has become both a model and a source of inspiration for historians and philosophers alike. The influence on contemporary historians of science is both explicitly acknowledged[1] and bordering on the classical, in the sense that his ideas and his approaches can now be freely borrowed and adapted and have become internalized in their professional sensibilities.

As regards philosophical influence, the view propounded in Thomas S. Kuhn's *Structure of Scientific Revolutions* is the one which in the last decade or two has caught the imagination, if not the intellect, of most scholars, scientists, and laymen alike. It would perhaps be an exaggeration to repeat the judgment I once heard from a philosopher, namely that, in Kuhn's book, what is new is not true, and what is true is not new, vis-à-vis Koyré's. Though exaggerated the judgment is not unfounded, and it does reflect the extent of Koyré's influence. At any rate one other philosopher, Joseph Agassi in *Towards an Historiography of Science,* has focused on one of Koyré's central techniques — error analysis, has articulated it, synthesized it with Karl Popper's philosophy of science, provided it with novel content and illustrations, and advocated it as the most fruitful approach to the philosophy of science. Some of Feyerabend's own critiques, examined in the last chapter, remind one of Koyré's technique of error analysis, though of course, Feyerabend's central concern is to explore the limitations of reason by studying the limitations of method and the power of rhetoric, whereas Koyré's concern was to extol the power of reason by exploring the rationalistic character of science, rationalist in the sense of a priori rationalism. This is all the more curious since they both use similar approaches and evidence. To what extent Koyré's rationalist apriorism is more adequate than Feyerabend's pseudo-irrationalistic anarchism will be examined presently.

I might pay my own homage to Koyré by interpreting his introduction of methods of intellectual history into the historiography of science as a move toward the technique of logical analysis, which is one whose powers I am interested in exploring. However, to do this would be insincere lip service on my part, for though it is true that Koyré's work is uniquely valuable as an introduction to the use of logical analysis in the interpretation of science, yet it needs such serious corrections, as it is shown below, that in their absence I feel it is bound to lead to abuse and to aprioristic and rationalistic excesses.

At any rate, now that Koyré "is the master of us all", to use words of historian Charles C. Gillispie,[2] the methodologically aware scholar, and most of all one interested in Galileo, ought to ask himself whether uncritical acceptance is any more justified here and now than it was at the time of Galileo vis-à-vis that supreme "master of those who know", Aristotle. To combat one authority with others, one might say that, though Koyré may be acknowledged as a founder of professional history of science in its present form, yet a discipline which hesitates to forget its founders is lost, a dictum coined by Alfred North Whitehead and popularized by sociologist Robert K. Merton.[3]

To be sure there have been critics of Koyré's work. For example, Koyré's erudition has been questioned by Eugenio Garin in 1957.[4] It concerns the character of Koyré's discussion of Galileo's alleged Platonism. Garin argues that Koyré's characterization of Platonism is abstract and unhistorical, and hence when the problem is discussed in such terms it is insoluble. He exhibits the inadequacies of Koyré's discussion of the actual historical situation of Platonism (that is, of Platonism as a historical entity) by pointing out a number of errors and confusions about relevant texts and persons. Garin argues that Koyré's distinction between Platonism as mathematicism and Platonism as mystical, magical numerology is a figment of Koyré's imagination in the sense that it has no historical basis in texts such as Ficino's works or even in Clavius's commentary to Euclid.[5] Another example given by Garin is Koyré's error in identifying the 'Lud. Buccaf.' mentioned in Bonamici's *De Motu* as a certain Lodovico Buccafiga[6] instead of Ludovico Boccadiferro; the latter is a nonnegligible figure, was a professor in various Italian universities including Bologna, and wrote several commentaries to Aristotle full of Platonic as well as Aristotelian doctrines;[7] moreover, he is mentioned in Galileo's own *Iuvenilia* (under the Latin name Buccaferrus).[8] Another example given by Garin is a reference to Crescas in Koyré's discussion of Descartes' notion (in *Le Monde*) of quantity of rest,[9] when, as Garin states, an educated person at the beginning of the seventeenth century could easily read Crescas's theses in the widely known *Examen vanitatis doctrinae gentium* by Francesco Pico.[10]

If Koyré's erudition can be so questioned, at the other end of the scholarly spectrum, the logic of his reasoning has been faulted in a significant way. It has been argued in my *History of Science as Explanation*[11] that an analysis of Koyré's central volume of *Études galiléennes* shows that the conclusions he himself draws are not supported by the evidence he himself gives.

Somewhere in between the extremes of these critiques lie two other equally serious problems. One is Koyré's persistent confusion of the context of scientific discovery and the context of scientific justification. There is nothing wrong with focusing on one or the other of these contexts, as long as one's conclusions are supported by evidence from the appropriate context. The error is to reach conclusions about one context on the basis of evidence from the other context. Koyré's study of Galileo's law of falling bodies examines evidence from the context of justification since he examines the various attempts found in Galileo's (published) writings to give a theoretical demonstration of it.[12] On the other hand, Koyré's conclusions[13] are about the method which enabled Galileo to succeed, where success can only be interpreted to pertain to his discovery of the law, since Koyré is at pains to point out the inadequacies of Galileo's various proofs of the law (and hence presumably there was no success in the context of justification), and since Koyré is contrasting that success to Descartes's failure, which was clearly a failure to arrive at the correct law of fall.[14]

It is interesting to point out that this type of criticism of Koyré has emanated from two very different quarters. On the one hand, it has been made in the terms just used in my work on historical method whose explicit and primary aim is to work out a philosophy of the historiography of science.[15] On the other hand, a similar criticism is made in the domain of pure Galileo scholarship, in Drake's attempt to emphasize 'Galileo Studies' as contrasted to the Koyré-type of 'Galilean Studies'.[16] Drake's distinction between the biographical context and the history-of-ideas context is similar to that between discovery and justification, and his judicious work in the former areas has led to some epoch-making results about Galileo's discovery of the law of fall.[17]

To this problem in Koyré's work which one may categorize as methodological or historiographical, we may add one that pertains to scholarship as such. Perhaps the single most striking feature that is apparent to the reader of Koyré's works, or at least of the book for which he is most famous, is the use of very long and very frequent quotations, interspersed with commentary. I am somewhat embarassed to report my having discovered that in the translation, citation, and arrangement of such quotations Koyré takes a number of impermissible liberties. It is almost as if these texts were being quoted so as to enable him to insinuate by distortion what could not be suggested by explicit argument.

For example, on p. 278 of *Études galiléennes* Koyré has a quotation for which he gives the reference '*Dialogo*, II, p. 423'. The quoted passage occurs neither on p. 423 (of the National Edition), nor in Day II, which at any rate does not include that page; since it is Day III that includes that page, one may look for the passage in Day III, but in vain. In fact, the passage is not from the *Dialogue* at all, but from Jacopo Mazzoni, as the interested scholar can discover by studying the rest of Koyré's footnote, which reads in part 'cf. Jacobi Mazzonii, . . . , *In Universam Platonis et Aristotelis Philosophiam* . . . , p. 187 sq. . . . ' Thus one concludes that Koyré's footnote should have read 'Jacobi Mazzonii, *In Universam* . . . , p. 187 sq.; cf. *Dialogo*, III, p. 423' rather than the other way around '*Dialogo*, p. 423; cf. Jacobi Mazzonii . . .' As it is, most readers will get the impression that Koyré is quoting from the *Dialogue*, when in fact he is not.[18]

However, the influence of Koyré's work persists in spite of such demonstrated inadequacies in erudition, logic (reasoning), methodology (historiography), and scholarship. So perhaps his appeal does not derive from the attraction of his approach but from the attraction of his substantive thesis, namely his apriorist-rationalist interpretation of science. For this reason, as well as because Garin's above criticism refers primarily to the first volume of Koyré's book, and because Drake's critique and mine above refer primarily to the second volume, it will be good to examine in detail the third which contains a statement of that thesis classic in its clarity and simplicity. It will emerge that Koyré confuses the activity of reasoning with the attitude of apriorist rationalism, and that his evidence supports an interpretation of science from the point of view of the former rather than one from the point of view of the latter.

The third volume deals with the law of inertia and contains a chapter entitled 'The *Dialogue on the Two Chief World Systems* and the Anti-Aristotelian Polemic' (K205–38).[19] It begins with an introductory discussion (K205–11) which emphasizes the connection between physics and cosmology in Galileo's work and in the rise of modern science in general; then it contains a brief discussion (K212–15) of the many aspects of the *Dialogue*: polemical, pedagogical, philosophical, and autobiographical. This is followed by an analysis of what Koyré takes to be the central core of the book, namely the physical arguments against the earth's motion. His analysis consists of three elements: a quotation of the main arguments (K215–19), an assessment of Galileo's criticism of these arguments (K219–20), and a justification of this assessment (K220–38).

The passage quoted by Koyré is that in which, first, Simplicio quotes from *De Caelo* Aristotle's original four arguments, and, then, Salviati gives a statement of several contemporary arguments inspired by the Aristotelian ones (F150–53).[20] The careful reader of Galileo's *Dialogue* will be struck by the last three lines of Koyré's quotation (K219) which read: "Moreover, the same thing would happen in *all* cases where one would shoot a cannon: the ball would pass above or below the mark according as one would shoot toward the east or toward the west . . . ." These lines constitute an excessively free translation[21] for the *initial segment* of the *last portion* of Salviati's speech on pp. 151–53 of the *Dialogue*. This last portion reads:

And not only the shots along the meridians, but also those toward the east or toward the west would not result right, the eastward ones resulting high, and the westward ones low whenever the shooting were point-blank; for, since the path of the ball in both shots would be along the tangent, namely along a line parallel to the horizon, and since if the diurnal motion belongs to the earth, the horizon would be constantly falling in the east and rising in the west (that's why eastern stars appear to be rising and western ones to be falling), therefore the eastern target would be falling under the shot, so that the shot would result high, and the rising of the western target would render the westward shot low. In this way one could not shoot right in any direction; and since our experience is different, we are forced to say that the earth is motionless (F153, my literal translation).

The liberties that Koyré takes are inadmissible because he thereby fails to see or to inform his readers that Galileo is here reporting a distinct problem with the earth's rotation, deriving specifically and exclusively from the point-blank nature of these gunshots. He is not merely repeating the east-west gunshot objection, stated earlier in Salviati's speech; in fact, the two problems require different solutions and these solutions are given by Galileo in two different passages (F193–97, and F205–09 respectively). Nor is Galileo merely summarizing the arguments he has just stated, which Koyré suggests by emphasizing the word 'all'. To excise this passage the way Koyré does, and to translate its beginning as he does, may be in accordance with his perception of the repetitive nature of the *Dialogue;* he argues that this repetitiveness has the important pedagogical and rhetorical function of familiarizing and accustoming its readers to the new concept of motion (K234, 237). However, if the alleged repetitiveness of the *Dialogue* is grounded on evidence like the present one, then it is an oversimplification at best, or perhaps an invention.

Another reason why it is improper for Koyré to quote the passage the way he does is that the point-blank objection and reply contain discussions of rectilinear motion along the tangent to the point of firing. The context of Koyré's discussion of the *Dialogue* is that of "Galileo and the Law of Inertia", namely, to what extent Galileo's conservation of motion involves rectilinear motion. Therefore, the point-blank objection would have been relevant, regardless of whether or not it would disconfirm Koyré's interpretation.

Let us now examine his assessment of Galileo's critiques. Here it is impossible to do Koyré any greater injustice than the one his own words do him. So let us quote this brief passage in full:

Let's now pass to the criticism. It is at once very profound and very simple. Galileo tells us that the arguments of the Aristotelians are nothing but paralogisms. They presuppose what must be shown. And, no doubt, it is true. But an Aristotelian could very well not accept the criticism, [which is] a consequence of the objection that Copernicus had already addressed to him: Aristotle does not reason, as he pretends, by starting from the facts, but on the contrary by starting from a theory. To this the Aristotelian could respond with good reason:
(a) that it is impossible to reason otherwise;
(b) that Galileo does the same.
In fact, the Aristotelian reasoning presupposes a theory, or if you prefer, a particular concept of motion, namely that of a process which affects the moving thing. It also presupposes that sense perception permits us to apprehend directly physical reality, that it is even the only means of apprehending it, and that, consequently, a physical theory can never cast doubt upon the immediate data of perception.
Now, Galileo expressly denies this. He starts from directly opposite assumptions:
(a) that physical *reality* is not given to the senses, but on the contrary apprehended by reason;
(b) that motion does not affect the moving thing, which remains indifferent to all motion that animates it, and that motion affects only the relations between a moving thing and one which does not move.
A paralogism from Galileo's point of view, the Aristotelian reasoning is in itself unobjectionable.
Nevertheless, dialectically speaking, Galileo no doubt has the right, at least within the *Dialogue,* to designate the Aristotelian reasoning as a paralogism. For, before having

stated the physical and mechanical proofs of the earth's immobility, Galileo has already laid down the double principle of the optical as well as mechanical relativity of motion (K219–20).

This is at best an oversimplification, and probably a disservice to the very rationalism that Koyré's interpretations are meant to support.

First, it is simply not true that "Galileo tells us that the arguments of the Aristotelians are nothing but paralogisms". In Koyré's own quotation, we have Aristotle's four original arguments (from violent motion, from double circular motion, from natural motion, and from vertical fall) plus five modern arguments (from the ship experiment, from vertical gunshots, from east-west gunshots, from north-south gunshots, and from point-blank gunshots). Of these nine arguments, Galileo claims that only three are paralogisms, namely the argument from violent motion (F159–62), the argument from double circular motion (F162–4) and the argument from vertical fall (F164–7). The problems with the other arguments, according to Galileo, are as follows. The argument from natural motion is the one to which most of the First Day is devoted (F164); it presupposes an untenable concept of natural motion, namely that straight and circular motion are two distinct instances of simple motion, whereas Galileo argues that they are two different stages of natural motion: straight motion can be acquired naturally but cannot naturally continue forever, whereas circular motion can naturally continue forever but cannot be acquired naturally without straight motion (F38–62; see Chapter 2). The ship experiment argument is simply based on a false premise, namely that on a moving ship a rock dropped from the top of the mast lands away from its foot (F169–75). The east-west gunshot argument involves a failure of hypothetical reasoning, namely a failure to take seriously the motion of the earth, even when examining the consequences of such motion (F193–7). The vertical gunshots argument, though in one place it is incidentally described as a paralogism just like the vertical fall argument (F200), in effect is shown to suffer from a failure to take into account both the relativity and the composition of motion, for this is what is actually discussed in the criticism (F197–203). The north-south gunshot argument is criticized as failing to take into account the conservation of motion (F203–5). Finally, the point-blank gunshot argument, which Koyré confuses with the one from east-west gunshots, is criticized as based on a phenomenon of such a small magnitude that it could not be detected even if it existed (F205–9), namely that on a moving earth the deviation from the horizontal would be of the order of a fraction of an inch.

Second, of the arguments that Galileo claims are paralogisms, it is not true that the paralogism is always that "they presupposed what must be shown". Only the vertical fall argument begs the question, the other two are fallacies of equivocation. The argument from violent motion misuses the ambiguity of the proposition "the parts of the earth would also move circularly", which can mean either that these parts would move around their own centers or that they would move around the earth's center (F159–62); Aristotle's second argument allegedly commits the fallacy of equivocation because its conclusion could mean either that the earth lacks the diurnal motion or that it lacks the annual motion (F162–4).[22]

Koyré's next error is to confuse begging the question (reasoning that presupposes what must be shown) with apriorism (reasoning based on theories rather than facts). For the reason that he gives why it must be admitted that the Aristotelian reasoning begs the question is that it starts from a theory rather than from facts. Koyré's confusion is a serious error. An argument presupposes what must be shown when one of its premises depends on its conclusion. Such reasoning is fallacious because an argument is an attempt to show that since you accept the premises you must accept the conclusion; if one of the premises depends on the conclusion, in the sense that it implicitly assumes the conclusion, then the argument is assuming what it is trying to prove, which is surely wrong. Nor would Simplicio or Aristotle deny this. Nowhere in the *Dialogue* does Simplicio take lightly the accusation of begging the question, which after all belongs to Aristotle's own list of fallacies.[23] On the other hand, to reason from a theory may or may not be correct, depending on the adequacy of the theory. But even if the presupposed theory is false or inadequate there is no *logical* error, no paralogism, unless the theory from which you are reasoning is the same as the theory (or alleged facts) *to* which you are reasoning; in this case you are involved in circular reasoning and are begging the question. In other words, not every instance of reasoning from a theory is an instance of presupposing what is being shown. Nor is every instance of begging the question an instance of apriori reasoning: it may be that what you are trying to prove is a fact; and it may be that you are proving it on the basis of other facts; then there would be no apriorism, but if one of the facts on the basis of which you are proving the conclusion, *depends on* the conclusion, then you are still begging the question. Circular reasoning does not become legitimate merely because it goes on within a domain of facts. In conclusion, then, apriorism is neither a sufficient nor a necessary condition for circular reasoning.

Let us illustrate such abstractions with discussions from the *Dialogue*. The

Aristotelian argument claimed by Galileo to be a paralogism in the sense of begging the question is the following (F164—6):

The earth cannot rotate because if it did bodies could not fall vertically; but they do since they can be seen to graze the edge of a tower when dropped from it. (My paraphrase.)

The final portion of this argument grounds the impossibility of the earth's rotation on the vertical fall of bodies, so that one premise of that final step is the proposition that "bodies fall vertically". Another portion of the argument bases this vertical fall of bodies on the apparent grazing; hence it is being assumed that this apparent grazing implies vertical fall. Now, Galileo argues, this implication can be questioned because it would not hold on a rotating earth (since *if* the earth rotated *and* bodies were seen to fall grazing the tower, *then* their actual path would not be vertical but slanted). Since the implication can be questioned, it is fair to ask for a justification of it: how do you know that the apparent grazing implies vertical fall? One abstractly possible justification would be the following: if the earth stands still then the implication holds; the earth does stand still; therefore the implication holds. There being no other means whereby an Aristotelian could justify the implication, he would have to use this abstractly possible argument. In so doing, though it is indeed true that the implication holds if the earth stands still, he is also assuming that the earth stands still, which is the final conclusion he wants to reach. Thus the Aristotelian argument from vertical fall begs the question because its premise that bodies fall vertically presupposes (i.e., would have to be justified by the proposition) that the earth stands still.

Feeling the force of Salviati's criticism, Simplicio with Sagredo's help states a new argument which can serve here as a good example of reasoning starting from a theory (F166—7):

If the earth rotated then bodies could not be seen to fall grazing the edge of a tower, since on a rotating earth this apparent grazing would imply that bodies would have two natural motions, toward and around the center, which is impossible. But bodies are seen to fall grazing the edge of a tower. Therefore, the earth can't rotate. (My paraphrase.)

The final portion of this argument grounds the impossibility of the earth's rotation directly on the apparent grazing, which is self-evident. But it also grounds the impossibility of rotation on the claim that if the earth rotated the apparent grazing would not occur. In another portion of the argument, this claim is grounded on two premises, a self-evident one and a theory. It is self-evident that on a rotating earth apparent grazing would entail simultaneous motions toward and around the center of the earth. But it is merely a "theory" to say that it is impossible for material bodies to have two spontaneous motions toward and around the center. This proposition may be merely a theory, but it is not the same as the conclusion of the argument. Hence this argument does not beg the question, though it is an instance of what Koyré would call a priori reasoning, reasoning from a theory.

Now, since Koyré is conflating the paralogism of presupposing what is being shown with the problem of a priori reasoning or reasoning from a theory, and since he claims that all the Aristotelian arguments are paralogisms, it follows that what he *means* is probably that these arguments are all reasoning from a theory. Let us ask whether he is right in so claiming. Unfortunately not, if for no other reason than because the ship experiment argument, which is quoted by him, is clearly not a piece of a priori reasoning. Koyré thinks that it is probably because he thinks that Galileo's criticism of it is a priori; however, even if this were so, it would not make the original ship experiment argument an a priori one. The simple truth is that someone may, on the basis of a theory, criticize an argument which is not itself based on a theory. Another reason Koyré might give for regarding the ship argument as a priori is to say that the experiment had never been made (K225). However, this may be a piece of a priori reasoning by Koyré himself; in fact Chiaramonti claims in his 1633 book answering Galileo's *Dialogue* that the experiment had been made by a certain Giovanni Cotunio of the University of Padua.[24] A final reason Koyré might give is his belief that "it is impossible to reason otherwise" than from a theory; however, he gives no justification for this allegation, unless it be the claim that the Aristotelians as well as Galileo did *in fact* reason from a theory; hence such an argument would beg the question.

Let us continue to place qualifications on Koyré's claims to see if we can find some truth in them. Though not all the Aristotelian arguments involve reasoning from a theory, some of them do. Concerning these, can we agree with Koyré that they are unobjectionable because it is impossible to reason otherwise and because Galileo does the same? It is true that if it were impossible to reason otherwise than from a theory, then an argument could not be faulted for doing so. However, it is obvious that reasoning from a theory is merely one type, and Koyré's claim that it is impossible to reason otherwise is an extravagant exaggeration at best. In fact, he gives no justification of this claim. Or perhaps his analysis of the Aristotelian and of Galileo's reasoning could be interpreted as a supporting argument, namely, that it is impossible to reason otherwise than from a theory because both the Aristotelians and Galileo did so. It would not be a very serious criticism of this argument to object that a generalization is being reached from two cases; its real problem is that these cases are not typical. In fact, the relevant Aristotelian reasoning consists of those arguments against the earth's motion which *happen* to involve reasoning from a theory; hence, these arguments would be prejudicially chosen to ground Koyré's generalization. On the other hand, the Galilean reasoning under consideration is that whereby he criticizes these and other Aristotelian arguments. Now, arguments criticizing a priori arguments are not typical since, whatever a priori element they might contain, it could be the result of the a priori arguments being criticized. Regarding Galileo's arguments that criticize Aristotelian reasoning that does not start from a theory, even if they did presuppose a priori elements, as Koyré tries to show, this very feature would make them atypical, since they would then be a priori criticism of empirical arguments. Since the question here is whether Koyré's alleged impossibility of reasoning other than from a theory has been reached by generalizing typical cases, we may dismiss Koyré's generalization.

Let us now examine Koyré's argument that the Aristotelian arguments which start from a theory are unobjectionable because Galileo is doing the same. Here it is important to see that, even if Galileo were reasoning from a theory, even if Koyré's account had shown this, his conclusion still would not follow: the Aristotelians could not justify their position with such a *tu quoque*. There are two reasons for this. First, as Koyré himself points out, an important element of the theory *from* which the Aristotelians argued was their emphasis on empiricism: that sense experience is the only means of apprehending reality; hence the Aristotelians could not consistently have admitted that their arguments against the motion of the earth were really reasoning from a theory. Second, much of Galileo's criticism consists of an analysis of the a priori elements of the Aristotelian reasoning, together with the argument that if one replaces these a priori elements with other, more plausible theories one cannot then draw the conclusion that the earth must stand still. Of course Koyré denies that these other theories, from which Galileo reasons, were more plausible than those from which the Aristotelians started; but even if Koyré is right about the equal plausibility of the respective presupposed theories, the situation is *not* otherwise symmetric. For, whereas the Aristotelians were trying to prove the impossibility of the earth's motion, Galileo in his criticism of these arguments is merely trying to prove that the Aristotelian reasoning is incorrect, not that the earth moves. In his criticism, Galileo never argues that, given his new concept of motion, it follows that the earth moves, but rather than given the new concept it follows the motionlessness of the earth is not proved by the Aristotelian arguments. In other words, the crucial difference is that Galileo's conclusion is that the Aristotelian reasoning is incorrect, whereas the Aristotelian conclusion is that the earth stands still; the Aristotelian point is that the earth stands still, Galileo's is that the Aristotelians haven't proved their point. That is why it is wrong for the Aristotelians to argue from a theory, but not for Galileo. Galileo argues from a theory in the context of showing that from a different theory one could not reach the conclusion the Aristotelians reach in their argument; since their argument presupposes a theory, they haven't proved their conclusion. For example, consider the above mentioned argument given as a good example of Aristotelian reasoning from a theory, the argument that based the impossibility of the earth's motion on the falling body's apparent grazing of the edge of a building, and ultimately, on denying the possibility of natural motion both toward and around the center. Given a theory that allows this possibility, one could not conclude the motionlessness of the earth from the apparent grazing. Galileo reminds us very frequently that his criticism is merely disproving the Aristotelian arguments, not proving the earth's motion, so much so that it would be tedious to collect the references.

Though Galileo's alleged reasoning based on a theory cannot justify the Aristotelian procedure, Koyré's claim is intrinsically interesting, and it is worth examining. In the passage quoted above, Koyré claims that Galileo's reasoning presupposes the principle that physical reality is apprehended by reason, not by the senses, and the principles of relativity and of conservation of motion. In the rest of his account (K220–38) Koyré then supports these claims by analyzing the passages where Galileo answers the Aristotelian arguments; special attention is given to the answer to the ship experiment argument which is the only passage used by Koyré as showing Galileo's presupposition of aprioristic rationalism. It should be noted first that, even if Koyré is otherwise correct in these interpretations, he would not be justified in claiming that the way in which Galileo presupposes the principle of apriorism is the same as that in which he presupposes the principles of relativity and of conservation, namely that all these principles are theories from which Galileo starts in his reasoning, so that his reasoning is no better grounded than that of the Aristotelians, who start from different theories. The fact that the principle of apriorism is an epistemological one concerning the nature of knowledge, and that the other principles are physical ones concerning the nature of motion, is perhaps the least of their differences. The more significant difference is that the physical principles are or could be used as *premises* in certain parts of Galileo's counterarguments, whereas the epistemological principle is not so used. For example, in the answer to the ship experiment it is clear that Galileo is or could be arguing as follows: given the principle of conservation, the rock will retain its horizontal motion even after it is dropped; and given the principle of relativity, the simultaneous downward motion will not constitute an interfering disturbance to this horizontal motion; therefore, on the moving ship the rock will end up at the foot of the mast. On the other hand, how would Galileo be committed to apriorism, assuming that he were? If Galileo can argue, by reasoning of the type just sketched, that the rock will land at the foot of the mast even on a moving ship, then it follows that one has to accept such a result on a moving ship unless one can show that there is something wrong with Galileo's argument. Now, to find something wrong with this argument one has to fault either some steps in reasoning or some premises being assumed in it; since presumably there is nothing wrong with the steps in reasoning, the only thing one could question is the assumed premises; but, if Koyré is right, these assumed premises are the principles of relativity and of conservation which cannot be faulted empirically but only by opposing to them another concept of motion such as the Aristotelian one; it follows that the only objection one could have to Galileo's reasoning is to produce the Aristotelian *argument* concluding that the rock must fall behind on a moving ship. Now, this may be in accordance with Koyré's desire of showing that both Galileo and the Aristotelians are reasoning from a theory, but it is not in accordance with his claim that Galileo is assuming apriorism as a principle *from* which he reasons to arrive at his physical conclusion, for it also follows from the above considerations that the apriorist principle is *implied by* Galileo's allegedly a priori argument. In other words, the apriorist principle is a *consequence* rather than an assumed premise of Galileo's ship experiment counterargument; the principle follows from the argument, rather than the other way around; Galileo does not need the epistemological principle to justify his physical argument, but rather his physical argument (the fact that such an argument can be given) justifies the epistemological principle. In short, *if* Galileo can answer the ship experiment argument in the a priori way that Koyré thinks, *then* his answer is *supporting, not assuming,* apriorism. Moreover, and conclusively for the present issue, if it is a fact that the Galilean and Aristotelian arguments are a priori, then apriorism is itself a fact and not a theory, for it would then be the consequence of a fact.

To explain further this difference between an assumption and an implication, and thus to reinforce this criticism of Koyré, I will call attention to an Aristotelian argument whose answer does involve Galileo in an epistemological assumption about the relationship between sense experience and reason similar (though not identical) with the apriorist rationalism here attributed to him by Koyré. The argument is that from the deception of the senses; it is explicitly discussed by Galileo in the Second Day, at the beginning of his critique of Chiaramonti's book on the three new stars (F272–81); it is also implicitly discussed at the beginning of the First Day (F56–62) and in the discussion of the gunshot arguments (F197–200). However, it is not one of the arguments quoted or mentioned by Koyré, though it would have provided him with a proper illustration of how an argument against the earth's motion and Galileo's answer to it assume an epistemological principle. The argument is that if the earth rotated then our senses would be deceived insofar as (a) we do not *feel* any motion, and (b) falling bodies would be appearing to move vertically but would in reality move in a slanted path. Obviously, in order to be able to conclude from this (by *modus tollens*) that the earth does not rotate, one needs another premise which would be a denial of the 'then' clause of the conditional premise; we could say that our senses cannot be deceived, or to use Koyré's words, that "sense perception permits us to apprehend physical reality directly" (K220). Galileo answers (F57–9, F166, F272–80) that our senses tell us merely that bodies *appear* to fall vertically (vision) and that the earth *feels* to be at rest (internal sense); to think from this that bodies really fall vertically and that the earth is really at rest is to draw a conclusion that involves reasoning; this reasoning happens to be incorrect since we cannot say without qualification that apparent vertical fall implies actual vertical fall (it does so only on a motionless earth), and since it is not true that we can feel all motion (we can only feel *changes* of motion); therefore, if the earth rotated it would not be our senses that would be deceived, but rather our reason, or at least the reason of those who would draw the unwarranted conclusion from the sense data; therefore the argument from the deception of the senses is incorrect. But in any case, it is by reasoning, not by the mere senses, that we can apprehend reality, or in Koyré's words, "physical reality is not given to the senses, but on the contrary apprehended by reason" (K220). From this 'theory' Galileo is refuting the Aristotelian argument, which was based on the opposite theory. However, I believe that, besides providing a good illustration for Koyré's point, this argument provides a good illustration of something he does not want to accept, namely that the two 'theories' are *not* on a par; at the very least Galileo's is the more plausible one; actually, in the sense in which they relate to the present argument, Galileo's 'theory' is true, whereas the Aristotelian one is false. Hence, though Galileo's critique of this argument illustrates Koyré's point, it does not support it; though it confirms the letter of his thesis, it does not confirm its spirit, since he attaches an apriorist meaning to Galileo's epistemological assumption, whereas in the present context it has a critical meaning. Galileo is being both a *critical* empiricist and a reasoning-oriented rationalist.

The same type of non-apriorist rationalism is presupposed by Galileo, though by way of implication rather than assumption, in the answer to the ship experiment argument examined by Koyré. Galileo is not a rationalist in the sense of an apriorist, but in the sense that he likes to use reasoning and arguments as much as possible, we might say a rationalist in the sense of a logician, a logician-in-action. The difference is that the logician will not limit himself to a priori reasoning, reasoning from a theory; some reasoning is reasoning based on facts, where by starting from facts one attempts to justify a conclusion; if the conclusion is a theory then we would have reasoning *to* a theory; or the conclusion may be itself factual, if only facts are used as premises and if all the inferences are strong. To be more specific, the thesis implied by the present passage (F169–80) could be formulated as follows: experiments are sometimes unnecessary to ascertain the results of a test, for sometimes it can be argued on the basis of known or more easily ascertainable facts, what these results must be. In fact, the passage can be reconstructed so as to become an illustration of this thesis, as was done in Chapter 5.

In his analysis Koyré neglects to take into account the very important fact that, though Galileo is justifying the conclusion of his own counterargument by reference to the principle of conservation, this is not being merely assumed but empirical evidence is given in its support. The principle of conservation corresponds to proposition (1) in the above reconstruction, which is justified by means of (a) and (b). (See Chapter 5.) Of course Koyré may say that the reasoning from (a) and (b) to (1) is not completely valid from a logical point of view, or perhaps that other principles are being assumed in this subargument. However, to the former it could be replied that it is one thing for a proposition to be incompletely supported and it is another for it to be completely unsupported; only the latter would be a 'theory' in Koyré's sense. To the second alternative one could reply that the other assumed principles would probably themselves be supportable when questioned, and though other principles might be assumed in these other arguments, perhaps they too could be supported if required, or if not perhaps one would be dealing with highly abstract, general, and universal metaphysical principles, whose assumption would present no practical or physical problem. Thus the impact of Koyré's point would end up being something to the effect that whenever one is reasoning one is justifying certain propositions on the basis of others which for the time being are not questioned; that is, in reasoning one can't prove everything simultaneously. But these statements about reasoning, far from presenting problems, constitute very elementary facts about the nature of reasoning; they amount to saying that reasoning is a step-by-step process. The only real question one could raise is whether such an activity is effective. To this the facts of life and of history provide the obvious answer.

Concerning the principle of relativity, the situation is more complicated. Koyré would argue as follows: though this principle is not directly present in the argument, the argument is assuming that "there is no way in which the horizontal impressed virtue could be disturbed by the vertically downward tendency due to weight", which is a form of the principle of composition; now, the composition of motion is a consequence of its relativity (K222–3). So presumably Galileo's answer is based on the principle of relativity in the sense that it would have to be used to justify one of the premises in his argument. However, Koyré's interpretation is not a faithful reconstruction of the argument actually given by Galileo. It is indeed true that if one assumes the principle of relativity of motion, and *if* one interprets it in such a way as to imply the composition of motion, *then* one could arrive at Galileo's result for the moving ship (having also used the principle of conservation). However, in the passage when Galileo gives his answer (F169–80), the relevant step (F175) is much more concrete: the vertical fall does not represent a disturbance to the horizontal motion because the body is indifferent to horizontal motion, and because the cause of the horizontal motion (the impressed 'virtue') is distinct from the cause of the vertical fall (gravity). Of course, Koyré could question the soundness of these inferences, though they are not questioned by Simplicio, who merely questions the truth of the two premises, that the body is indifferent to horizontal motion, and that the cause of this motion is an impressed 'virtue'. But even if the soundness of the inferences is questionable, Koyré's substitute argument is not unquestionable either, for though it may contain no problem of the soundness of the inferences, it contains the problem that the crucial premises are just being assumed. I am inclined to believe that the inferences in Galileo's actual answer are contextually sound. It follows that this answer is not assuming, but proving (supporting) the principle of relativity. This can be seen even more clearly from Galileo's answers to the vertical fall argument (F164–6) and to the east-west gunshot argument (F193–7), as we will see below.

Koyré's evidence that the principle of relativity is being presupposed in the sense of assumed consists of (1) the fact that Galileo obviously holds the principle, (2) the fact that by *postulating* the principle one *could* answer the objections to the earth's motion being considered by Koyré, and (3) the alleged fact that Galileo gives several statements of the principle before he gives his answers to the objections (K220, n.3; K221, n.1; K222, n.1; K237). (1) is indeed true, but by itself it does not support Koyré's conclusion, since it might be, as I think is the case, that Galileo holds the principle because it is supported by his critiques, rather than because he needs to assume it in order to make those critiques. (2) is irrelevant, even if true, for the fact that the logic of the situation could be rearranged differently, does not change the facts of the *actual* logic of the situation, as we have just seen for the case of the ship experiment.

As for Galileo's alleged previous references, they are misinterpreted by Koyré (K220, n.3: F57, F101, F139, F141; K221, n.1: F139 ff.; K237). The first reference alleged by Koyré occurs at the beginning of the First Day where Simplicio objects to Salviati's principle of circular motion, by saying that it conflicts with the clear evidence of the senses that bodies are seen to fall in a straight line (F57). Salviati answers, among other things, that it is questionable, as he promises to show later, whether falling bodies really follow a straight path. However, there is no statement of the principle of relativity; there is only a reference to a future discussion, hence even if that discussion were the proper one, the present passage could not be regarded as evidence of any antecedent commitment to relativity. At any rate the subsequent passages being referred to can only be those containing the critique of the argument from vertical fall (F164–66) and the one containing the suggestion that the real path of a falling body is circular (F188–93); and in these passages the principle of relativity is not in sight.

Koyré's second reference is to p. 101 where I find the general topic to be the roughness of the lunar surface. More specifically, on that page we find the ending of the discussion of the experiment comparing reflections onto a wall from a flat and from a spherical mirror, and the beginning of the discussion where Salviati attempts to explain the just-observed experimental facts in terms of eye irradiation and of the microscopic irregularities of the reflecting surface. Perhaps, Koyré meant a different page.

Koyré's other references (F139, F141, F139 ff., and F141 ff.) are to the beginning of the Second Day, after the preliminary discussion criticizing the slavish followers of Aristotle's authority. The passage (F139–43) begins with Salviati saying:

Then let the beginning of our reflections be the consideration that whatever motion comes to be attributed to the earth must necessarily remain imperceptible to us and as if nonexistent, so long as we look only at terrestrial objects; for as inhabitants of the earth, we consequently participate in the same motion. But on the other hand it is indeed just as necessary that it display itself very generally in all other visible bodies and objects which, being separated from the earth, do not take part in this movement. So the true method of investigating whether any motion can be attributed to the earth, and if so what it may be, is to observe and consider whether bodies separated from the earth exhibit some appearance of motion which belongs equally to all.[25]

Here Koyré not only takes this passage out of context, but he manages to perpetrate an equivocation with the Italian word *principio*. This can mean either 'beginning' or 'principle'. The text suggests 'beginning' as the meaning, and both Drake and Salusbury-Santillana[26] translate it this way. Koyré *interprets* the meaning as being, "Let the principle of our reflections be . . . ." More importantly, the context of the present passage is as follows. The discussion mentioned by Koyré is part of a passage containing several arguments favorable to the earth's rotation (F139–50); this is followed by another passage containing statements of the arguments against (F150–9); and then come the critiques of these arguments (F159–244). The favorable arguments are explicitly labeled by Galileo as being merely probable (because based on the principle of simplicity); as requiring the removal of the apparently conclusive counterevidence, since a single conclusive objection would overcome all probable arguments (F148); and as being based on a principle of relativity of motion which both parties agree is good old Peripatetic doctrine, and perhaps even older than Aristotle (F142). Therefore, the principle being explicitly laid down here must be that of optical, not mechanical, relativity. The reasons for this are the presentation of it as Aristotelian doctrine, just mentioned, and the explicit requirement that it remains to be shown that there is no conclusive counterevidence (F148–50). For Galileo is very clear that on the basis of the kind of relativity mentioned here, because of the principle of simplicity, it only follows that, *other things being equal,* the earth's rotation is more probable. At this stage Simplicio believes that the other things are *not* equal, since he thinks there are conclusive arguments against the earth's motion; also at this stage Salviati is confident that he will be able to show that the other things *are* equal, though it is obvious that he has not yet done so. The other things that have to be equal are all the terrestrial phenomena usually mentioned as proof of the earth's rest: these phenomena have to be equally explicable by the motion and by the rest of the earth. So Salviati knows that his task is to show this; but this is what is usually called the principle of mechanical relativity, which is very different

from that of optical relativity, and which everyone including Koyré agrees to
be one of Galileo's great innovations. But this mechanical relativity is not the
principle already stated in this passage but the one that remains to be shown.
In Salviati's words:

All inconveniences will be removed as you propound them. Up to this point, only the
first and most general reasons have been mentioned which render it not entirely improbable that the daily rotation belongs to the earth rather than to the rest of the
universe. Nor do I set these forth to you as inviolable laws, but merely as plausible
reasons. For I understand very well that one single experiment or conclusive proof to the
contrary would suffice to overthrow both these and a great many other probable arguments. So there is no need to stop here; rather let us proceed ahead and hear what
Simplicio answers, and what greater probabilities or firmer arguments he adduces on the
other side . . .
    Finding out whether both positions satisfy us equally well will be included in the
detailed examination of the appearances which they have to satisfy. For we have argued
ex hypothesi up to now, and will continue to argue so, assuming that both positions are
equally adapted to the fulfillment of all the appearances.[27]

In fact, when the time comes to answer the objections, Galileo does not
answer them in the way portrayed by Koyré, namely by appealing to the
principle of mechanical relativity, on the basis of which the answers would be
trivial, and then by trying to familiarize and accustom Simplicio to this new
concept (K222–3, K231–6). Instead Galileo engages in a detailed critical
analysis showing that the evidence adduced by the objections cannot be used
to decide whether or not we are on a moving earth. Let us take the vertical
fall argument (F164–6). Of course, given the principle of mechanical
relativity, "if the rock and the tower share the same motion of the earth, then
this motion will be for them as nonexistent, and everything will happen as if
it did not really exist, that is, as if the earth were motionless", as Koyré puts
it (K223). And, of course, "that's exactly what the Aristotelian cannot
admit", as Koyré also says (K223). But that is why Galileo attempts to prove
his point without previously assuming the principle of mechanical relativity.
As we saw above, Galileo argues, by very concrete and basically correct considerations, that the argument from vertical fall begs the question, because we
can't know that bodies really fall vertically unless we know that the earth
stands still. Though Galileo here does not argue explicitly for the principle of
mechanical relativity, it is obvious what the next few steps of the argument
would be. Given Galileo's detailed critique of the objection from vertical fall,
it follows that the phenomenon of vertical fall, that is fall which is really
perpendicular to the earth's surface, cannot be used to prove the motionlessness of the earth. But since we can't know that vertical fall is a fact unless

we know that the earth stands still, it follows that we can't know that vertical fall is a fact unless we know that the earth moves; and from the latter it follows that vertical fall can't be used to prove the motion of the earth either. Therefore, vertical fall cannot be used to prove either that the earth moves or that it doesn't. So here we have one crucial phenomenon that has been shown to obey the principle of mechanical relativity.

However, the fact that this is true for one phenomenon doesn't mean that it will be true for other phenomena; and so Galileo criticizes the east-west-gunshot argument and ends up showing that the range of east-west gunshots also obeys the principle. Thus Galileo is not being repetitious for the sake of familiarizing Simplicio with a new concept by recourse to experience, as Koyré seems to think. Moreover, there is no real need for him to make it look as if it were Simplicio who was requiring the recourse to experience. Koyré so claims when he says that "the reader contemporary to Galileo . . . through Simplicio's mouth, once again asks for a recourse to experience: 'I should like, he tells us, to find some means of making an experiment concerning these projectiles . . .'" (K234). A footnote refers us to p. 194 of the *Dialogue*. When this is consulted one finds that the speaker is Salviati, not Simplicio![28] Finally, the specific reasons, not based on a previous assumption of mechanical relativity, why the range would be the same on a moving earth, involve a comparison to shooting arrows from a moving cart (F193–7), and a comparison between this and shooting them by throwing them with different speeds from a motionless cart. Galileo argues that the arrows can be made to move through correspondingly different distances either by shooting with equal force from a moving cart or by imparting them different initial speeds from a motionless one, for the equal forces from a moving cart generate more speed in the arrow in the direction of the cart and less in the opposite direction than the same forces do on a motionless cart; the different traveled distances, in turn, generate equal ranges from a moving cart. Galileo's argument hinges on the analogy between the earth and the cart, and on the difference in effective force produced by applying the same force in different directions on a moving cart. It follows, that the range of east-west gunshots is a phenomenon that would occur in the same way whether the earth moves or not. To that extent the principle of mechanical relativity is thereby justified.

In conclusion, we may say that Koyré's account of the logic of the Aristotelian objections to the earth's motion and of Galileo's counterarguments is mistaken both in its fundamentals and in its details. The basic problems seem to be superficiality in logical analysis, oversimplification, injudicious exaggerations, and questionable manipulation of the text by means of excessive quotations, of taking passages out of context, and of not infrequent scholarly carelessness. Nevertheless, Koyré does deserve the credit for having called attention to the logical structure and validity of Galileo's arguments and to his rationalism, even though he misunderstands the former as circular and misinterprets the latter as apriorism.

Finally, it would be unhistorical to deny that the study of the history of science made great progress with Koyré; to turn the clock backwards is simply unthinkable. Nevertheless, even historical sensibility is not an absolute methodological requirement; indeed scientific method (as conceived here, Chapters 5, 6, and 8) tells us that the historical understanding of the Koyré case would presently be rather injudicious. For at a time when the technique of error analysis which he pioneered is undergoing a 'cancerous growth',[29] and when the apriorist rationalism which he defended is acquiring the status of a dogma, the 'scientific' thing to do is to be critical of Koyré. The fact that the logical analysis advocated here may be seen to stem from his technique of error analysis, and that our emphasis on reasoning may be taken to resemble his emphasis on rationalism, ought to serve as a warning of how far our 'science of history' is from the *scientism* of history or any other mechanical rule or oversimplified panacea.

NOTES

[1] Professors Marshall Clagett and I. Bernard Cohen have dedicated to him their major works, and Professor Charles C. Gillispie states that "he is the master of us all" (*The Edge of Objectivity*, p. 523).

[2] *The Edge of Objectivity*, p. 523.

[3] Quoted from Whitehead's *The Organization of Thought* in Merton's *On Theoretical Sociology*, p. 1, as the epigraph to Chapter I.

[4] 'Chi legga di A. Koyré . . . ', *pp. 406–408*.

[5] *Ibid.*, pp. 406–407.

[6] A. Koyré, *Études galiléennes*, p. 44.

[7] Garin, pp. 408–409.

[8] Galilei, edited by Favaro, *Opere* 1, 134, 167, 172.

[9] Koyré, p. 340, n. 1.

[10] Garin, p. 408.

[11] M. A. Finocchiaro, *History of Science as Explanation*, pp. 86–116.

[12] Koyré, pp. 83–107 and 136–155.

[13] *Ibid.*, pp. 155–158.

[14] *Ibid.*, pp. 107–136.

[15] *History of Science as Explanation*, especially pp. 234–238.

[16] S. Drake, *Galileo Studies*, p. 14.

[17] S. Drake, 'Galileo's Discovery of the Law of Free Fall'.

[18] This reference has been corrected in the English translation of *Études galiléennes*, just published under the title *Galileo Studies* (p. 233, n. 285).

[19] Hereafter references to Koyré's *Études galiléennes* will be made in the text, by prefixing page numbers with 'K', short for 'Koyré'.

[20] G. Galilei, edited by Favaro, *Opere* 7, 150–53. Hereafter, references to this volume of Galileo's *Opere*, which contains the *Dialogue*, will be made in the text by prefixing page numbers with 'F', short for 'Favaro', the main editor of the 'National Edition' of the *Opere*.

[21] The English translation of Koyré's book does not read this way for this passage since, instead of translating Koyré's words, it quotes from Drake's translation of the *Dialogue*. Cf. Koyré, *Galileo Studies*, p. 162.

[22] See Chapters 2 and 16.

[23] *Topics*, 162b34; *Prior Analytics*, 65a10, 64b33; cf. C. L. Hamblin, *Fallacies*, pp. 50–58.

[24] S. Chiaramonti, *Difesa al suo Anti-Ticone e Libro delle Tre Stelle Nuove*, p. 339; cf. G. Barenghi, *Considerazioni sopra il Dialogo*, p. 183.

[25] G. Galilei, *Dialogue Concerning the Two Chief World Systems*, tr. Drake, p. 114; cf. K139–40.

[26] G. Galilei, *Dialogue on the Great World Systems*, tr. Salusbury-Santillana, p. 127.

[27] Galilei, *Dialogue*, tr. Drake, pp. 122 and 124.

[28] The English translation of Koyré's book corrects his mistake; cf. Koyré, *Galileo Studies*, p. 172, and p. 226, n. 181. What effect this correction has on the Galilean-Aristotelian drama which constitutes the tenor of Koyré's text, the reader can judge for himself.

[29] To give an ironical twist to this phrase which Agassi used to characterize the pre-Koyré period in the historiography of science; see J. Agassi, *Towards an Historiography of Science*, pp. 33–40.

CHAPTER 10

THE ERUDITION OF LOGIC AND
THE LOGIC OF ERUDITION:
CRITIQUE OF GALILEO SCHOLARSHIP

The critiques of the last several chapters, undertaken after the analysis of various aspects of Galileo's *Dialogue* of the earlier chapters, have been doubly instructive. In part, we have gained new or deeper insights into his work; but we have also derived useful theoretical and methodological lessons concerning scientific rationality, the history and the nature of philosophy, and the historiography of science. Let us continue our two-faceted critiques by exploiting the opportunities offered by classic Galileo scholarship from the point of view of the logic of erudition.

Historians have seldom been good logicians, nor have logicians often excelled in historical sensibility. Is this fact a historical accident or a logical necessity? One way to explore this problem would be to undertake a methodological examination of the history of logic; this could be done partly by analyzing works by historians of logic with an eye toward the formulation of methodological lessons, and partly by a philosophical, more or less speculative reflection on the nature of logic, history, the history of logic, and the historiography of logic. One useful distinction here would be between logical theorizing and logical practice. Almost all works in the history of logic deal, of course, with the development of logical theories. There is no reason, however, to neglect unduly the history of logical practice; in fact, given the preponderance of the former, now would seem an appropriate time to do some work in the latter. This would give us a second way to explore the above mentioned problem. But what is meant by logical practice? It is the more or less self-conscious practice of reasoning combined with some explicit use of (basic) logical concepts and terminology. Important works for the historian of logical practice to consider would thus be Euclid's *Elements*, Plato's *Dialogues*, Aquinas's *Summa Theologica*, Machiavelli's *Discourses*, Galileo's *Dialogue Concerning the Two Chief World Systems*, Hume's *Dialogues Concerning Natural Religion*, perhaps Marx's *Capital*. It seems that the study of such works would require both logical acumen and historical sensibility and would therefore be very fruitful and challenging.

A third way to explore the problem would be to study the methodology of the history of scientific thought, for though one may not want to equate scientific thought with logic, few would deny that reasoning plays a central and effective role in science. I have carried out some such studies elsewhere.[1]

Fourth, one could study the logic of history, partly in the sense of the nature of historical development, but primarily in the sense of the reasoning of historians. Here one would discover whether or not the difficulties are insurmountable.

I believe I can combine several of these approaches by a critical examination of selected writings by historians of science dealing with Galileo's *Dialogue*. Most directly I will be dealing with the logic of historiography, attempting to derive methodological lessons by analyzing the reasoning of historians. Indirectly I will be dealing with the history and the historiography of logical practice. Because of the place of Galileo and the *Dialogue* in the history of science, this will be relevant to the historirgraphy of science. Finally, I will be dealing with the historiography of logical theorizing insofar as some of the subjects of my discussion will be remarks by historians about a number of logical concepts used by Galileo.

The choice for my analysis is Emil Strauss's Notes to his 1891 German translation of Galileo's *Dialogue*[2] because they are universally regarded as the most informative and erudite ever compiled; Antonio Favaro's text of the *Dialogue* in the 'National Edition' of Galileo's complete works, since this is the standard and definitive critical edition;[3] Stillman Drake's translation of the book since Drake is perhaps the greatest Galilean translator of all times and the leading Galilean scholar of our time;[4] and Maurice Clavelin's *Natural Philosophy of Galileo*[5] because it is the most widely acclaimed recent work on the topic.

Though no general solution to our problem will emerge, it will become apparent that, when the subject of historical investigation is itself logic and reasoning, even the erudition and scholarship of the best historians are not enough; at least as important are logical skill and sensitivity.

ERUDITE COMMENTARY ON REASONING: STRAUSS

The erudition of Strauss's notes is impressive; it has been recognized by Drake, who explicitly expresses his indebtedness to him,[6] and by Giorgio de Santillana, whose notes to his revision of the Salusbury translation are full of references to Strauss. Though these notes can still be read with profit and perhaps still remain unsurpassed, the careful reader of the *Dialogue* can detect a certain weakness in one specific area, namely logic and reasoning. It is useful, then, to examine carefully Strauss's comments on these topics.

1.   Let us begin with the note Strauss has to the passage where Galileo discusses Aristotle's authority as a logician (F59–60).[7] Strauss comments:

The merits of Aristotle in logic are also emphasized elsewhere. His writings pertaining to this part of philosophy are as a whole traditionally called *Organon* (instrument, tool). [S502, n. 42][8]

The other passage to which Strauss refers (F157) need not concern us here. This note contains two points: the interpretative claim that Galileo is here emphasizing Aristotle's logical merits, and the informative remark that his logical treatises are entitled *Organon*.

These comments are beset by the following problems. Though it is true that Simplicio is emphasizing Aristotle's logical merits, the general tone of the discussion, including Salviati's speech, is critical of Aristotle's authority as a logician. Strauss's misinterpretation is due to his failure to note the distinction between logic-as-theory and logic-as-practice, which is contained in Salviati's remark that one "can be a great logician, but not too skillful in knowing how to use logic" (F59–60); hence Strauss does not see that Galileo's emphasis on Aristotle's merits in logic refers only to his merits as a logical theorist, whereas regarding his skill as a logical practitioner Galileo is emphasizing his deficiency, at least in reasoning about the topics at hand.

The second problem is that one wonders about the relevance of Strauss's remark concerning the title of Aristotle's logical treatises. The only connection seems to be purely verbal, in the sense that the passage contains some remarks about organs, indeed almost a pun about the definition of logic as the organ of philosophizing, and the difference between organ-making and organ-playing, which suggests a difference between the making and the use of the organ of philosophizing. But Strauss's reference to Aristotle's *Organon* misses these subtleties, both the pun and the substance of Galileo's point, which is the distinction between logical theory and practice.

2.   Galileo's criticism of the contrariety argument includes a discussion of the liar's paradox (F66), in connection with the question of whether the Peripatetics are obliged to argue that because heaven is inalterable, and the earth alterable, heaven has a contrary and hence is alterable after all. Strauss comments (S503, n. 50):

The well-known fallacy of the Cretan here repeated does not really belong to the class which one calls 'sorites' in the narrow sense. The typical example of this is rather the supposed argument of whether one can form a pile by adding one piece to a quantity of things. I have translated the odd expression *'argomento cornuto'* by 'pseudo-argument'.

I am inclined to agree that the argument about the lying Cretan is not a sorites, as Galileo calls it, and that it is strange to call it an *'argomento cornuto'*. But why mention this and not the fact that Galileo's version of the paradox is incorrectly stated? This is so because Galileo's statement is as follows (F66):

(1) A Cretan said that all Cretans are liars.
(2) Therefore, he is lying in saying that all Cretans are liars.
(3) Therefore, all Cretans tell the truth.
(4) Therefore, he is telling the truth.
(5) Therefore, it is true that all Cretans are liars.
(6) Therefore, he is lying.
(7) Etc.

The problem with this is that (3) does not follow from (2). What does follow is that *not all* Cretans are liars, which does not entail that *he* is lying. What Galileo should have said is that the Cretan said that *he* was lying; then it would follow, from *his* lying in saying that *he* is a liar, that he is telling the truth.

3. At the beginning of the Second Day several of Aristotle's arguments against the motion of the earth are examined. In the course of his critique, Galileo gives what look like verbatim quotations from Aristotle's works. This gives Strauss the opportunity for some very informative erudite commentary. For example, when Galileo finds it useful to quote the Latin text of Aristotle's second argument (F162–3), Galileo refers to it as 'Text 97'. Strauss explains (S520, n. 33):

Which Aristotelian text Galileo used can perhaps be determined from the reference here given, namely *'testo* 97', which I have translated as 'Paragraph 97'. In our Aristotle editions the paragraphs are commonly not numbered with consecutive numbers, but rather their numbering begins anew with each chapter. Thus the place here cited is determined as being in the first paragraph of the 14th chapter. In 'Dido's' edition the preceding 13 chapters comprise altogether 95 paragraphs, so that the arrangement of the text used by Galileo does not agree exactly but only approximately with ours.

Such information is obviously of little relevance to the understanding of the content of Galileo's text; moreover, in this case it is even superfluous for the identification of the corresponding passage in Aristotle since, in addition to saying *'testo* 97', Galileo also gives the reference *"On the Heavens,* second book, Chapter 14", which is sufficient.

In a previous note to the passage where Galileo first quotes Aristotle's second argument in Italian (F150), Strauss does have a word of explanation (S515, n. 23):

By the apparent lagging behind is meant the fact that the planets require somewhat longer than 24 hours to complete a revolution in the heavens; this longer period is precisely the result of their particular movement (partly apparent, partly real).

This remark is certainly helpful, but it explains the easiest part of the argument. How problematic the rest is can be glimpsed from de Santillana's note to this passage, which note, however, is itself very difficult to understand;[9] its being almost unreadable is perhaps not surprising in the light of the fact that it is an attempt to understand the meaning of Galileo's passage by reference to other non-Galilean texts, rather than by analyzing the other remarks that Galileo makes about the argument where he criticizes it (F162–4). An interpretation of this whole passage will be given below (Chapter 16).

4.   Galileo's criticism of the argument from vertical fall (F165–6) begins by showing the sense in which the argument begs the question, or is a *petitio principii*. Strauss comments that (S520, n. 35):

petitio principii is a well known logical technical term by which one designates an invalid argument whose conclusion is substantiated by means of a premise dependent on the conclusion. The Aristotelian syllogism, brought into canonical form, would read:
  (A) If the earth moves, the body cannot fall vertically.
  (B) But the body falls vertically.
  (C) Therefore the earth does not move.
But in this case the second premise is questionable as long as the conclusion is questionable. – Here this second premise itself is called the 'middle term', whereas it is the middle concept which is more usually referred to by this designation. – *Ignotum per aegue ignotum*: the unknown by means of the equally unknown.

This is the full text of Strauss's note 35 to the Second Day, of which I have included the third part (which translates Galileo's Latin term) in order to point out that Strauss himself commits this fallacy in the second part of his note, concerning 'middle term'. The reason is that someone who doesn't already know what is the more usual meaning of 'middle term', will not know it after he is told that it more usually refers to the 'middle concept', rather than to a premise like (B) above, which admittedly is what Galileo is referring to. Ordinarily the middle term is a term common to both premises of a syllogism, and not present in its conclusion; for example, 'Y' is the middle term in the following:

All *X* are *Y*.
All *Y* are *Z*.
Therefore, all *X* are *Z*.

However, it is the first part of Strauss's note that requires comment. His definition of *petitio principii* is correct. His canonical formulation of the argument is also correct, insofar as it goes. However, it is not clear what he means in saying that (B) is questionable as long as (C) is. In accordance with his own definition of *petitio principii* this should mean that (B) is dependent on (C). But in what sense is (B) dependent on (C)? The answer is in the sense that (B) is justified by assuming (C). To explain this answer Strauss should have reconstructed Galileo's reasoning, which is the following:[10]

(1) *If* the earth moves *and* the body grazes the tower, *then* the body is not really falling vertically. (This is obvious on reflection.)

(2) Therefore, *if* the earth moves, then the grazing of the tower would *not* imply real vertical fall.

(3) But, if the earth does *not* move, then the grazing of the tower *does* imply real vertical fall. (This is obvious on reflection.)

(4) Therefore, the grazing of the tower implies real vertical fall if and only if the earth does not move.

(5) But the argument is assuming that the grazing of the tower implies real vertical fall, since (B) is justified on the basis of the grazing of the tower.

(6) Therefore, the argument is assuming that the earth does not move, namely its conclusion (C).

5.   One of the objections to the earth's motion criticized by Galileo was that a simple body like the earth could not have multiple natural motions (F281–8). In the course of his criticism Galileo argues briefly that there is no more incompatibility between the Copernican straight-downward motion of heavy bodies and circular rotation of the earth than there is between the Peripatetic straight-downward motion of heavy bodies and motionlessness of the earth (F288). On this argument Strauss comments that (S542, n. 148):

the following consideration by Sagredo is somewhat sophistical, a special principle implying rest is not necessary from the Peripatetic point of view. Galileo obviously does not attach a very high scientific value to the whole discussion, as shown among other things from the fact that he puts it in Sagredo's mouth.

There are several problems with this remark. First, it is not clear why Sagredo's argument (F288) should be regarded as sophistical and not merely invalid or incorrect. Second, it is questionable whether it is even incorrect.

Sagredo's point would be incorrect if he were asserting that the peripatetics do in fact need two distinct principles, one for downward motion and the other for rest for the whole earth; Strauss is right to point out that the rest of the whole earth is a consequence of the motion of its parts toward the center of the universe. However, Sagredo's (and Galileo's) point is that the mixing of downward and circular motions by the Copernicans is less implausible than the mixing of downward motion and rest by the Peripatetics; that just as the two prima facie imcompatible states of motion and of rest do not really conflict, there may not be any conflict either between the two motions, straight and circular. It should also be noted that Salviati intervenes to say that *another* reason why there may not be any conflict between terrestial straight and circular motions, is that the apparently straight motion may be really circular, which he had suggested in a previous digression (F188–93). The 'fanciful' character of this digression and its scientific error do not invalidate its use in the present context, namely to suggest a plausible way to resolve the straight-circular conflict.

6.   At the beginning of the Third Day, after giving a sketch of the heliocentric system, Salviati complains about the existence of people who make silly objections to the earth's motion, for example, that it is impossible for the heavy earth to go up above the sun. These people provide Galileo with the pretext for the following incredibly subtle and humorous witticism:

One should not pay attention to such people, whose number is infinite, nor take into account their fooleries; nor should one try to win the sympathies of men whose definition contains only the genus but not the difference, in order to have them as companions in very subtle and delicate discussions. [F355, my translation]

Strauss comments that (S551, n. 34)

according to orthodox logic every definition of a concept consists of the specification of a 'genus' and a 'specific difference'. The conventional definition of man, which at the same time supplied one of the most common examples to illustrate the nature of definition, correspondingly read 'rational mortal animal'; if one then disregarded the 'specific difference' which is expressed by the term 'rational', then the mere genus 'animal' is left.

Here Strauss must be given the credit for getting almost right something completely missed by other otherwise competent scholars. However, his comment is not completely adequate. For if we take the definition of man to be 'rational mortal animal', and if we delete from it the specific difference 'rational', then we are left not with 'animal' but with 'mortal animal'; in this case the joke does not get off the ground since Galileo would be labeling these people 'mortal animals'. Instead Galileo must be thinking of the definition of man as 'rational animal', so that when we take out the specific difference 'rational' from this definition, we are left with the genus 'animal' and thus with the clearly intended vituperative though unprosaic description of these men as 'animals'.

7.   There is a passage in the Third Day where Galileo asks whether Tycho or his followers have searched for any evidence from stellar appearances which might confirm or disconfirm the earth's motion (F399). Galileo answers that they haven't since they argue in an *ad hominem* manner, namely by deriving anti-Copernican conclusions from Copernicus's admission that there are no changes in stellar appearances. This gives Strauss the occasion to comment (S561, n. 70):

Ad hominem, 'against the man', a term in dialectics by which one understands an argument which is not based on established or generally accepted premises but rather on those held as correct by the opponent; such an argument has therefore of course no scientific worth but can serve for the persuasion of an opponent.

The first thing to point out is that Strauss is, correctly, not equating Galileo's *'ad hominem* argument' with its modern meaning, which is the *fallacy* committed when a case is argued not on its own merits but merely on the basis of the motives or background of its supporters or opponents. Nevertheless, there are several problems with this note. First Strauss is not explaining Galileo's meaning of *ad hominem* argument because Galileo regards as *ad hominem* the Tychonic argument based on the premise that there are no annual changes in stellar appearances, which was a 'generally accepted' premise, rather than one merely held by Copernicans. In other words, it is not true to say about the Tychonic argument being considered by Galileo that it was "not based on established or generally accepted premises but rather on those held as correct by the opponent". Of course, it is indeed true that the argument was based on premises which were 'not established', but that is another matter, and that is precisely Galileo's point, that the lack of changes in stellar appearances was generally accepted, but not 'established' as a result of search and research.

Second, Strauss claims that 'of course' such an argument has 'no scientific worth'. Of course this is more nearly correct of the type of argument that Strauss is talking about. But is it true of the type that Galileo has in mind? In this passage (F399) what Galileo finds wrong with Tycho and his followers is not that they argue *ad hominem,* but rather that they do so "more in defense of another man than out of any great desire to get at the truth",[11] and hence that what they do "may suffice to refute the man, but certainly not to clear up the fact".[12] This presupposes that *ad hominem* arguments *can* be used to get at the truth, though they can also be used for merely rhetorical or dialectical purposes.

All of this suggests that by *ad hominem* argument Galileo means an argument which derives a conclusion unacceptable to an opponent from premises accepted by him, but *not necessarily* by the arguer, or generally accepted, or established. Hence such arguments (if otherwise correct) provide conditional knowledge, namely that if one accepts the premises in question then the conclusion undesirable to the opponent follows. In this passage Galileo correctly points out that such arguments are improper to deal with the question of the factual correctness of the premises; however there is no evidence to attribute to him the erroneous claim that such arguments have no scientific value. On the contrary, there is evidence from other works that supports my interpretation of Galileo's conception of *ad hominem* arguments.[13]

EDITING A TEXT OF REASONING: FAVARO

The name Favaro certainly needs no introduction for historians of science interested in Galileo. What I wish to do here is merely to call attention to a few examples where the text of the National Edition could be improved. Since these imperfections center around one definite area, namely logic, it is of some importance to mention them.

1.   One difference between the Favaro text and that of the 1632 edition is the modernized spelling of the former. Usually no problem results, but the following passage is an exception. On F57 Salviati has just completed his argument that circular motion is the only type of natural motion, and that straight motion is the simplest means of acquiring the natural state of rest at the proper place or of circular motion. Simplicio's reply to this is in fact an argument designed to show that the natural motion of the earth (in the sense of terrestrial globe) is straight, which is a direct refutation of the first part of the conclusion just reached by Salviati; the 1632 edition (EP24)[14] makes this clear by using a capital 'T' in the spelling of '*Terra*' (Earth). Here, as in most other occurrences of this word, Favaro uses a lower case 't'; the net effect is that one gets the impression that Simplicio's reply is merely a claim that Salviati's conclusion conflicts with the sense-experience of straight-downward

motion of terrestrial (heavy) bodies, and of straight-upward motion of light bodies, which constitutes a misunderstanding of Simplicio's own speech and, more importantly, leads one to misconstrue Salviati's natural motion argument. I am not saying, of course, that Favaro's text necessarily gives such a wrong impression, but only that it facilitates this misinterpretation, and renders more difficult the correct one.

2.   The logical significance of punctuation is implicitly admitted by Favaro when he says, "We have however departed from the original edition (as we did in preceding volumes) for what pertains to punctuation, which we have rendered more rational" (F12–13). I would agree that on the whole Favaro's changes in punctuation are a great improvement of the 1632 edition. However this is not *always* the case, as the following example shows. The passage is such that it would be impossible to understand it from Favaro's text, whereas the original is less misleading by being more amorphous. The passage is part of Galileo's answer to the objection from birds, and the original punctuation may be seen in the following literal, and line-by-line translation from the 1632 edition (EP 180):

1   Your having more difficulty for this, than for the other
2   objections, seems to me, to depend on birds being animate,
3   and being able therefore to use force at will against the primary
4   motion innate in earthly things; in such a way precisely, that
5   we see them while they are alive flying even upwards, a motion
6   impossible to them as heavy bodies; whereas dead they cannot, but
7   fall downwards; and therefore you judge, that the reasons, which take
8   place for all sorts of projectiles mentioned above, cannot
9   take place for birds; and this is very true, and because it is true,
10 therefore we do not see Mr. Sagredo those projectiles doing, what
11 birds do; for if from the top of a tower you let
12 fall a dead bird, and a live one, the dead one will do the same, that
13 a stone does; that is it will follow first the general diurnal motion, and
14 then the downward motion, as a heavy body; but if the released bird is
15 alive, who will forbid it, the diurnal motion always staying with it,
16 from going by a beating of wings toward whatever part of the horizon
17 it likes most? and this new motion, being its own in particular, and
18 not shared by us, must become perceptible to us; and if it had
19 by its flight moved toward the west, who is to forbid it,
20 from returning to the tower by the same beating of wings?
21 Because finally to fly up toward the west,

22   was nothing other, than to subtract from the diurnal motion, which has
      for example ten degrees
23   of speed, a single degree, so that it was left with nine while
24   it was flying, and if it had alighted to earth, it would have regained
25   the ten common degrees; to which by flying toward the east, it
26   could have added one, and with the eleven return to the tower.
27   And in sum, if we will consider, and more intimately
28   contemplate the effects of the flying of birds, they do not
29   differ in anything from projectiles thrown in all directions,
30   except that these are moved by an external projector, and those
31   by an internal principle.

After many fruitless attempts to arrive at an accurate reconstruction, the only interpretation that made sense became one consisting of two distinct comments about the birds argument (see Chapter 16): a favorable one on lines 1–9, and an unfavorable one on lines 9–26, with lines 27–31 constituting a remark extraneous to these two comments but summarizing the entire discussion of the birds argument. Thus it was misleading for Favaro to eliminate the period from line 26 (cf. F212.26),[15] and to introduce one in line 18 to replace the semicolon (cf. F212.19). Favaro makes the biggest break in lines 1–18 on line 11 by replacing the semicolon by a colon; since there is no way of combining the first 11 lines, he should not have left the semicolon of line 9 (F212.10) but replaced it by a period. There is no question that Galileo's punctuation was odd, and his language in this passage very colloquial; moreover, the question mark in line 20 and the capital letter in line 21 are misleading. Nevertheless, Galileo's original passage is easier than Favaro's, because Favaro gives the wrong clues.

3.   The Favaro text introduces a number of corrections in the text of the 1632 edition (F10–13). Favaro states that most of these corrections are grounded on (1) the "Errata" page of the 1632 edition (F10); (2) Galileo's own corrections handwritten in his copy of the 1632 edition (F10); (3) a manuscript that exists for the discussion of the 1572 nova at the beginning of the Third Day, on pp. F301.34–346.26 (F10); (4) the text of Galileo's *Discorso del flusso e reflusso del mare,* many passages of which he transcribed verbatim in his composition of the Fourth Day (F12, n. 1); and (5) the two versions of the wording of the marginal postils, one alongside the text and the other in a table at the end of the 1632 edition (F11–12). Besides these groups of corrections, Favaro corrected a number of obvious typographical errors of the 1632 edition which had not been noted by Galileo in his own

copy (F10–11). Favaro also states that he made a few other corrections on the grounds that "the correction was necessary, otherwise the meaning would have been seriously affected" (F10, n. 5) and that they fall into the pattern of Galileo's habit of exchanging one term for one opposite in meaning (*Ibid.*). These corrections are Favaro's replacement of Galileo's *retto* (straight) by *circolare* (circular) on p. 45, in the second postil; the replacement of Galileo's *minori* (minor) by *maggiori* (major) on F55.31; the replacement of Galileo's *occidentali . . . bassi, e alti . . . orientali* (western . . . low, and eastern . . . high) by *occidentali . . . alti, e bassi . . . orientali* (western . . . high, and eastern . . . low) on F208.27; the replacement of *segante* (secant) by *tangente* (tangent) on F228.14; and the replacement of *quoziente* (quotient) by *prodotto* (product) on F324.2.

To these I believe one should add at least the following two corrections, though they were not made by Favaro. On p. F193.31 I think that the word should be *annuale* (annual) instead of *diurno* (diurnal). The sentence involved may be translated literally as follows:

But this one, taken from bodies falling perpendicularly, I do not regard as one of the strongest arguments for the immobility of the earth, and I don't know what will happen about gunshots, and especially about those against diurnal motion. [F193]

The speaker is Simplicio and the remark is made at the end of the criticism of the objection from falling bodies and the beginning of the discussion of the gunshot objections. I believe that the 'those' of the last clause refers to 'arguments' rather than to 'gunshots', and hence the last clause should be 'and especially about those against annual motion', so that the last sentence would mean that Simplicio doesn't know what will happen about the arguments from gunshots, and especially about the arguments against annual motion. There are two reasons. First, logically speaking, if 'those' referred to gunshots and 'diurnal' is not corrected, then we would have an implicit statement of the opinion that the arguments from cannons shot in a direction against the diurnal motion (i.e. gunshots in a westward direction) are the strongest. However, in the list of geostatic arguments, there are none from gunshots toward the west, though of course we have the argument from east-west gunshots, that is the argument involving considerations about cannons shot toward the west *and* toward the east. The logic of *this* argument is such that no argument would remain if one considered only gunshots toward the west, since the argument attempts to derive from the earth's rotation the consequence that there would be a *difference* in the range of gunshots toward the east and toward the west. Moreover, the east-west gunshots argument is nowhere in the book regarded as particularly strong, whereas one of the arguments against the annual motion (the parallax argument) is so regarded, so much so that Galileo in fact is unable to give a conclusive refutation of it, but rather criticizes it by suggesting a research program to discover new evidence that would refute it and confirm the earth's annual motion (F164, F406, and F407—16).

The second reason for my interpretation is linguistic. If the arguments referred to in the last clause are arguments from cannon shots in a direction against the diurnal motion, then there would be no need of the 'and' with which the last clause begins. The use of this 'and' indicates that the structure of the whole sentence is: But [(this ... earth), and (I don't ... gunshots), and (especially ... motion)], where the third conjunct in this series introduces a relatively different claim, such as one about the special strength of objections to the annual motion. However, if the term 'diurnal' were correct, then the structure of the sentence would be: But {[this ... earth], and [I don't ... about (gunshots, and especially ... motion)]}, where the last 'and' would be superfluous.

4.   My second example of overlooked correction involves Galileo's exchange of *perchè* (because) and *però* or *perciò* (therefore) on F216.37—217.1. The relevant passage may be translated literally as follows:

... all these are arguments for the truth of the conclusion, namely that whirling gives to the moving body an impetus toward the circumference, if the motion is fast; and because, if the earth turned on itself, the motion of the surface, especially near the equator, being incomparably faster than those mentioned above, would extrude everything toward the sky. [F216.35—217.3]

The speaker is Salviati who in the sentences just preceding describes a number of easily performable experiments supporting the generalization about the extruding power of whirling. Here he is strengthening this objection to the earth's rotation, before criticizing it. It is obvious that in order to make sense the 'because' must be replaced by 'therefore', a good example of Galileo's slips involving exchange of opposites; the slip of the pen is all the more understandable here in the light of the similarity of spelling in Italian.

5.   An example of a marginal postil uncorrected by Favaro is the second one on F413: "If annual variations in the fixed stars were observed, the earth's motion would not be contradicted". Since it is contextually obvious that if annual stellar variations were observed, that would be a very strong (almost

decisive) confirmation of the earth's motion, logical intuition tells us that the antecedent clause should be a negative one: "if *no* annual variations in the fixed stars were observed . . . ". In fact, an examination of the text to which the postil refers shows the need for such a correction, for the text has Salviati affirm that, though if variations were seen that would prove the earth's motion, if variations were not seen that would not disprove it, since the failure to see them might be due to immense stellar distances or to the failure to look for them or to the failure to look carefully enough.

6.   The passage on centrifugal force (F214–44) is generally considered as one of the scientifically most important ones, hence it is important here to be especially careful about what Galileo is saying and what he is not saying. On F238.12–238.22 we have a passage which may be translated literally as follows:

. . . no one will think that the cause of extrusion in the big wheel grows as the proportion of the speed of its circumference to the speed of the circumference of the smaller wheel, because this is most false, as for now a very easy experiment can show us somewhat roughly: because with a stick one cubit long we could throw such a stone that we could not with one six cubits long, even if the motion of the end of the long stick, namely of the stone held therein, were more than twice as fast as the motion of the end of the shorter stick; this would happen if the speed were such that in the time of a whole rotation by the bigger stick, the smaller one would rotate three times.

The problem here is with the reasoning indicator 'because' (*chè*) after the colon. It makes little sense to our logical intuition. In fact, a check with the 1632 edition reveals that Galileo used a 'che' without the accent, which is no reasoning indicator but rather means 'that', as in 'the experiment that'.

THE TRANSLATION OF REASONING: DRAKE

English-speaking people are fortunate that there exists a scholar like Stillman Drake whose efforts have made available to them in their own tongue almost all of Galileo's major works.[16] Add to this the fact that Drake's devotion has enabled him to internalize the spirit of Galileo to an unprecedented degree, and we get a truly unique situation. Not content with merely translating Galileo, Drake has also formulated with increasing frequency, a number of explanatory interpretations which cannot be easily excelled for their textual accuracy or biographical well-foundedness, whatever their alleged faults from other points of view.[17] As if this were not enough, he has in recent years undertaken the epoch-making task of deciphering and chronologically arranging Galileo's unpublished and previously unanalyzed notes on motion, and this effort has always led either to new insights into Galileo's work or to novel evidence for some old beliefs, or to new counter-evidence against recent popular historical interpretations.[18]

Having said this much, I must add that in my own investigations I have not always been satisfied with Drake's translations.[19] If these inadequacies were those of an ordinary Galileo scholar, or if they pertained to random topics, then it would not be worth discussing them explicitly. However, they do center around one definite area, namely logic and reasoning; moreover, this topic is relatively controversial since the practice of reasoning can be easily confused with commitment to apriorist rationalism (as we saw in the case of Koyré), and since the art of reasoning could be easily opposed to the art of experimenting (though there is no opposition, since experiment is either an activity distinct from reasoning or else it is observation guided by reasoning).

Before discussing a number of passages from the *Dialogue* to illustrate these imperfections of translation, I want to give one example to illustrate that Drake's translation does indeed capture and convey the spirit of Galileo very well, so that my criticism is perhaps as much of a comment on the nature of translation in general as on the adequacy of Drake's, or at least a comment on the problems of translating a multi-faceted work like the *Dialogue*. The example is a sentence from Galileo's preface "To the Discerning Reader" in which Galileo is referring to his deceased friends Sagredo and Salviati. Translated literally the sentence would read:

Now, since bitter death has deprived Venice and Florence of these two great lights while they were in the brightest part of their years, I have wanted, as far as my meager abilities can, to prolong the life of their reputation on these pages of mine by using them as interlocutors in the present controversy. [F31]

Drake's translation reads, "Now, since bitter death has deprived Venice and Florence of those two great luminaries in the very meridian of their years, I have . . . ." (D7).[20] The important phrase here is "in the very meridian of their years", which is a more Galilean phrase than Galileo's own "*nel più bel sereno degli anni loro*". In the light of the astronomical dimension of the book, Drake's metaphor is very much of the type used by Galileo, who is constantly finding appropriate metaphorical uses for terms which literally apply to the natural phenomena he is discussing.[21]

1.   Let us now pass to criticism. The very first sentence with which the dialogue begins may be translated as follows:

SALV. It was yesterday's conclusion and plan that today we should discuss, as distinctly and particularly as we can, of the natural reasons and their efficacy which by one side and by the other have so far been produced by the proponents of the Aristotelian and Ptolemaic position and by the followers of the Copernican system. [F33, my translation]

Instead of "the natural reasons and their efficacy", Drake has "the character and the efficacy of those laws of nature" (D9). The problem here is with the 'laws of nature', which is supposed to be a translation of the Italian *'ragioni naturali'*, which means natural reasons in the sense of physical arguments. Drake's rendition is misleading because it obscures the logical dimension of the work, which is bound to be present in a discussion of arguments and their strength. In other words, whereas the first sentence promises that the book will be a work in applied logic (if we define logic as the study of the nature and validity of reasoning), Drake's translation makes it a work of physics (if we define the latter as the study of the laws of nature).

2. In the course of the discussion of natural motion in the First Day, Simplicio gives the following argument, as it is translated by Drake:

Who is there so blind as not to see that earthy and watery parts, as heavy things, move naturally downward — that is to say toward the center of the universe, assigned by nature itself as the end and terminus of straight motion *deorsum*? Who does not likewise see fire and air move directly upward toward the arc of the moon's orbit, as the natural end of motion *sursum*? This being so obviously seen, and it being certain that *eadem est ratio totius et partium,* why should he not call it a true and evident proposition that the natural motion of the earth is straight motion *ad medium,* and that of fire, straight *a medio.* [D32–3]

In a note, Drake tells us that the Latin *eadem est ratio totius et partium* means that "the reasoning which applies to the whole applies also to the part" (D471). However, this makes nonsense of Simplicio's argument in a way certainly not intended by Galileo. The argument is that since the natural motion of the parts of the element earth is (visibly) straight down and that of the parts of the element fire is (visibly) straight up, therefore the natural motion of the whole earth is straight down and of fire straight up, in virtue of the principle expressed in Latin. To be meaningful, the argument must be appealing to the principle that the reasoning that applies to the *parts* applies also to the *whole,* which is the reverse of Drake's translation. In fact, the Latin sentence means literally "the reasoning applying to the whole and to the parts is the same", which refers both to reasoning from parts to whole and from whole to parts; hence, Drake's translation is not incorrect from an abstract point of view, but only in the context of the passage in which the

Latin sentence occurs. However, in this context we have an inaccurate representation of the logic expressed in the text.

3.   Among the objections to the earth's diurnal motion, several dealt with gunshots. Though interrelated to one another, and as a group to the various versions of the falling bodies argument, these gunshot objections should be distinguished from each other if they are to be fully understood. However, this is difficult to do from Drake's translation where the same term 'point-blank' is used both to refer correctly to shots which are '*di punto in bianco*' (F153.9, D127) or '*di punto bianco*' (F205.33, D180), as well as to refer incorrectly to shots which are '*di volata*' (F152.20, D126). The latter type of shot instead is one where the artillery piece is given a great elevation.[22] The confusion is compounded by the fact that the second time the latter phrase occurs (F194.12) Drake drops it from his translation (D168).

4.   In the critique of Aristotle's objection from falling bodies, there is a speech by Salviati which Drake translates as follows:

SALV. So Aristotle's defense consists in its being impossible, or at least in his having considered it impossible, that the rock might move with a motion mixed of straight and circular. For if he had not held it to be impossible that the stone might move both toward and around the center at the same time, he would have understood how it could happen that the falling rock might go grazing the tower whether that was moving or was standing still, and consequently he would have been able to perceive that this grazing could imply nothing as to the motion or rest of the earth.

Nevertheless this does not excuse Aristotle, not only because if he did have this idea he ought to have said so, it being such an important point in the argument, but also, and more so, because it cannot be said either that such an effect is impossible or that Aristotle considered it impossible. The former cannot be said because, as I shall shortly prove to you, this is not only possible but necessary; and the latter cannot be said either, because Aristotle himself admits that fire moves naturally upward in a straight line and also turns in the diurnal motion which is imparted by the sky to all the element of fire and to the greater part of the air. Therefore if he saw no impossibility in the mixing of straight-upward with circular motion, as communicated to fire and to the air up as far as the moon's orbit, no more should he deem this impossible with regard to the rock's straight-downward motion and the circular motion natural to the entire globe of the earth, of which the rock is a part. [D141]

The logical impression being conveyed here is that the final and/or main conclusion in this passage is the last sentence, "If he saw no impossibility in . . . , no more should he deem this impossible . . . ". Presumably this consequence is being drawn from the preceding statements; perhaps the basis is the immediately preceding clause, i.e., that "Aristotle himself admits that . . . to the greater part of the air". But such an inference would make no sense, for it would be like arguing that because Aristotle admits A, therefore if Aristotle admits A then he should admit B. In fact, no other preceding sentence can be used to ground the alleged conclusion; moreover, one will soon realize that this alleged conclusion is intrinsically plausible in the context, and therefore it would make more sense merely to assert it. If one checks the Italian text (F167) that is exactly what one finds; there is nothing in the last sentence corresponding to 'therefore if . . . ' but rather a phrase (*se dunque*) corresponding to 'if then . . . '. Moreover, the punctuation, which serves as a logical indicator, is clearer in the Italian. In short, Drake's last twelve lines are better translated as follows: "The former cannot be said . . . necessary. The latter cannot . . . greater part of air; if then he saw no impossibility . . . of which the rock is a part". When so modified, the main conclusion in Drake's second paragraph is that "this does not excuse Aristotle", i.e., that the defence of Aristotle mentioned in the first paragraph does not really excuse him, for two reasons. The first is that he ought to have said so, i.e., given explicitly that defense; the second reason is that neither one of two things can be said. The first of these things cannot be said because so will Salviati shortly prove; and the second thing cannot be said because of an argument having the form: Aristotle admits A, *and* if he admits A then he should admit B.

Drake's inaccuracy is significantly misleading because it prevents appreciation of the geostatic objection being discussed and of Galileo's critique of it. In fact, Simplicio's response is, "It does not look that way to me at all" (D141), and then he says why; therefore it is important to know what 'it' is, what Simplicio is denying. This must be the main or final conclusion in Salviati's argument, and if one gets it wrong (as one is bound to do from Drake's translation), then one does not understand what is going on; in particular, Simplicio's response would not make sense, and then one would get the oversimplification that Aristotelianism made no sense, but that only Galileo's theories did; whereas it is more plausible to hold that the former made *some* sense, but that Galileo's made *greater* sense.

Simplicio justifies his denial by saying, in Drake's translation (which may be accepted here):

If the element of fire goes around together with the air, this is a very easy and even a necessary thing for a particle of fire, which, rising high from the earth, receives that very motion in passing through the moving air, being so tenuous and light a body and so easily moved. But it is quite incredible that a very heavy rock or a cannon ball which is dropping without restraint should let itself be budged by the air or by anything else. Besides which, there is the very appropriate experiment of the stone dropped from the

top of the mast of a ship, which falls to the foot of the mast when the ship is standing
still, but falls as far from that same point when the ship is sailing as the ship is perceived
to have advanced during the time of the fall, this being several yards when the ship's
course is rapid. [D141]

This would make no sense as a reply to Salviati's thesis alleged in Drake's
text, namely the proposition we have symbolized as "if Aristotle admits A,
then he should admit B"; for the last sentence about the ship experiment
sounds like Simplicio is just changing topic, while the first two sentences
seem to constitute a somewhat irrelevant counterargument of the form "if C
holds, then A follows but B does not", where C is a statement of conditions
obtaining on a rotating earth. Such a counterargument does not lead to a
denial of Aristotle's inconsistency in admitting A but not B, not only because
it says nothing about Aristotle, but also because the C is a proposition that he
obviously did not hold. However, in reality, such a counterargument does
plausibly lead to a denial of Salviati's claim about the character of "Aristotle's
defense". Salviati had said that Aristotle's defense seemed to consist in
holding the impossibility of mixed motion, toward and around the center.
Aristotle's defense in turn was the following 'defensive' argument contained
in Sagredo and Simplicio's speeches, just preceding Salviati's:

SAGR. On behalf of Simplicio I should like, if possible, to defend Aristotle . . . . I reply
that if the tower were moving, it would be impossible for the rock to fall grazing it;
therefore from the scraping fall is inferred the stability of the earth.
SIMP. So it is. For to expect the rock to go grazing the tower if that were carried along
by the earth would be requiring the rock to have two natural motions; that is, a straight
one toward the center, and a circular one about the center, which is impossible.
[D140–141]

This may be called the apparent vertical fall argument, as distinct from the
actual vertical fall argument, which in the preceding dialogue has just been
shown to be circular. Salviati is showing how the apparent vertical fall argument assumes the impossibility of mixed, down-and-around motion, and he is
arguing that this assumption is unjustified because Aristotle admits mixed up-
and-around motion, and if he admits this he should admit mixed, down-and-
around. Simplicio is implicitly agreeing with Salviati that the apparent vertical
fall argument assumes the impossibility of mixed, down-and-around motion,
and, explicitly, he is denying that this assumption is unjustified, for two
reasons: (1) that on a rotating earth the mixed, up-and-around motion of fire
would easily result, but the mixed, down-and-around motion of heavy bodies
would not, and (2) the ship experiment. If there is a problem with Simplicio's
reply it is his failure in (1) to take seriously the idea (expressed in Salviati's

last sentence) that on a rotating earth heavy bodies would have a natural tendency to follow the rotation, for Simplicio expects that the cannon ball should be carried around forcefully by the rotating air. Regarding (2), Salviati explicitly argues that there is a disanalogy between what might happen on a ship and on a rotating earth (D141–143).

To conclude, Drake's above-mentioned inaccuracy is likely to lead one to miss (a) the existence of the apparent vertical fall argument (as a distinct objection to the earth's motion), (b) the fact that one function of the ship experiment is to be a part of this argument, and (c) Galileo's argument for the disanalogy between the ship and the earth.[23]

5. One of the objections to the earth's motion was an argument based on a ship experiment. In Drake's translation, the argument says that

since when the ship stands still the rock falls to the foot of the mast, and when the ship is in motion it falls apart from there, then conversely, from the falling of the rock at the foot it is inferred that the ship stands still, and from its falling away it may be deduced that the ship is moving. And since what happens on the ship must likewise happen on the land, from the falling of the rock at the foot of the tower one necessarily infers the immobility of the terrestrial globe. [D144; cf. F169–70]

Everything is acceptable here, except that the sentence "what happens on the ship must likewise happen on the land" should read "what is true of the ship must likewise happen of the Earth".[24] This is significant because it is clear that the objection examined by Galileo is an argument from analogy, the two analogues being the ship and the Earth; this fits very well with the critique that follows, which amounts to the argument that, *even if* the analogy were sound (cf. F167–9, D141–3) the objection is inconclusive because it is based on the *false* premise that on a moving ship the rock falls *away* from the foot of the mast (F169–75, D143–9). On the other hand, as translated by Drake, the objection seems to be a generalization from sea to land, and this may mislead the reader into interpreting the ship experiment as a thought-experiment on a rotating earth, and Galileo's counterargument as an attempted proof of the earth's motion to the effect that since the rock would fall at the foot of the mast on a moving earth, it would also fall at the foot of the tower, which is what happens, thereby confirming the earth's motion.[25]

6. Between the end of the discussion of the falling bodies objection and the beginning of the discussion of the gunshot arguments, there is a brief digression about the flying of birds. Sagredo exclaims, "If only the flying of birds gave me as much trouble as the difficulties caused by cannons and by all

other experiments mentioned above!" (F193, my translation). In the rest of
his speech, and in Salviati's response, the implication is that the objection
from birds is much stronger than all others. This emphasis tends to be lost in
Drake's translation, where Sagredo's exclamation reads, "If only the flying
of birds didn't give me as much trouble as . . ." (D167–8), which implies
that the flying of birds gives *as much* trouble as the other alleged counterevidence.

7.   In the discussion of the heliocentrism of planetary motions, Galileo
makes the following comment about certain people who give very silly arguments against the earth's motion:

One need not take into account these people, whose number is infinite, nor notice their
fooleries; neither should we try to acquire as companions in very subtle and delicate
opinions, *men in whose definition only the genus enters but the difference is missing.*
[F355, my translation, my italics.]

Drake's translation reads:

There is no need to bother about such men as these, whose name is legion, or to take
notice of their fooleries. Neither need we try to convert *men who define by generalizing
and cannot make room for distinctions,* just in order to have such fellows for our company in very subtle and delicate doctrines. [D327, my italics]

The problematic phrase here is *"uomini nella cui difinizione entra solo il
genere e manca la differenza"*, which corresponds to the italicized expressions,
and which refers indirectly to the definition by genus and difference of man
as a rational animal, as we saw in our discussion of Strauss.

8.   The next passage is Salviati's disclaimer after the discussion of sunspots
in the Third Day. When translated literally the passage reads:

I will attribute to them neither the label of conclusive nor of inconclusive since, as I have
said other times, my intention was not to resolve anything about such a lofty matter, but
only to propose those natural and astronomical reasons which can be adduced by me for
the one and for the other position. I leave to others the determination, which ultimately
must not be ambiguous since, it being fitting for one of the two arrangements to be
necessarily true and for the other to be necessarily false, it is impossible that (staying
within the terms of human doctrines) the reasons adduced for the true side should not
manifest themselves as conclusive as the contrary ones vain and ineffective. [F383, my
translation]

In Drake's translation, the last clause of the first sentence reads: "but merely to set forth those physical and astronomical reasons which the two sides can give me to set forth" (D356). Moreover, he drops the qualification "necessarily" in the first half of the second sentence, and the crucially important comparison "as conclusive as . . ." in the last clause:

I leave to others the decision, which ultimately should not be ambiguous, since one of the arrangements must be true and the other false. Hence it is not possible within the bounds of human learning that the reasons adopted by the right side should be anything but clearly conclusive, and those opposed to them, vain and ineffective. [D356]

Finally, Drake misrepresents the structure of the claim in the second half of the passage. Salviati is claiming primarily that ultimately the determination must not be ambiguous; the reason he gives for this claim is that it is impossible that the reasons adduced for the true side should not appear as conclusive as the reasons for the false side appear inconclusive; and, in turn, the reason for this impossibility is that it is fitting for one of the two arrangements to be necessarily true and the other necessarily false. In Drake's translation we find instead the respective truth and falsity of the two arrangements as the direct reason for the ultimate unambiguousness of the decision, which is less plausible, besides being inaccurate; while the impossibility in question becomes the apparent final conclusion in the passage. In the light of the book's rhetorical and philosophical import, it is important to understand such passages correctly. The present passage is particularly significant because it is explicitly used in its mistranslated version by a well known scholar to reach unwarranted interpretations.[26]

9.   In his discussion of stellar dimensions one of the things that Galileo shows is that the actual size of stars and their annual parallax required by Copernicanism are much smaller than was believed by some of its opponents. At one point he gives a rough calculation to show that the annual parallax of a sixth magnitude star would be about the same as the solar parallax due to the earth's radius. Sagredo comments: "As a first step, this makes a big drop" (F387, my translation). Salviati replies: "Indeed it does, since . . ." (F387, my translation); and then he continues his argument, arguing next that the anti-Copernicans make such stars 10 million times bigger than necessary. In Drake's translation, Sagredo's comment reads, "For a first step, this is a bad fall" (D360), and Salviati's response, "It is indeed wrong, since . . ." (*Ibid.*); this misrepresents a favorable judgment by Sagredo about Salviati's argument, as a negative remark.

CONTEXT AND THE EPISTEMOLOGICAL CONTENT
OF REASONING: CLAVELIN

Maurice Clavelin's *Natural Philosophy of Galileo* has been deservedly praised
ever since its original French edition in 1968. For example, Drake finds it
"a book that deserves the most careful reading in its entirety. Ultimately it
will prove to be as stimulating and as illuminating as the classic work of
Koyré himself."[27] Comparing the relation between Koyré and Clavelin to
that between Galileo and Newton, Drake concludes that Clavelin has achieved
"a substantial advance. No longer, one may hope, will it be considered naive
or rash to disagree with Alexandre Koyré."[28] Howard Stein judges it to be
"by far the best account known to the reviewer of the *oeuvre* of Galileo; and
an examplar of a type of study in the history of thought that is, in the
reviewer's opinion, both highly desirable and all too rare".[29] A. R. Hall calls
it "dominatingly the best study of its subject yet written";[30] characterized by
"wide range, depth, and subtlety . . . solidity and detail . . . Galilean scholars
will need to turn to this book again and again, and never without rich
rewards".[31] Michael Mahoney finds that "if it is Galileo's thought that one is
interested in, one can hardly find a better guide than Clavelin's study"[32] and
that it is "rewarding at every turn".[33] Even less favorable reviewers find it to
possess 'obvious merits'[34] and to be "a classical *exposition de textes*".[35]

If, however, we examine carefully the evidential adequacy of Clavelin's
interpretations dealing with logic and reasoning, we discover that they are
unsatisfactory, and that the root-cause of this deficiency is his practice of
interpreting passages out of context. This means that the practice is generally
widespread,[36] since its problems have not been detected by his reviewers,
even those who wrote article-length studies. This in turn means that the
scholarship of logical practice needs fundamental reforms.

Clavelin's interpretations are part of his account of Galileo's philosophy of
science, which is alleged by him to have three elements: (1) a concept of
explanation as rational reconstruction of phenomena (C404–9);[37] (2) an
identification of physical with mathematical intelligibility (C409–21); and
(3) the principle of simplicity (C421–4). Presumably (1) and (2) are part of
what Galileo means by explanation, whereas (3) is something like an a priori
necessary condition for the possibility of explanation in this sense. (1)
reminds us of the hypothetico-deductive model of explanation, (2) is the
mathematicism widely attributed to Galileo, and (3) is his supposedly well-
known commitment to simplicity.

The idea of explanation as rational reconstruction is interesting and intrinsically plausible, and in some sense it would be 'nice' if it was Galileo's idea. Unfortunately, the way in which Clavelin supports this point is inadequate. He does so by referring to the discussion of the apparent motion of sunspots in the *Dialogue*.[38] Clavelin assumes that in this passage Galileo is giving an explanation of this apparent motion (C404) and then proceeds to determine the structure inherent in the explanation (C405–9). Galileo is portrayed as first knowing the basic facts about the slanted and curved paths of sunspots (C404), as being initially puzzled and ignorant of how and why these paths result (C404), then as having put forth his explanatory hypothesis that the sun's axis of rotation is inclined to the ecliptic (C405–8), and finally as having made and confirmed the prediction that the slant and the curvature of the paths should alternate upwards and downwards at six-month intervals (C408–9). In other words, Clavelin interprets this hypothesis as having the function of (1) rendering comprehensible the previously known but not understood fact that the path was sometimes slanted but straight and sometimes curved but not slanted, and (2) predicting the annual period of the direction of these paths. However, if one examines the relevant text in the *Dialogue*, one finds that the hypothesis about the inclination of the sun's axis is presented by Galileo as having had the function of leading him to the discovery of the details of the motion of sunspots, that is, the function not of explaining known facts, but of leading him to previously unknown ones. Galileo states that he originally thought the sun's axis was perpendicular to the ecliptic (F373, D345), but then, having accidentally observed only once merely that a particular path was curved (F374, D346), he immediately got the idea of axial inclination (F374, D347) which led him to observe further and to test the reality of its consequences (F374–5, D347–8); then it so happened that all the predictions were confirmed (F379, D352). I am not saying that Galileo's account in this passage is an accurate chronological report; the chronological question is here irrelevant. What I am saying is that Galileo is discussing logic of discovery and prediction, rather than logic of explanation, as Clavelin thinks. So I conclude that the evidence presented by Clavelin does not support the first element which he attributes to Galileo's philosophy of science, namely explanation as rational reconstruction with predictive consequences, though this concept is intrinsically plausible and though there may be other Galilean evidence in its favor. Finally, Clavelin is to be commended for calling attention to the philosophical significance of the passage, especially since a related nearby passage is used by him to support another claim about Galileo's philosophy of science, namely his commitment to the principle of simplicity.

The passage in question (F379–83, D352–6) is the one immediately following the discussion of the details of the motion of sunspots and is used by Clavelin to argue that, for Galileo, "the principle of simplicity became a superior criterion of deciding between physical truth and falsehood" (C421). Clavelin's argument is that in this passage Galileo admits that a geostatic explanation of the motion of sunspots is possible, but rejects it as being inferior in simplicity since one would have to attribute to the sun two new motions besides the annual and the diurnal, namely a monthly axial rotation and an annual rotation of this axis of rotation. Now it is true that on one occasion in this passage Galileo describes as 'very simple' (F382, D355) the one motion attributed to the sun in the geokinetic theory; but he also describes the geostatic explanation as 'innovating' (F382, D355) of solar motions, and the geokinetic explanation as predictively fruitful (F379, D352); and he calls attention to the inexplicability by the geostatic theory of the annual period of the rotation of the sun's axis, which it requires. Moreover, he does not elaborate his simplicity remark in the way he does with the others: the predictive fruitfulness of the geokinetic theory is supported by the whole account of how he supposedly discovered the factual details (F372–9, D345–52); the 'innovating' character of the geostatic explanation is supported by the fact its statement emphasizes how the sun's fourth motion is somehow extra (F380–82, D353–5); and the geostatic inexplicability of the annual period of this fourth motion is a very decisive point in favor of the greater explanatory power of the geokinetic explanation. Finally, if *we* (the readers of this passage) try to elaborate on the simplicity remark, the following would be the most plausible elaboration: using as clue Galileo's remark that the four 'incongruous' (F382, D355) motions attributed to the sun in the geostatic explanation are reduced to "a single and very simple one" (F382, D355) in the geokinetic theory, it seems that the latter is simpler because in it there are two congruous motions by the earth and a single one by the sun, whereas in the geostatic model there are four incongruous ones by the sun. Now, since Galileo says nothing about the incongruity of these four motions, we may conclude he is referring primarily to the opposite directions of the sun's annual and diurnal motions which he discusses elsewhere in the *Dialogue*; it is not obvious that there is any additional incongruity for the sun's third and fourth motions, either between themselves, or between them on the one hand and the annual and diurnal ones on the other. Hence the comparison in terms of simplicity of the two explanations seems to involve merely a comparison of the *number* of motions involved; that is, the geokinetic explanation is simpler in the sense of requiring one motion less than the geostatic explanation. But if this is the meaning, then the point is implicit in two of the other philosophically significant remarks, mentioned above: to describe the geostatic explanation as 'innovating' of motions means that it requires at least one extra motion; and to emphasize the geostatic inexplicability of the annual period of the sun's fourth motion calls attention, among other things, to the fact that there is this extra motion in the geostatic model. I conclude that Galileo's simplicity remark in this passage is incidental, because it is not elaborated in the text, and relatively superfluous, because its information content is reducible to the other remarks made and elaborated by Galileo. Hence it may be disregarded if we can give an interpretation of the passage which takes into account the philosophical remarks that predominate in it.

It is possible to give such an interpretation for the entire passage. It is the following: Logically speaking, whenever certain observable facts are shown to follow from a theory, this does not constitute a conclusive argument for the truth of the theory, unless one shows that there is no other way of explaining those facts (F379–80, D352–3); nevertheless, methodologically speaking, given two different theories each of which explains the relevant facts, the theory which has greater explanatory power and which is less ad hoc is the better and more probable one. This makes sense of the scientific discussions in the passage, as well as of all the philosophical remarks in it; for the whole passage can be reconstructed as being an illustration of the philosophical theses just formulated, which must therefore be attributed to Galileo himself.[39]

The discussion about explanation and simplicity so far has shown primarily that the passage under consideration does not support Clavelin's interpretation, but some other thesis about Galileo's philosophy of science, attributing to him a different philosphical claim; at the same time it may be that Clavelin's theses may be supported with other evidence. Next, in order to justify further the incidental character of the simplicity remark in this passage, and in order to *refute* another of Clavelin's theses about simplicity in Galileo, I shall discuss a passage where simplicity is a main topic. Clavelin's thesis is the following: "Nor did Galileo consider this appeal to the principle of simplicity a mere expedient ... the greater simplicity of a theory must be clear proof of its closer agreement with reality" (C423). The passage is the one at the beginning of the Second Day in the *Dialogue* where Galileo states a number of arguments favorable to the earth's diurnal motion (F139–50, D266–83). The passage stresses the connection between simplicity and probability, both by means of explicit philosophical remarks, and by the fact that the concrete scientific topics discussed can be reconstructed as illustrations of these philosophical theses. Note that this interpretation is of the same type as the one given above for the passage on the motion of sunspots; the similarity consists in the fact that the interpretation is formulated in such a way as to do justice both to the philosophical and to the scientific content of the passage, in such a way that we have an integration of both this double aspect of the passage and also of all the main philosophical remarks made in it. The reconstruction is the following: Arguments based on the principle of simplicity are neither worthless nor conclusive, but rather probable; they are not worthless because, in the absence of conclusive arguments, the simplest idea is the most acceptable; they are not conclusive because a single piece of counterevidence or conclusive counterargument is sufficient to refute an idea based on simplicity; and they are probable because nature acts by means of the fewest possible operations.[40]

Next let us look at the third element which Clavelin finds in Galileo's philosophy of science, his identification of physical with mathematical intelligibility. Clavelin's evidence consists of "several arguments in the *Dialogue,* all the more convincing in that they were not yet couched in the language of geometry" (C412). The first is the "proof that if the earth did indeed spin east[ward] on its axis every twenty-four hours, then a stone dropped from a tower would not be deflected toward the west" (C412). This proof is reconstructed by Clavelin as follows: first Galileo shows that on a moving ship a stone dropped from the top of the mast lands at the foot; next Galileo claims that "the case of a tower carried along by the earth's diurnal motion is in all respects *identical* with that of the ship" (C413); hence he concludes that the same thing will happen on the earth. Now, in this argument, according to Clavelin, Galileo is identifying physical and mathematical intelligibility in the sense that this physical argument and a mathematical proof "both involve the reduction of a statement to one or several previously established propositions with the help of successive identifications" (C413). This is a very weak sense indeed; the interpretation is ingenious but has no basis in the text; perhaps that is the reason why in his discussion (C412–13) Clavelin gives no references to the *Dialogue.* I suggest, in fact, that we cannot find in this work the proof constructed by Clavelin. The relevant passage (F164–75, D138–49) has three parts and is a critique of three different versions of the falling bodies objection to the earth's motion. The first version says that the earth can't move because bodies fall vertically; Galileo shows that this argument is circular because we can know that vertical fall is actual (as distinct from merely apparent) if and only if we know that the earth stands still (D138–40).

The second version grounds the earth's immobility on the undisputed fact that bodies *appear* to fall vertically; Galileo argues that this objection depends on the impossibility of mixed motion, namely motion around and straight-toward the center, which is *invalidly* justified by the alleged fact that bodies dropped from the mast of a moving ship fall behind; he then argues that the justification is invalid because the motion of the ship is violent, whereas the earth's diurnal motion would be natural, and because the relevant portion of the air does not move along with the ship but would do so near the earth's surface if the earth moved (F167–9, D140–43). The third version is an argument from analogy, comparing a tower on the earth with a mast on a ship, and inferring that just as the rock falls behind when dropped on the moving ship, so it would fall behind on a rotating earth; Galileo criticizes this argument by arguing that it is based on the false premise that the rock falls behind when dropped on a moving ship (F169–75, D143–9).

It is another question, of course, whether the proof constructed by Clavelin, though not present in the text, is suggested by it. However, even this is doubtful because Galileo's critique of the second version of the objection, the argument from *apparent* vertical fall, amounts to a criticism of the soundness of the analogy between the rotating earth and the moving ship. This would prevent one from using in the proof the proposition (crucial in Clavelin's construction) that "the case of a tower carried along by the earth's diurnal motion is in all respects *identical* with that of the ship" (C413).

Once again, however, I wish to make my own criticism more than merely negative. In fact, I can give a positive philosophical interpretation of the passages that Clavelin is talking about. These interpretations, like the ones I have given above, take into account both the various philosophical remarks present in them and the concrete scientific discussions in the context of which they occur. Following this approach, Galileo's discussion of the ship analogy argument (F169–80, D143–54) is an illustration of the following thesis: Experiments are sometimes unnecessary to ascertain the results of a test, for sometimes it can be argued on the basis of known or more easily ascertainable facts, what these results must be. The details of Galileo's discussion support this interpretation.[41]

The other relevant passage, namely Galileo's critique of the earth-ship analogy (F167–9, D141–3), is part of a longer passage (F159–69, D133–43) which, when analyzed in the same way, can be shown to be illustrating the value and usefulness of logical analysis.[42]

Let us now look at a second piece of evidence given by Clavelin to support his claim that Galileo's philosophy of science identifies physical and mathematical intelligibility. It is the passage (F129–30, D103) at the end of the First Day in the *Dialogue* where Galileo discusses the powers and limitations of human understanding. Clavelin thinks it shows that Galileo "presented mathematics as the most perfect knowledge to which man can aspire – knowledge so perfect, in fact, that it can be compared intensively if not extensively with divine understanding" (C414). However, Clavelin's argument looks plausible only because he neglects to quote the part of the passage where Galileo indicates one important *qualitative* difference between divine understanding and human understanding even for the case of mathematical reason, namely God knows the infinitely many mathematical truths he knows without using a step-by-step reasoning process. In other words, though it is true that Galileo is making a philosophical point in this passage, this point is much more complex than even Clavelin realizes. When the specific passage mentioned by him is examined in context (F124–31, D98–105), and when the examination is carried out, as before, by integrating all philosophical remarks with one another and the philosophical theory with the scientific practice, we get a thesis to the effect that the powers of human understanding are likely to disappoint both the optimists and the pessimists.[43]

Clavelin ends his discussion of Galileo's identification of physical and mathematical intelligibility by stating that "Galileo reversed the traditional roles of logic and mathematics" (C416). That is, Galileo gives logic the function of analyzing given arguments and mathematics the function of constructing new ones. Though the relevance of this alleged fact is not readily apparent, let us disregard this problem and consider some of the evidence that leads Clavelin to allege it as a fact. Clavelin quotes Galileo as saying that the art of proof is acquired by the "reading of books filled with demonstrations, and these are exclusively mathematical works, not logical ones" (C417, quoted from F60). Once more, by taking the remark out of context, Clavelin misses its main point, which is that the art of proof is acquired by reading books which contain actual proofs, rather than books which contain theories of proof. For in the context in which the remark is made (F59–60, D35), Galileo is discussing how logical theory and logical practice are distinct; the meaning of the phrase 'logical books' in the sentence quoted by Clavelin is that of 'books in the *theory* of reasoning'; and the meaning of the term 'demonstrations' is that of 'logical practice' or 'actual reasoning'. Finally, Galileo's equation of logical practice exclusively with mathematics must be regarded as merely a rhetorical excess by Galileo, since it does not tie in at all with the rest of the passage in which it occurs. When this entire passage (F59–71, D35–47) is analyzed by the same technique used before, we get the following results: The logical authority of Aristotle is not sacrosanct and at any rate pertains primarily to the domain of logical theory rather than logical practice; his authority as a logical practitioner cannot be grounded on his undisputed authority as a logical theorist, but only on the soundness of his concrete reasoning; the latter can only be tested by the logical analysis of his actual arguments, which analysis one should therefore be free to carry out; if you do this you will discover that his logical practice leaves much to be desired.[44]

To conclude, our analysis of some concrete problems in the study of the historical record of reasoning suggests that more than mere erudition and scholarship is needed when the subject matter is the kind of logical practice found in Galileo's *Dialogue*; thus the logical theorist must become his own scholar.

NOTES

[1] In my *History of Science as Explanation*.

[2] G. Galilei, *Dialog Über Die Bieden Hauptsachlichsten Weltsysteme*, tr. Emil Strauss.

[3] G. Galilei, *Opere*, edited by Favaro, Vol. 7.

[4] G. Galilei, *Dialogue Concerning the Two Chief World Systems*, tr. S. Drake.

[5] This is a translation of Clavelin's *La philosophie naturelle de Galilée*.

[6] Galilei, *Dialogue*, tr. Drake, p. xxvi.

[7] Galilei, edited by Favaro, *Opere* 7, 59–60. Subsequent references to this book will be made only in parenthesis in the text, by prefixing the page number(s) with an 'F', as a reminder that Favaro's edition is meant.

[8] Galilei, *Dialog*, tr. Strauss, p. 502, note 42. Subsequent references to this book will be made only in parenthesis in the text, by prefixing the page number(s) with an 'S', as a reminder that Strauss's edition is meant.

[9] G. Galilei, *Dialogue on the Great World Systems*, tr. Salusbury-Santillana, p. 138, note 22.

[10] For more details see Chapters 8 and 12.

[11] Galilei, *Dialogue*, tr. Drake, p. 372.

[12] *Ibid.*

[13] Galilei, edited by Favaro, *Opere* 6, 316, 317, 319, and 321, translated in *The Controversy on the Comets of 1618*, edited and tr. S. Drake and C. D. O'Malley, pp. 276, 279, and 280; also Galilei, *Opere* 8, 105–6, and 5, 351–63. Cf. my 'The Concept of *Ad Hominem* Argument in Galileo and Locke'.

[14] G. Galilei, *Dialogo di Galileo Galilei Linceo . . .*, p. 24. Subsequent references to this book will be made only in parenthesis in the text, by prefixing the page numbers with 'EP', as a reminder that the *editio princeps* is meant.

[15] This notation refers to line 26 on page F212; similarly for subsequent such references.

[16] Besides the *Dialogue* and the *Controversy on the Comets* already cited, we have: *Two New Sciences, Discoveries and Opinions of Galileo, Galileo Against the Philosophers*.

Other Galilean works can be found in parts of the following volumes, translated by Drake in conjunction with other scholars: *On Motion and on Mechanics,* and *Mechanics in Sixteenth Century Italy.* Finally, his *Galileo at Work* contains translations of many important letters, papers, and documents.

[17] Most of his interpretations up to 1970 are contained in his *Galileo Studies.*

[18] Besides the work already cited, see for example, 'The Evolution of *De Motu*', 'Galileo and the First Mechanical Computing Device', 'Galileo's "Platonic" Cosmogony and Kepler's *Prodromus*', 'Impetus Theory and Quanta of Speed Before and After Galileo', and 'The Uniform Speed Equivalent to a Uniformly Accelerated Motion from Rest'.

[19] See my 'Vires Acquirit Eundo: The Passage Where Galileo Renounces Space-Acceleration and Causal Investigation', my 'Galileo's Space-Proportionality Argument: A Role for Logic in Historiography', and Drake's 'Velocity and Eudoxian Proportion Theory'.

[20] Subsequent references to G. Galilei, *Dialogue,* tr. Drake, will be given only in parenthesis in the text, by prefixing the page number(s) with a 'D', as a reminder that Drake's translation is meant.

[21] See Chapter 3, dealing with the literary aspects of Galileo's book. It should be mentioned here that the purpose of Drake's translation was to popularize the *Dialogue,* and that it was with deliberation that he chose "reasonably easy reading in preference to strict literalness, even at the price of taking certain liberties with the text" (D, p. XXV), so that my criticism is aimed primarily at scholars who make uncritical use of Drake's translation. On the other hand I am not sure that the problems mentioned below are above the layman's head, and so I am inclined to think they could have been avoided without falling into pedantry. It should also be noted that Santillana's revision of Salusbury's translation is no substitute for Drake's translation, primarily because the Santillana edition is incomplete. See Appendix, note 3 for details.

[22] Cf. the glossary in L. Sosio's edition of the *Dialogue*: G. Galilei, *Dialogo sopra i due massimi sistemi,* edited by Sosio, p. 574. See also the *Vocabolario Illustrato della Lingua Italiana* 2, 1541. Santillana seems to commit the same oversight as Drake: cf. Galilei, *Dialogue,* edited by Santillana, pp. 141, 183.

[23] Additional support for my interpretation is the use of the subjunctive mood in Simplicio's reply (F167) to Salviati, within the sentence that I have structured as being of the form "if *C* holds, then *A* follows but *B* does not" (*quando l'elemento del fuoco vadia in giro insieme con l'aria . . .*).

[24] "*quello che occorre della nave deve parimente accadere della Terra*" (F170).

[25] Feyerabend is a notable example of such a misinterpretation; cf. his *Against Method,* pp. 69–98, esp. pp. 83–87. Some of his critics disagree with the philosophical consequences to be drawn from his analysis but seem to share his misinterpretation, though they ought to know better; for example, P. K. Machamer ('Feyerabend and Galileo: The Interaction of Theories and the Reinterpretation of Experience', pp. 27–28), in criticizing Feyerabend on this point, states the geostatic objection analyzed by him as the argument that "if the earth were moving, then when a stone is dropped from the top of a mast (on a ship) it would fall behind the mast (because the earth will have carried the ship beyond the point where the stone was dropped)". Feyerabend's interpretation also conflicts with Galileo's explicit clarification at the end of his critique that "I have not claimed to prove it yet, but only to show that nothing can be deduced from the experiments offered by its adversaries as one argument for its motionlessness" (D154).

[26] See E. McMullin, 'Introduction: Galileo, Man of Science', pp. 31–32.

[27] S. Drake, Review of *La philosophie naturelle de Galilée*, p. 277.

[28] *Ibid.*

[29] H. Stein, 'Maurice Clavelin on Galileo's Natural Philosophy', p. 375.

[30] A. R. Hall, Essay review of *La philosophie naturelle de Galilée*, p. 80.

[31] *Ibid.*, p. 84.

[32] M. S. Mahoney, 'Galileo's Thought', p. 944.

[33] *Ibid.*, p. 945.

[34] W. R. Shea, Review of *La philosophie naturelle de Galilée*, p. 125.

[35] J. L. Heilbron, Review of *La philosophie naturelle de Galilée*, p. 342.

[36] See, for example, W. Wisan, 'Galileo's Scientific Method: A Reexamination'; cf. my review of the collection where this paper appears, *New Perspectives on Galileo*, edited by R. E. Butts and J. C. Pitt. For the sake of perspective, it must be said that Wisan's work represents the best and latest study of the mathematical aspects of Galileo.

[37] M. Clavelin, *The Natural Philosophy of Galileo*, pp. 404–409. Subsequent references to this book will be given in the text in parenthesis, by prefixing the page number(s) by 'C', as a reminder that Clavelin's book is meant.

[38] Galilei, edited by Favaro, *Opere* 7, 372–79; idem, *Dialogue*, tr. Drake, pp. 345–52. As before page references to these books will be denoted respectively by 'F' and by 'D'.

[39] For more details, see Chapter 5, section dealing with passage F372–83.

[40] For more details, see Chapter 5, section dealing with passage F139–50.

[41] For more details, see Chapter 5, section dealing with passage F169–80.

[42] For more details, see Chapter 5, section dealing with passage F159–69.

[43] For more details, see Chapter 5, section dealing with passage F124–31.

[44] For more details, see Chapter 5, section dealing with passage F59–71.

CHAPTER 11

THE PSYCHOLOGY OF LOGIC AND
THE LOGIC OF PSYCHOLOGY:
CRITIQUE OF THE PSYCHOLOGY OF REASONING

The argument so far has been that Galileo's *Dialogue* has classic significance for science, and important implications for the philosophy of science, for the historiography of science, and for the history and historiography of philosophy. It has also emerged that reasoning is the book's central feature, both at the level of scientific practice, and at the level of philosophical reflection. It follows that reasoning, which of course is not to be confused with rationalism, is an essential feature of scientifc rationality. However, though we have thereby exhibited the structure of scientific rationality, we have not yet examined the structure of this structure. That is, so far we have examined the macro-structure of scientific rationality, but not its micro-structure. In order to do the latter, we need to study the finer details of the reasoning in Galileo's book. There is, however, an additional independent reason why it is desirable to undertake this study, namely that it provides materials for and examples of a concrete approach to the theory of reasoning, and to the foundations of a reformed science of logic. This I now proceed to show.

In recent years we have seen the emergence of new expressions of the old critiques of formal logic. Some have questioned its generality, claiming that it represents basically a theory of mathematical reasoning and completely neglects the very wide and important area of so-called rhetorical or dialectical reasoning which includes most of the arguments found in law, ethics, philosophy, the human sciences, and everyday life.[1] Others have questioned its empirical import, claiming that the empirical study of the psychology of reasoning shows that formal logic is inaccurate as a description of human reasoning, and misleading and useless for its theoretical explanation.[2] Still others have objected to its artificiality (or formalistic character), claiming that it pays very little attention to informal argumentation in natural language, the study of which was and ought to be its *raison d'être*.[3] There are some who question its usefulness, arguing that the rigor and precision of its concepts present an intrinsic limitation for its practical application to ordinary logical problems, which need to be encoded before and decoded after the formal apparatus is used; and unfortunately the encoding step is usually as debatable as the original problem was when discussed in its own terms.[4] Finally there are those who object to the abstractness of formal logic, claiming that what is needed is not more theorizing but more analysis, not the construction of other abstract schemas, but the critical understanding of concrete instances of reasoning.[5] It is, of course, easy to defend logic from such criticism by regarding it as a branch of pure mathematics; however, I know of no philosophical logician who is willing to accept the full consequences and practice logic as mathematicians do; the pretension always lingers that he is dealing with arguments and reasoning rather than with purely abstract entities (such as truth functions and sets).[6] At any rate, just as there is a place for logic practiced as mathematics (though it is up to mathematicians to determine it), so there ought to be a place for the kind of general, empirical, informal (natural), useful, and critical-concrete logic suggested by the above mentioned criticism. However, qualifications and elaborations need to be made to such critiques in order to get a better idea of what this type of investigation should be like and a better understanding of the relation between actual reasoning and formal logic. This is especially true since none of the positive results of the scholars from whom such criticism has been taken combine all the mentioned virtues.

Let us explore the sense in which a logic of reasoning could be empirical by examining some recent work in the psychology of reasoning. Some of the general questions to keep in mind are whether 'empirical' here ought to mean experimental or historical, and whether introspective experience is an effective empirical guide in this case.

The most extensive recent investigations into the psychology of reasoning are those of Wason and Johnson-Laird.[7] Three main conclusions seem to emerge from their studies. First they argue that the propositional calculus does not adequately characterize propositional reasoning because that calculus is two-valued, truth-functional, and formal in the sense of ignoring the specific content of propositions (p. 93). And they present evidence that the content of a proposition has a significant effect on how people interpret its logical form (pp. 54–85), that people do not attach a material conditional meaning to the 'if then' (pp. 54–65 and 86–96), and that many statements in everyday language presuppose a state of affairs which when unfulfilled renders the statement neither true nor false but irrelevant or inapplicable (pp. 90–1); as regards this last point they investigate in great detail the role of negations and show that "a denial generally (not invariably) functions in language to correct the preconception which it denies" (p. 30), so that "the affirmative preconception has to be recovered before the meaning of the negative can be grasped" (p. 39), and in turn "when a negative is itself denied by an affirmative, it becomes difficult to keep track of the argument" (p. 53), that is, one finds people not reasoning in accordance with what one would expect from formal logic.

Second, Wason and Johnson-Laird exhibit the empirical inadequacy of predicate calculus by presenting evidence concerning the interpretation of universal affirmative and of particular propositions. Concerning the former there is a tendency to treat the subject term of 'All S are P' as the whole universe of discourse, which would justify its conversion to 'All P are S'; and there is also a tendency to interpret 'Some S are P' as equivalent to 'Some S are not P' (pp. 157–8).

These conclusions seem well founded, and the logician with a sensitivity for actual reasoning will feel that they correspond to what he has known all along. In fact, Wason and Johnson-Laird occasionally are explicit in expressing their debt to a philosopher like Strawson.[8] Strawson's method of ordinary language analysis may be regarded as the introspective psychology counterpart of Wason and Johnson-Laird's experimental approach, hence as a type of a fundamentally empirical approach, and a proper one at that.

Wason and Johnson-Laird's most original work is perhaps that dealing with the testing of general hypotheses. It is here that emerge the limitations of their approach as well as a point in favor of formal logic. The main conclusion they reach in these investigations is one about the effect of content on reasoning, namely that the testing of a general hypothesis dealing with a real life situation is in accordance with what one would expect from logic, whereas when the hypothesis deals with abstract material the subjects' performance diverges from the supposedly logical one (p. 193). This conclusion will be shown below to be based on an inadequate analysis of their own evidence; however, it should be pointed out first that the authors seem to be right in their general conclusion about the effects of content on the way people reason, namely that the nature of the subject matter (concrete or abstract) affects logical performance, sometimes facilitating it, sometimes hindering it.

Let us look at their experimental results. The first group of experiments involves the testing of the hypothesis "If a card has a vowel on one side, then it has an even number on the other side" (p. 173) by reference to four cards whose visible, up sides show respectively the symbols E, K, 4, and 7; the subjects are informed that each card has a letter on one of its sides and a number on its other side, and their task is to name all and only those cards which need to be turned over in order to determine whether the hypothesis is true or false (p. 173); for a total of 128 subjects in the course of four experiments the choices are as follows (p. 182):

Cards 1 and 3:	59 subjects
Card 1:	42 subjects
Cards 1, 3, and 4:	9 subjects
Cards 1 and 4:	5 subjects
Other combinations:	13 subjects

Since the right answer is 'Cards 1 and 4', this means that only 5 out of 128 subjects got the right answer.

The second experiment relevant to our discussion (pp. 190–1) involves the testing of the hypothesis "Every time I go to Manchester I travel by train". The four cards show respectively the words Manchester, Leeds, Train, and Car. The subjects are informed that each card has the name of a city on one side and the name of a means of transportation on the other; moreover, they are asked to imagine that each card represents a journey made by the speaker and that the hypothesis represents a generalization about these journeys. The task was once again to select all and only those cards that needed to be turned over to check whether the generalization was true or false. Ten out of sixteen subjects made the correct selection of cards 1 and 4 (as contrasted to two out of sixteen in a control group tested with abstract material).

The third relevant experiment (pp. 191–192) involves both the ('concrete') hypothesis "If a letter is sealed, then it has a 5d stamp on it" and the ('abstract') hypothesis "If a letter has a D on one side, then it has a 5 on the other side". Instead of cards, two sets of four *envelopes* were used; one ('concrete') set was arranged as follows: the back of a sealed envelope, the back of an unsealed envelope, the front of an envelope with an address and a 5d stamp on it, and the front of an envelope with an address and a 4d stamp on it; the other ('abstract') set of envelopes was arranged as follows: the front of an envelope with a D written in the middle of it, the front of an envelope with a C written in the middle of it, the back of an envelope with a 5 written in the middle, and the back of an envelope with a 4 on it. There were two tasks to be performed: testing of the abstract hypothesis by reference to the abstract set of envelopes, and testing of the concrete hypothesis by reference to the concrete set of envelopes. (For testing of the concrete hypothesis subjects were told to imagine that they were Post Office workers engaged in sorting letters.) There were two groups of subjects, and each group performed both tasks in the opposite order. The results were that 21 out of 24 subjects were correct in the concrete task, but only 2 out of 24 in the abstract task.

The difference in logical performance is certainly striking, but the real problem is to account for it. The explanation favored by the authors is the following (pp. 182–187). First they interpret the logical structure of the hypothesis being tested as that of the conditional "If p, then q"; p would refer, respectively, to the propositions that a card has a vowel on one side, or that I go to Manchester, or that a letter is sealed, or that a letter has a D on one side; and q would refer, respectively, to the proposition that the card has an even number on the other side, or that I travel by train, or that the letter has a $5d$ stamp on it, or that the letter has a 5 on the other side. Second, in the light of this interpretation, they refer to card 1 as the p card, card 2 as the not-p card, card 3 as the q card, and card 4 as the not-q card. Third, their reconstruction of the subjects' reasoning is as follows:

It assumes that the subject will initially focus only on items mentioned in the rule. From this hypothetical 'list', only cards which could verify the rule will be selected, as a function of whether he assumes that the converse holds (selection of p and q), or does not hold (selection of just p). Two levels of insight have been retained, but unlike those in the preliminary model, they are no longer independent. *Partial insight* consists in realizing that all cards should be tested, and that those which could verify, and those which could falsify, should be selected, i.e., p, q and not-q. Even if q had not been selected initially, it will now be selected because it could verify the rule. *Complete insight* consists in realising that only cards which could falsify should be selected, i.e. p and not-q. [P. 185]

Fourth, the greater insight into the concrete task is attributed to the coincidence in it of a logical and of a causal analysis of the situation, the assumption being that humans have a tendency toward causal analysis of problems (p. 193).

It is unnecessary to examine in detail the arguments and evidence they give in support of this explanation. One reason is that their reconstruction of the subjects' reasoning is somewhat unclear. For example, it is not clear how and why the subjects with partial insight chose the q card (#3), whether it is because they think that a p on the down side would verify the hypothesis 'if p then q', or whether they think that this discovery would verify its converse 'if q then p' which they regard as equivalent to the hypothesis itself. If the former then the question would arise how any subject could choose only p, that is, why the subjects who choose only p do not regard that they would have a verification if the third (q) card had p on the other side; if the latter then how can the subjects who chose p, q, and not-q fail to choose the second card (not-p) which would yield a falsification if q were on the other side (since this would falsify 'if q then p', and hence its converse allegedly equivalent to it).

However, the main reason for the inadequacy of the authors' explanation is their analysis of the hypotheses being tested as conditionals of the propositional calculus, whereas they are universal generalizations of conditionals involving quantifiers. This initial oversimplification in turn leads them to attribute equal complexity to the concrete as to the abstract propositions, whereas the latter are much more complex than the former. I now turn to showing this.

The hypothesis in the first group of experiments reads: "If a card has a vowel on one side, then it has an even number on the other side". This should be symbolized as follows:

$$(x)\,(y)\,(z)\,(Cx\;\&\;Pxy\;\&\;Syx\;\&\;y\neq z\;\&\;Szx \longrightarrow Qxz)$$

where

Cx: x is a card;
Pxy: x has a vowel on side y;
Syx: y is a side of card x;
Qxz: x has an even number on side z.

The generalization refers to four cards and eight sides. Let us use the following individual constants and denote the given cards and their sides:

c_1 :   card #1, with an 'E' on its up side;
c_2 :   card #2, with a 'K' on its up side;
c_3 :   card #3, with a '4' on its up side;
c_4 :   card #4, with a '7' on its up side;
s_{11}: 'up' side of card 1, showing 'E';
s_{21}: 'down' side of card 1;
s_{12}: 'up' side of card 2, showing 'K';
s_{22}: 'down' side of card 2;
s_{13}: 'up' side of card 3, showing '4';
s_{23}: 'down' side of card 3;
s_{14}: 'up' side of card 4, showing '7';
s_{24}: 'down' side of card 4.

The generalization is true if and only if it is true in each of the eight cases to which it refers, that is if and only if all of the following sentences are true.

$$\sigma_1: \quad Cc_1 \ \& \ Pc_1s_{11} \ \& \ Ss_{11}c_1 \ \& \ s_{11} \neq s_{21} \ \& \ Ss_{21}c_1 \longrightarrow Qc_1s_{21};$$

$$\sigma_2: \quad Cc_2 \ \& \ Pc_2s_{12} \ \& \ Ss_{12}c_2 \ \& \ s_{12} \neq s_{22} \ \& \ Ss_{22}c_2 \longrightarrow Qc_2s_{22};$$

$$\sigma_3: \quad Cc_3 \ \& \ Pc_3s_{13} \ \& \ Ss_{13}c_3 \ \& \ s_{13} \neq s_{23} \ \& \ Ss_{23}c_3 \longrightarrow Qc_3s_{23};$$

$$\sigma_4: \quad Cc_4 \ \& \ Pc_4s_{14} \ \& \ Ss_{14}c_4 \ \& \ s_{14} \neq s_{24} \ \& \ Ss_{24}c_4 \longrightarrow Qc_4s_{24};$$

$$\sigma_5: \quad Cc_1 \ \& \ Pc_1s_{21} \ \& \ Ss_{21}c_1 \ \& \ s_{21} \neq s_{11} \ \& \ Ss_{11}c_1 \longrightarrow Qc_1s_{11};$$

$$\sigma_6: \quad Cc_2 \ \& \ Pc_2s_{22} \ \& \ Ss_{22}c_2 \ \& \ s_{22} \neq s_{12} \ \& \ Ss_{12}c_2 \longrightarrow Qc_2s_{12};$$

$$\sigma_7: \quad Cc_3 \ \& \ Pc_3s_{23} \ \& \ Ss_{23}c_3 \ \& \ s_{23} \neq s_{13} \ \& \ Ss_{13}c_3 \longrightarrow Qc_3s_{13};$$

$$\sigma_8: \quad Cc_4 \ \& \ Pc_4s_{24} \ \& \ Ss_{24}c_4 \ \& \ s_{24} \neq s_{14} \ \& \ Ss_{14}c_4 \longrightarrow Qc_4s_{14}.$$

In each case the first, third, fourth, and fifth conjuncts of the antecedent are true (because the symbols have been so chosen as to insure their truth). So the truth value of each sentence depends on the truth values of the second conjunct of the antecedent (which asserts that a given card has a vowel on a given side), and on the truth value of the consequent (which asserts that the corresponding card has an even number on the other side).

Now, for σ_1, Pc_1s_{11} is true since card 1 does have a vowel on its up side, but the truth value of Qc_1s_{21} is unknown since we don't know what number is on the down side of card 1. Hence we need to turn over card 1 in order to determine the truth value of σ_1.

For σ_2, Pc_2s_{12} is false since card 2 has a consonant on its up side. Hence, without looking at the consequent of σ_2, we can conclude that σ_2 is true.

For σ_3, Pc_3s_{13} is false since card 3 does not have a vowel on its up side; hence σ_3 is true.

For σ_4, Pc_4s_{14} is false since card 4 does not have a vowel on its up side either, hence σ_4 is true.

For σ_5, Pc_1s_{21} is false since we have been informed that every card has a letter on one side and a number on the other, and hence the down side of card 1 does *not* have a vowel; hence σ_5 is automatically true, by falsity of antecedent.

For σ_6, the same applies as for σ_5; so σ_6 is true.

For σ_7, its consequent Qc_3s_{13} is true because card 3 has an even number on its up side; hence, regardless of the truth value of Pc_3s_{23}, σ_7 is true.

For σ_8, its consequent Qc_4s_{14} is false because card 4 does not have an even number on its up side; hence σ_8 would be false if Pc_4s_{24} were true; but we do not know the truth value of Pc_4s_{24} since we don't know what is the letter on the down side of card 4; hence we need to turn over card 4 to determine the truth value of σ_8.

To summarize, the generalization is demonstrably true in every relevant case except the first and the eighth, which involve cards 1 and 4 respectively.

These are all and the only cards that need to be turned over for an exhaustive test.

This analysis shows that the answer which Wason and Johnson-Laird regard as correct is indeed one in accordance with a formal-logical interpretation of the hypothesis in question. The analysis can be simplified somewhat by considering a more abstract symbolization of the hypothesis; such a simplified analysis will be closer to the one given by Wason and Johnson-Laird, without, however, diluting away the quantificational aspect of the problem, as they do. The hypothesis in question could also be symbolized as follows:

$$(x)\,(y)\,[Dxy \longrightarrow (Px \longrightarrow Qy)]\,,$$

where

Dxy: x and y are different sides of the same card;
Px: side x has a vowel on it;
Qy: side y has an even number on it.

If the individual constants denoting the sides of the cards are the same as before (s_{nm}), then we would have again eight cases corresponding to the following ordered pairs as abstract entities, where the truth-values have been written under the relevant portions of the formulas:

(1) $<s_{11}, s_{21}>$: $Ds_{11}s_{21} \longrightarrow (Ps_{11} \longrightarrow Qs_{21})$;
                        T $?$ T $?$ $?$

(2) $<s_{12}, s_{22}>$: $Ds_{12}s_{22} \longrightarrow (Ps_{12} \longrightarrow Qs_{22})$;
                        T T F T $?$

(3) $<s_{13}, s_{23}>$: $Ds_{13}s_{23} \longrightarrow (Ps_{13} \longrightarrow Qs_{23})$;
                        T T F T F

(4) $<s_{14}, s_{24}>$: $Ds_{14}s_{24} \longrightarrow (Ps_{14} \longrightarrow Qs_{24})$;
                        T T F T F

(5) $<s_{21}, s_{11}>$: $Ds_{21}s_{11} \longrightarrow (Ps_{21} \longrightarrow Qs_{11})$;
                        T T F T F

(6) $<s_{22}, s_{12}>$: $Ds_{22}s_{12} \longrightarrow (Ps_{22} \longrightarrow Qs_{12})$;
                        T T F T F

(7) $<s_{23}, s_{13}>$: $Ds_{23}s_{13} \longrightarrow (Ps_{23} \longrightarrow Qs_{13})$;
                        T T $?$ T T

(8) $<s_{24}, s_{14}>: Ds_{24}s_{14} \longrightarrow (Ps_{24} \longrightarrow Qs_{14}).$
                      T   ?   ?   ?   F

Once again all the truth values are determinable except for the case of the ordered pair of the up and down sides of the first card and the case of the ordered pair of the down and up sides of the fourth card; so that cards 1 and 4 need to be turned over. From the point of view of formal logic it is immaterial whether the denotations of the predicate symbols D, P, and Q are those just given, which correspond to the first group of experiments, or whether the following interpretations hold, which would correspond to the 'abstract' case in the third group of experiments mentioned above:

Dxy: x and y are sides of the same envelope;
Px: side x has a D;
Qy: side y has a 5.

Hence the structure of the 'abstract' hypothesis of the third group of experiments is identical to that of the hypothesis in the first group, and we would expect analogous experimental results.

However the concrete hypotheses involve simpler formulas. In a second group of experiments the generalization was:
"Every time I go to Manchester, I travel by train".
This can be symbolized as follows:

$$(x)(Mx \longrightarrow Tx),$$
where
Mx: x is a journey to Manchester;
Tx: x is a journey by train.

This is the more obvious and natural interpretation, even though a more complex one could be given, namely

$$(x)(y)[Jxy \longrightarrow (Mx \longrightarrow Ty)],$$
where

Jxy: x is the destination in a journey where the means of transportation is y;
Mx: the destination of journey x is Manchester;
Ty: the means of transportation of journey y is train.

In the simpler interpretation the universe of discourse is the set of journeys, whereas in the more complex interpretation the universe of discourse is the union of the set of journey destinations and the set of journey means of transportation. Because of the simpler interpretation, we may ignore the more complex one.

Could the structure of the generalization in the abstract experiments be simplified further than it was done above? Our simplified analysis resulted in a formula which was a doubly universally quantified generalization of a conditional whose antecedent was an atomic formula with a binary predicate and whose consequent was itself a conditional consisting of two atomic monadic formulas:

$$(x)\,(y)\,[Dxy \longrightarrow (Px \longrightarrow Qy)].$$

Could we reduce this to something like

$$(x)\,(Vx \longrightarrow Ex),$$

where

Vx: x is a card with a vowel on it;
Ex: x is a card with an even number on it?

This would not have the intended meaning because this sentence would be true for cards which had an even number *and* a vowel on one side and nothing on the other, whereas the generalization being tested would be false in this case. I do not think that the structure of this generalization can be simplified any further. We do need a universe of discourse of *sides* of cards, and we need to consider such sides in ordered pairs.

Let us now look at the concrete generalization in the third set of experiments: "If a letter is sealed, then it has a 5d stamp on it", which can be symbolized as

$$(x)\,(Sx \longrightarrow Fx),$$

where

Sx: x is a sealed letter;
Fx: x is a letter with a 5d stamp on it.

The universe of discourse is simply the set of letters.

Thus both of the concrete experiments involve universal generalization with a single quantifier and with monadic predicates only:

$$(x)\,(Px \longrightarrow Qx).$$

In either experiment we have only four cases, in each of which only one individual constant is involved:

JOURNEY EXPERIMENT

j_1:   journey described by the first card, namely to Manchester;
j_2:   journey described by the second card, namely to Leeds;
j_3:   journey described by the third card, namely by train;
j_4:   journey described by the fourth card, namely by car.

The four sentences involved are the following, where the truth values have been written under the appropriate portion of the sentence:

(1) $Pj_1 \longrightarrow Qj_1$; (2) $Pj_2 \longrightarrow Qj_2$; (3) $Pj_3 \longrightarrow Qj_3$; (4) $Pj_4 \longrightarrow Qj_4$.
     T ?   ? F T ?   ? T T ?   ? F

POST OFFICE EXPERIMENT

ℓ_1:   first letter, sealed;
ℓ_2:   second letter, unsealed;
ℓ_3:   third letter, properly stamped;
ℓ_4:   fourth letter, improperly stamped.

(1) $P\ell_1 \longrightarrow Q\ell_1$; (2) $P\ell_2 \longrightarrow Q\ell_2$; (3) $P\ell_3 \longrightarrow Q\ell_3$; (4) $P\ell_4 \longrightarrow Q\ell_4$.
     T ?   ? F T ?   ? T T ?   ? F

In both experiments items 1 and 4 need to be turned over.

The differences in complexity between the concrete and the abstract hypotheses may be summarized as follows.[8] In the concrete case, the universe of discourse is a set referring to a single relatively natural class of objects; in the abstract case the universe is a set of relatively abstract entities, like sides of cards. Second, in the concrete case only monadic predicates occur, in the abstract there occur binary as well as monadic predicates. Finally, in the concrete case we have only one universal quantifier; in the abstract case we have two. Because of these differences, the possibility emerges that the great difference in logical performance by the experimental subjects could be accounted thereby. This possibility becomes likely in view of the existence of certain disturbing factors, which are mentioned for other purposes by Wason and Johnson-Laird, but which can be shown to have just the required effect. These features are the natural ('non-material') interpretation of the 'if-then' and the relative ambiguity of the hypothesis in the abstract experiments.

Let us turn to the latter first. Wason and Johnson-Laird present considerable evidence to show that a widespread and persistent feature of the abstract experiments was "the failure to appreciate that the cards are *reversible,* i.e.

the significance of a card is the same regardless of which part of it is exposed" (p. 195). They call attention to this fact in their discussion of what they regard as a more speculative and less plausible explanation of the poor performance, according to which the novelty and abstractness of the task induces a temporary regression to less mature forms of cognitive functioning, such as failing to appreciate the importance of a reversible operation (pp. 193–201). However, we need not worry about the authors' reasons for mentioning the fact. Instead we may use it as evidence that the abstract hypothesis as initially stated, without qualification, comment, or context, could mean either one of two things:

(H1)   If a card has a vowel on one *of the presently visible* sides, then it has an even number on the other side.

(H2)   If a card has a vowel on either one of its two sides, then it has an even number on the other side.

The difference between these two hypotheses is their scope as generalizations: the first refers to four entities, the second to eight. Earlier, in analyzing the hypothesis in the experiment I interpreted it to mean (H2). However, the overwhelming evidence from the experimental subjects is that they took it to mean (H1). Let us examine the consequences of this (more restricted) interpretation. We could symbolize it as

$$(x)\,(y)\,[Uxy \longrightarrow (Px \longrightarrow Qy)],$$

where Uxy: x is the up side of a card whose down side is y, and the other symbols are interpreted as before. For this formula there are only four relevant cases that would need to be checked, namely those corresponding to the first four in the list above, referring respectively to $<s_{11}, s_{21}>$, $<s_{12}, s_{22}>$, $<s_{13}, s_{23}>$, $<s_{14}, s_{24}>$. When the hypothesis is so interpreted, the only card that needs to be turned over is the first one. This would explain the choice of card 1 by a large percentage of the subjects in the abstract experiments (42 out of 128, in one series).

In this manner, however, we could not explain the other responses. To explain them we need to call attention to the fact that the natural interpretation of 'if-then' is not as a material conditional. The authors themselves present evidence (pp. 61–65, 89–93) that the 'if-then' does not have the truth table of formal logic. It seems clear, therefore, that in reconstructing the subjects' reasoning, we should use their nonformal truth table for 'if-then'. The authors themselves are aware that this should be done, however they erroneously state without argument that the subjects' responses are inconsistent with their own interpretation of 'if-then' (p. 183). Let us see what really emerges when we take this into account.

Wason and Johnson-Laird present evidence that one common interpretation of the 'if-then' is in accordance with the following truth table, as contrasted to the one used in formal logic:

p	q	if p then q (actual reasoning)	$p \longrightarrow q$ (formal logic)
true	true	true	true
true	false	false	false
false	true	void	true
false	false	void	true

If we look at the list above, we see that cases 2 through 6 would be evaluated as void or irrelevant, rather than true, because of the falsity of the 'P' formula. The value of case 1 would be true *or* false depending on the truth value of Qs_{21}; hence card 1 would have to be turned over. The value of case 7 would be true or void (irrelevant), though we could be sure that it's not false; hence we would have to turn over card 3. The value of case 8 would be false or void (irrelevant), though we would be sure that it can't be true; hence card 4 would have to be turned over. What follows here is that, if we prescind from the ambiguity of the hypothesis, and we interpret it in a 'reversible' manner (H2), then the response would be card 1, 3, and 4. A significant group of subjects, in fact, made this response.

Another common interpretation of the 'if-then' is such that a conditional is taken to imply its converse. Wason and Johnson-Laird have discovered that this is especially true when the conditional is some sort of generalization (p. 61) and when the material is abstract (p. 62). These conditions are precisely those occurring in the hypothesis being tested in the abstract experiment. According to this interpretation, 'if p then q' would be an abbreviation of 'if p then q, and if q then p', where the 'if-then' would obey the 'defective' truth table mentioned above. It is not easy, and perhaps unnecessary to formalize such a concept of 'if-then'. Nevertheless, I believe it can be applied to the present situation as follows. The second and fourth card would be ruled out as irrelevant, the second because it falsifies the antecedent of the hypothesis, the fourth because it falsifies the antecedent of its converse. Card 1 needs to be turned over to determine the truth value of the hypothesis as stated, card 3 in order to determine the truth value of its converse. These are the choices made by the largest percentage of subjects (59 out of 128) in the original group of experiments.

To summarize this explanation of the experimental results using abstract material, we may say that it is possible to give an empirically supported rational reconstruction of the subjects' reasoning, both those who get answers in accordance with formal logic and those that don't. In the case of one common response (card 1), it is possible to give a reconstruction completely in accordance with formal logic, based on an interpretation of the original hypothesis different from the one intended by the experimenter, but at least as plausible. In the case of another common response (cards 1, 3, and 4), it is possible to give a reconstruction which is formalistic in the sense of being in the spirit of formal logic, but using a nonstandard truth table. In the case of the most common response, a formalistic reconstruction is not easy or perhaps not possible. The few subjects whose response is labeled 'correct' by Wason and Johnson-Laird are merely those who happened by accident to choose the meaning of the hypothesis intended by the experimenters and who then reasoned in accordance with formal logic. This is the most that can be said in favor of the approach espoused by the authors. For one finds it simply incredible that the subjects could have actually been engaged in reasoning as complex as that outlined above. In that case the root of the problem would be the experimenter's choice of the original abstract hypothesis, which is simply *deceptively simple*. It is really very complicated, from the point of view of formal logic, and hence we may say it was a bad choice on the part of the experimenters. Could a better choice have been made? Since the hypothesis complexity derives from its abstractness, as we saw above, it seems that the only better choice would have been a more concrete hypothesis and situation, which they did use in another series of experiments. But when an experiment is so 'concretized', how significant does it become? Doesn't it perhaps become trivial? Let us ask for example, what one can learn from the concrete experiments mentioned above? One cannot deny that when the concrete experiments are contrasted to the abstract ones they are very interesting and significant; in fact, such a contrast leads Wason and Johnson-Laird to conclusions about the importance of content in reasoning and hence to question the empirical import of formal logic; or the contrast may lead one, as it led us in the above critique, to conclude that the use of abstract material in experimental situations is tricky since the very abstractness of the material makes it very complex, and what's worse, gives it the *appearance* of being simple, from the point of view of formal logic. Thus, if one rejects the usefulness of experiments with abstract material, then the question remains whether experiments with concrete material have any value *per se*.

Let us then examine these experiments by themselves. They certainly show such things as that the subjects would be able to do post office work involving such simple tasks as determining whether sealed letters have a 5d stamp on them. But what does it show about their reasoning? In particular what does it show about whether or not they reason in accordance to formal logic? The authors themselves confess that the almost trivial simplicity of the concrete task makes it extremely difficult to determine exactly what steps the subjects are following in their reasoning (pp. 190–1, 192–3). Of course, one could ask them to give their reasons, as the authors report that has been done in the case of the responses to the abstract material. But to do this is really to have them express *arguments*. This is certainly a step in the right direction, but then it would have been better to ask the subjects to *write out arguments,* rather than merely give answers to pointed questions. But if one is going to have experimental subjects write out arguments, it would be better to examine examples of already written out arguments, existing in various kinds of literature, produced by people in more realistic and concrete circumstances, when they had the time to be relatively explicit in the steps of their reasoning. In short, these problems in Wason and Johnson-Laird's account of reasoning suggest that a better approach to the study of human reasoning is the study of written works containing relatively explicit argumentation. Now, one of the best examples of such work is Galileo's *Dialogue Concerning the Two Chief World Systems,* which is full of explicit statements and analyses of arguments for and against the earth's motion, a very concrete and realistic issue at the beginning of the seventeenth century when the book was written. It is in part in this spirit that our investigation into Galileo's *Dialogue* should be viewed.

We may conclude this chapter by saying that on the one hand formal logic has more explanatory power than these psychologists believe since it turned out to be possible to give full (or almost) formal-logical reconstructions of the subjects' reasoning in ways unnoticed by the authors. On the other hand, when formal logic is combined with an experimental approach by devising experiments in its terms, we seem to get the situation that formal reasoning is either not formal or not reasoning, in the sense that we get either intuitive and trivial reasoning or formal operations that are likely to be psychologically unreal because of their complexity.

NOTES

[1] C. Perelman and L. Olbrechts-Tyteca, *The New Rhetoric;* C. Perelman, *The Idea of*

Justice and the Problem of Argument; idem, 'The New Rhetoric'; idem, 'A Reply to Henry W. Johnstone, Jr.'; S. Toulmin, *The Uses of Argument.*

[2] P. C. Wason and P. N. Johnson-Laird, *Psychology of Reasoning*; and D. Osherson, 'Models of Logical Thinking', in *Reasoning: Representation and Process in Children and Adults,* edited by R. J. Falmagne.

[3] Y. Bar-Hillel, 'A Neglected Recent Trend in Logic'; idem, 'Comments', in J. F. Staal (ed.), 'Formal Logic and Natural Languages (A Symposium)'; idem, 'Argumentation in Natural Languages'; idem, 'Argumentation in Pragmatic Languages'.

[4] M. Scriven, 'Definitions, Explanations, and Theories', p. 100; idem, *Reasoning*, p. xv.

[5] S. Toulmin, *The Uses of Argument,* pp. 252–59.

[6] See, for example, D. Kalish and R. Montague, *Logic*, p. 3; Benson Mates, *Elementary Logic*, pp. 3–4; W. Salmon, *Logic*, pp. 1–17; I. Copi, *Introduction to Logic*, pp. 3–5.

[7] *Psychology of Reasoning*; in this chapter, subsequent references to this book will be made within the text by indicating the page number(s) in parenthesis.

[8] It might be objected that the second concrete example is more similar to the abstract one and rather unlike the first concrete one, since the second concrete example refers to sides of envelopes which allow for permutations similar to the cards, whereas for the journeys to Manchester of the first concrete example there corresponds no 'opposite'. This objection suggests that perhaps the structure of the post office problem is reflected in the formula

$$(x)\,(y)\,[Dxy \longrightarrow (Px \longrightarrow Qy)]\,,$$

where perhaps

Dxy: x and y are different sides of an envelope;

Px: side x is sealed;

Qy: side y has a 5d stamp on it.

The meaning of this would be that if an envelope is sealed on one side then it has a 5d stamp on the other side.

In reply I should begin by pointing out that the first concrete example (travel to Manchester) *can* be given a more complex structure, in spite of the fact that it makes no sense to speak of the 'opposite side' of a journey. I suggested such a more complex structure in my discussion above. My reason for focusing on the simpler formula was the mere possibility of the simpler interpretation, whereas I argued that a simpler formula would have misinterpreted the abstract situation. In short, whereas in the two concrete examples the simpler formulas do not do any injustice to the situation, in the abstract case the simpler formula does. Thus the relevant similarity between the two concrete cases remains, together with their common difference from the abstract example.

Ultimately, of course, this is due to the physical difference between the abstract and concrete situations. On the one hand, in the post office example, the physical fact is that an envelope can be sealed only one one side, and this reduces to four the number of real possibilities covered by the generalization "if an envelope is sealed on one side then it has a 5d stamp on the other side." (Moreover, there is no problem in the original hypothesis "if a letter is sealed, then it has a 5d stamp on it", deriving from the possibility that the stamp and the sealing might be on the same side.) Similarly, in the travel example, let the generalization be expressed by saying: "for any given journey accomplished by a given means of transportation, if the destination is Manchester then the transportation is by train". The apparent possibilities involve the following pairs:

<Manchester, train>, <Manchester, car>, <Leeds, train>, <Leeds, car>, <train, Manchester>, <car, Manchester>, <train, Leeds>, and <car, Leeds>. Here the first four entities are not really (physically) different from the last four, respectively.

This answer to the initial objection might now lead to another, namely that the difference between the two concrete and the one abstract hypotheses is not one of logical structure, but one of background assumptions, relating to the the physical facts of the respective situations. This new objection is true in one sense, but not in the sense required to make it valid. For I am not claiming that the difference between the two types of hypotheses is merely one of *syntactical* structure, but rather that the syntactical complexity of the *relevant* symbolic formulas is one way of discussing the logical differences that has explanatory power; moreover, I argue below that another important element in explaining the experimental differences is that the meaning of the abstract hypothesis is contextually ambiguous in a way in which that of the concrete one is not.

CHAPTER 12

THE RHETORIC OF LOGIC AND
THE LOGIC OF RHETORIC:
CRITIQUE OF THE NEW RHETORIC

If our critique of the psychology of reasoning is correct, works and texts like the Galilean ones examined earlier provide more adequate *data* for a theory of reasoning than that provided by the usual controlled experiments done by psychologists. Moreover, since what we want is a *scientific* theory of reasoning, there is no reason why we should not avail ourselves of the scientific methodology worked out earlier, when we discussed the methodological content of Galileo's science (Chapter 5), the essential characteristics of his methodology (Chapter 6), and the problem of scientific rationality (Chapter 8). Though that methodology does not provide us with simple recipes and rules to follow, it does provide us with a model to emulate and a basic idea to follow. The idea is to isolate or define a number of relevant concepts or categories, and while thinking in their terms to practically exercise our judgment on the concrete materials being studied. For a theory of reasoning, the material is of course arguments, and the relevant categories are such potentially conflicting polarities as: logic and history, logic and psychology, the conceptual and the empirical, logic and rhetoric, the abstract and the concrete, the formal and the contextual, the theoretical and the practical, the descriptive-explanatory and the prescriptive-evaluative, the general and the particular. Having already examined the first few of these, let us go on to logic and rhetoric.

LOGIC AND RHETORIC

Formal logic is primarily the study of mathematical reasoning, or to be more exact, of those aspects of reasoning which predominate and acquire their clearest expression in mathematics. This is a fact that has always been explicitly recognized by the best logicians.[1] It is also something which is usually presupposed in an implicit manner by those who claim to want to deal with everyday reasoning.[2] The fact has also provided the basis for a very cogent though nonconstructive diatribe against formal logic.[3] The creators and followers of the so-called new rhetoric[4] have followed a more constructive approach by working out the details of an alternative theory of reasoning, centering around those aspects which predominate in legal arguments.

It is not clear what the relation between formal logic and the new rhetoric is or ought to be. Perhaps the two are parallel studies of different particular domains of reasoning, formal logic dealing with mathematical arguments, the new rhetoric dealing with legal arguments (and/or everyday arguments, and/or evaluative arguments). Perhaps the two are complementary studies of distinct aspects of all reasoning, formal logic dealing with the techniques of establishing the *truth* of propositions, the new rhetoric dealing with the linguistic techniques of securing the acceptance of propositions (either inducing or increasing adherence to them). A third possibility is that formal logic is the special case of the new rhetoric where the only way of securing acceptance is to establish the truth of the proposition.

In spite of this ambiguity of purpose one cannot but applaud the efforts of Perelman's 'new rhetoric' and of his associates and followers. They are correct in emphasizing the danger that the exclusive reliance on the formal logic of reasoning has the irrationalistic tendency of leaving most of the human sciences and of human affairs in the realm of the arbitrary and capricious. They are also correct in mentioning the humanistic value of the emphasis by the new rhetoric on the noncompelling, nonnecessitating (but not non-existing) force of argumentation, which leaves the door open for a sound notion of 'free will'. Third, very commendable is their desire to take the study of rhetorical problems out of the hands of propaganda ministers, Madison Avenue men, public relation experts, and preachers from the pulpit, who are the only groups who have in recent times studied these problems.

Nevertheless an attempt must be made to resolve the ambiguity for the following reasons. If we have two separate, parallel fields then in order to decide whether to use formal logic or the new rhetoric for the understanding of a concrete instance of reasoning, one would have to decide whether one is faced with a demonstration (deduction) or with an informal argumentation. In other words we would have to apply whatever concepts of demonstration and of argumentation had been arrived at as a result of the two different studies; ultimately such applications would reduce to determining whether the reasoning under investigation was more similar to mathematical or to legal reasoning. No doubt, in many cases there would be no problem with such a determination; however, if rhetoric and logic are essentially points of view which though distinct can be applied to the same material (as the second alternative above suggests), then the determination could be expected to run into trouble, because in this case the demonstration/argumentation question would have to be decided in terms of the *propriety* of adopting one of the two points of view (with the understanding that the other point of view would *also* be possible), rather than in terms of which properties are present in the material. If the third of the above alternatives is the correct one, then the question would have to be decided in a correspondingly different way. Thus the application of the new rhetoric depends on how its relation to formal logic is conceived since most actual reasoning contains a mixture of elements of deduction and argumentation (and other things besides).

A culturally, scientifically, and historically important illustration of these problems relates to Galileo's *Dialogue Concerning the Two Chief World Systems*. Was the book an attempt to establish the truth of the earth's motion, or an attempt to induce or increase the acceptability of this idea in the minds of Galileo's contemporaries? If the former, the book was a failure, if the latter a success. And to which activity, demonstration or argumentation (or both?) did the Church's condemnation of Copernicanism of 1616 refer? At any rate, does not this controversy show that the propriety of treating something as the subject of demonstration or argumentation is historically dependent, since what can only be treated rhetorically at one time may later become the subject of demonstration? Moreover, since argumentation is proper, indeed necessary, when something is incapable of proof, it would seem that the *theory* of argumentation presupposes (at least in these cases) the *theory* of proof. In other words, for mixed cases of reasoning, which are the typical cases, rhetorical analysis can only come after logical analysis. Let me explore this idea by discussing the rhetorically all-important concept of begging the question.

The concept of begging the question (*petitio principii*) is a very important one in the new rhetoric because it is the best example of a procedure which is unobjectionable from the point of view of formal logic, but necessarily faulty from the point of view of inducing or increasing assent.[5] In view of this importance, it is unfortunate that Perelman does not give a better example of begging the question than the following one taken from Navarre, who in turn was following Blass, involving a passage in Antiphon's speech on the murder of Herodes: "I would have you know that I am much more deserving of your pity than of punishment. Punishment is indeed the due of the guilty, while pity is the due of those who are the object of an unjust accusation."[6] Perelman interpets this[7] as an argument with the first sentence as conclusion, the second sentence as major premise, and the missing minor premise "I am the object of an unjust accusation", which cannot be accepted by the judges before deciding the case. This is rather inadequate because the passage could also be interpreted as consisting of two different ways of stating the same conclusion, stated one way and explicitly in the first sentence, and stated with a different emphasis and implicitly in the second sentence; and there is no direct evidence in the passage (such as reasoning indicators) to favor either interpretation. Moreover, even if we accept Perelman's interpretation, it should be noted that the missing premise is not identical with the conclusion; hence if Antiphon proceeds to give an argument supporting that missing premise, and if this argument is independent of the conclusion in Perelman's reconstruction, then there would be no begging of the question. In other words, the passage quoted by Perelman could be the last step of a long argument which does not assume the proposition that "I am much more deserving of your pity than of punishment". Of course, in order to determine this condition we would have to look at the rest of the speech; so I am not concluding that Perelman's interpretation is wrong, but rather that it is not justified by the passage he presents, and hence that the passage is not a good example of what he is trying to illustrate. Unfortunately, this is a general problem characterizing Perelman's most important book (*The New Rhetoric*): though a great merit of this book is that the examples and materials mentioned are both more numerous, substantive, and realistic than the artificial trivialities found in most logic books, nevertheless the passages are usually taken out of context and the propriety and strength of their illustrative power depends on assumptions which are not made explicit and which may or may not be accurate. It follows from this that Perelman's approach would be improved if his discussions were centered around different portions of one and the same work. Of course it is not easy to find a single book rich enough to provide illustrations for and ground discussion of all the concepts in the 'new rhetoric'. However, there exists such a book, and it is Galileo's *Dialogue*.

Besides being excessively abstract in the sense just specified, the new rhetoric may be improperly abstract in the sense of suggesting that rhetorical analysis is independent of logical analysis. If we again consider Perelman's example of begging the question, it seems obvious that the rhetorical judgment that a begging of the question is being perpetrated presupposes some kind of logical analysis, either the one given by Perelman himself, or an expanded one along the lines I suggested. So, in any case, in order to understand the general concept of begging the question, or in order to explain a given instance of it, one needs beforehand to rely on concepts of formal logic, such as premises implying a conclusion. In other words, the very definition of begging the question is full of formal-logical ideas: an argument whose (formal) validity requires it to have as a premise a proposition which is either identical with its conclusion or supportable only by means of an argument having that conclusion as a premise. This is not to deny that the importance of such arguments is a rhetorical one; but what this means is that only when they are examined from a point of view different from formal logic, can one see their ineffectiveness and why they are to be avoided. Thus, if the aim is to induce or intensify adherence to a certain proposition, then one cannot accomplish this by means of any process which at some stage before the last assumes that the adherence has taken place.

In summary then, we may agree with Perelman that the concept of begging the question is *the* place to start to begin to understand the differences between formal logic and the new rhetoric, as well as the importance of the latter. At the same time I feel that it is also the place to start to see the limitations of the new rhetoric, and how these limitations may be overcome; the limitations center around its abstractness in the double sense of separateness of rhetorical analysis from logical analysis and of the analyzed material (reasoning) from its context. Far from being separable from logical analysis, rhetorical analysis presupposes it though the two are (conceptually) distinguishable and distinct; moreover, it is precisely in rhetorical analysis that the (linguistic) context of reasoning acquires paramount importance. A beautiful illustration of these ideas can be given with an argument contained in Galileo's *Dialogue,* and the illustration will suggest the importance of this book as material for theorists of reasoning.

A QUESTION-BEGGING ARGUMENT

The argument is Galileo's tower argument, discussed in previous chapters for its relevance to the understanding of scientific rationality and other matters. Those discussions were relatively cursory since it was sufficient then to suggest the broad outline of Galileo's reasoning. Here, however, we shall need a more detailed analysis since we are attempting to develop a rigorous, empirical, historical approach to the study of reasoning, and so we need to experience and discover what it is really like to do this sort of investigation.

The general context of Galileo's situation is the question of whether or not the earth rotates. The argument from vertical fall was one of the classical objections to the earth's motion. The portion of dialogue relevant to our purposes here is the passage consisting of the quotations cited in Chapter 8. Referring to that passage, we find that the first step of the original Aristotelian argument can be reconstructed as follows:

(1)   If the earth moved, the place of ejection of a body thrown vertically upward would move along with it and the body would fall some distance west of that place.

(2)   Therefore, if the earth moved, bodies would not fall vertically.
(3)   But bodies do fall vertically.
(4)   Therefore, the earth does not move.

Next Salviati asks Simplicio how he knows that bodies fall vertically. Simplicio answers that we know it because they are *seen* to fall vertically. This argument by Simplicio is the following:

(5)   Bodies are seen to fall vertically.
(3)   Therefore, bodies do fall vertically.

Now comes the crucial step. By what some would call a thought experiment, but is really a brilliant piece of hypothetical reasoning, Salviati gets Simplicio to recognize that

(6)   *if* the earth moved *and* bodies were *seen* to fall vertically, then they would (actually) *not* be falling vertically.

From this Salviati, with Simplicio's approval, infers that

(7)   if one does not assume that the earth stands still, then one cannot conclude that bodies would actually fall vertically from the mere fact that they are seen to fall vertically.

This justification of (7) with (6) is repeated later in Sagredo's speech. Hence we must regard as a partial restatement of (6), the reason for (7) given by Simplicio:

(8)   if the earth moved, then bodies would (actually) *not* be falling vertically.

In making Simplicio say this, Galileo means that if the earth moved, then (since bodies are seen to fall vertically) they would actually not be falling vertically. The following argument elucidates the sense in which (8) is a partial restatement of (6) and must be what Galileo has in mind.
    Letting

M = the earth *moves*
S = bodies are *seen* to fall vertically, and
V = bodies do actually fall *vertically*,

the reasoning could by symbolized as follows:

(6)   If M and S, then not-V.
(6a)   Therefore, if S and M, then not-V.

(6b)   Therefore, if S, then if M then not-V.
(5)   S. (This is obvious, since bodies are seen to fall vertically.)
(8)   Therefore, if M, then not-V.

Transformations of this sort also shed some light on what may have been in Galileo's mind in inferring (7) from (6). Using the same abbreviations:

(6)   If M and S, then not-V.
(6.1)   Therefore, if M, then if S then not-V.
(6.2)   Therefore, if M, then not-(if S then V).[8]
(6.3)   Therefore, if not-(not-M), then not-(if S then V).
(6.4)   Therefore, if one does not assume not-M, then one cannot conclude V from S.

Finally equating, 'not-M' with 'the earth stands still' and substituting for V and S, (6.4) gives (7). The crucial and problematic step is that from (6.3) to (6.4). For the moment, suffice it to say that (6.4) *expresses in rhetorical language what (6.3) says in logical language.*

It is clear how from (7) Salviati reaches his next conclusion to the effect that

(9)   the Aristotelian argument begs the question by assuming the very same thing it is trying to prove.

In inferring (3) from (5), the Aristotelian argument is concluding that bodies would actually fall vertically from the mere fact that they are seen to fall vertically. In the light of (7), it follows that it is being assumed that the earth stands still (doesn't move), which is the argument's conclusion (4).

What is not so clear is why Salviati does not give this simple explanation to Simplicio when he asks for a justification of (8). The way Salviati expresses himself is the following:

(10)   The conclusion of the argument – that the earth doesn't move – ought to be unknown, for otherwise it would be superfluous to give the argument.
(11)   The middle term premise – that bodies fall vertically – ought to be known, for otherwise one would be proving the unknown by means of the equally unknown.
(12)   (But it has already been argued that) if one does not first know that the earth doesn't move, one cannot know that bodies fall vertically.
(13)   Therefore, the certainty of the premise derives from the uncertainty of the conclusion.

(14)   Therefore, the argument is a *petitio principii* and seriously faulty.

The problem with this argument is (13); one might be inclined to be very literal and claim that it neither implies (14) nor is implied by (10), (11), and (12). But the real problem is its obscurity. In this context the word 'certainty' means 'knowledge', so it might seem that (13) means that knowledge of the premise derives from *lack* of knowledge of the conclusion. Though this does not make sense, it is a step toward the correct interpretation. For what seems to be happening is that Galileo is carried away by the stylistic brilliance of the expression, "the certainty of the premise derives from the uncertainty of the conclusion". The logic of the situation is such that the term 'certainty' cannot be performing a function analogous to that of the term 'uncertainty'. The sentence is best interpreted as saying that knowledge of the premise derives from the conclusion, which is supposedly unknown. This follows plausibly from (10), (11), and (12), and in turn we can see that, by the definition of *petitio principii,* the argument is one because the premise depends on the conclusion. And we can see that one fault with *petitio principii* lies in its contradictory epistemological presuppositions, in our case that *knowledge of* one proposition derives from another *unknown* proposition.

After this explanation of the nature and seriousness of the paralogism in the Aristotelian argument, Sagredo, with the help of Simplicio, objects to this criticism, especially to the justification of the crucial (7), which is practically synonymous with (12). The objection is basically that (6), the reason for (7), hides a contradiction in its antecedent, and that a new argument for the motionlessness of the earth can be formulated:

(15)   If the earth moved, then it would be impossible that bodies should be seen to fall vertically.

(5)   But bodies are seen to fall vertically.

(4)   Therefore, the earth doesn't move.

And the justification for (15) is that

(16)   if the earth moved, then if bodies were seen to fall vertically, they would simultaneously have two motions — toward and around the center; and

(17)   it is impossible to have two such motions.

As Salviati implicitly admits, this revised argument is not open to the same objection as the original one. In particular, (16) is an immediate consequence of (6), indeed another way of expressing the same thought. What the

Aristotelian is doing at this point is to turn the tables around. Whereas, Galileo had used (6) to show the question-begging character of the original Aristotelian argument, the Aristotelian is using (6) to formulate a revised argument not beset by the same problem. Of course, his having done so is an implicit admission that the original argument was indeed question-begging.

As Salviati notes, the worth of the revised argument depends wholly on the alleged impossibility of motion being both toward and around the center (17). He finds three things wrong with this premise:

(18)   If Aristotle had (17) in mind when giving the original argument, he should have explicitly said so (since it is such an important part of the argument);

(19)   the mixed motion mentioned in (17) is not only not impossible, but necessary (something which Galileo promises to demonstrate later);

(20)   Aristotle himself would not accept (17), since he allows that kind of mixed motion when he admits that fire and some of the air move up by nature and rotate by participation.

(18) is obviously true, but it is not clear in what sense, if any, it makes the *revised* Aristotelian argument faulty. This question will be pursued below.

In the case of (19), it is clear how the argument would be affected: one of its premises would be false and it would be unsound (though formally valid). It is also clear that (19) is physically true, in the sense that such mixed motion is possible, though not necessarily 'necessary'. However, the relevant critical question would be whether (19) is true *in the context of Galileo's discussion,* that is, whether he does give the promised argument, and whether it is contextually sound. He does fulfill the promise, at least in part, in his argument dealing with the falling of bodies on a moving ship, and I believe the argument is basically sound. But these matters will not be pursued here.

(20) is partly true, because Aristotle does allow mixed motion for fire (*Meteor.,* 344all). But whether it is completely true, i.e. whether Aristotle would consequently allow mixed motion for falling bodies is questioned by Simplicio on the ground that

(21)   there is an important difference between the two situations: whereas particles of fire are so light that they could easily be carried along by the rotating air, falling rocks are so heavy that they would not be.

This objection is misconceived first because the question is the possibility, not the likelihood of such mixed motion, and second because the Aristotelian fire particles were supposed to be carried along not by the rotating air but by the rotating lunar sphere, which compared with them was as tenuous, or even more so, as air is compared to rocks and cannon balls. Thus Salviati's third criticism of the revised Aristotelian argument seems correct. Its effect is to make this argument faulty in the sense of un-Aristotelian.

The statement of (21) is followed by a discussion of the ship experiment as possible evidence for (17). This further discussion need not concern us here.[9]

EVALUATIONS

So far the main concern has been with the reasoning as found in the text. The evaluation of that reasoning involves primarily two questions. Is the revised Aristotelian argument faulty, and if so how, given that it is a revision of the original question-begging argument? And is the following argument valid and if so how: if one does not assume that the earth stands still, then one cannot conclude that bodies actually fall vertically from the mere fact that they are seen to fall vertically, because if it is not true that the earth stands still, then it's not true that if bodies are seen to fall vertically they actually fall vertically.

Let us pursue this second question first. It was noted above that the conclusion is simply a rhetorical formulation of the premise. When this is noted, the first thing that comes to mind is that Galileo might be using two principles relating the rhetorical and the logical situation:

Principle	Logic	Rhetoric
I	If p, then q.	If one assumes p, one can conclude q. I.e., one can conclude q from p.
II	If p is not true, then q is not true.	If one does not assume p, then one cannot conclude q.

Then, using the earlier abbreviations, one could start with

(6.3)   if not-M is not true, then it is not true that if S then V;

apply Principle II and get

(6.31)   if one does not assume not-M, then one cannot conclude that if S then V;

from which, by applying Principle I, we get

(6.32)   if one does not assume not-M, then one cannot conclude that one can conclude V from S;

which means the same as

(6.33)   if one does not assume not-M, then one cannot conclude V from S;

which is the desired conclusion (7).

The problem with this interpretation is that whereas Principle I is intrinsically plausible, II is not. Moreover, II can be shown to conflict with I. In fact, 'if p is not true, then q is not true' is equivalent to 'if q, then p'; and if we apply I to the latter we get 'one can conclude p from q', which differs from the result of applying II to the former, namely from 'if one does not assume p, then one cannot conclude q', since this is equivalent to 'one can conclude q only if one assumes p'.

Fortunately, there is a simpler explanation of Galileo's argument, using only Principle I. Start with

(6.2)   If M, then it's not true that if S then V.

Apply I to the main conditional to get

(6.21)   If one assumes M, then one can conclude that it's not true that if S then V.

Now apply I to the conditional in the consequent of (6.21), to get

(6.22)   If one assumes M, then one can conclude that it's not true that one can conclude V from S;

that is, combining the two occurrences of 'conclude',

(6.23)   If one assumes M, then one cannot conclude V from S.

Now it is obvious in the context that

(6.24)   If one does not assume not-M (that the earth stands still), then one is assuming M (that the earth moves).

Hence we get

(6.25)   If one does not assume not-M, then one cannot conclude V from S;

which is the desired conclusion (7). Thus the step from (6) to (7) in Galileo's reasoning seems valid.

Let us now consider the second evaluative problem, which is to determine whether the revised Aristotelian argument is somewhat faulty because it is a revision of the old one. The revised argument grounds the immobility of the earth on the alleged impossibility of motion being both around and toward the center; it is formally valid though unsound insofar as this premise is false and unjustified. The question is, what does it matter if Aristotle did not have the revised argument in mind when he gave the original one. Is it really improper to revise the original Aristotelian argument? The impropriety, if any, is clearly not a 'formal' one, any more than question begging is. It has to do with the way the argument originates. The Aristotelian (Sagredo acting as one) thinks of it when he clearly sees that if one infers actual vertical fall from seen vertical fall, as he is, then one is assuming that the earth stands still. Now, this was implied by the crucial hypothetical that if the earth moves, then if bodies are seen to fall vertically they do not actually fall vertically. So we might say that the Aristotelian makes the switch when he realizes the damaging consequences of that crucial hypothetical. What he does at this point is to draw from it a consequence that suits him, namely

(16)   if the earth moved, then if bodies were seen to fall vertically they would be moving with 'mixed' motion.

How does this consequence suit him? In the sense that according to it, the motion of the earth implies mixed motion for falling bodies, which he then quickly denies. The propriety of the revision then depends on the propriety of this denial, more specifically on whether or not the denial is arbitrary. Not whether it is true (absolutely or contextually), which relates to the second possible deficiency of the revised argument which Galileo discusses; and not directly whether the denial is un-Aristotelian, which relates to its third possible deficiency discussed above.

That denial would be arbitrary if it were arbitrarily introduced. And it would be arbitrarily introduced if it were un-Aristotelian, which it is in one sense. Hence the premise on which the argument depends seems arbitrary. The concept that this deficiency suggests is that of an *ad hoc* argument. This could be regarded as a kind of question-begging argument, one where a conclusion is supported by means of a premise which is as much in need of support as the conclusion. This kind of argument is referred to by Galileo in the passage, and it is described as "the attempt to prove *ignotum per aeque ignotum*", i.e. to prove the unknown by means of the equally unknown. The reference is made when it is taken for granted that the original Aristotelian argument is *not* an instance of this kind of question begging. Hence the original argument must be question-begging in a different, and presumably more serious, sense.

RECONSTRUCTIONS AND THEORETICAL CONSIDERATIONS

We shall now explore this other concept of question begging and simultaneously seek a deeper understanding of the original Aristotelian argument and of Galileo's criticism of it. With the above discussion behind us, we shall now feel more free to engage in abstractions and reconstructions. In addition to the previous abbreviations, we shall use the following:

W = bodies ejected vertically upward land some distance *west* of the place of ejection;

X = falling bodies move with mixed motion which is simultaneously toward and around the center.

One version of the Aristotelian argument can be reconstructed as follows:[10]

(1−2221)		If not-M and S, then V.
(1−222)	∴	If not-M, then if S then V.
(1−221)		Not-M.
(1−22)	∴	If S, then V.
(1−21)		S.
(1−2)	∴	V.
(1−11)		If M, then W.
(1−1)	∴	If M, then not-V.
(1−0)	∴	Not-M; from (1−2) and (1−1).

The last four lines would constitute the initial segment of the argument. (1−21) and (1−22) constitute the justification of V. Since (1−22) is not true if M is, it needs a justification, which can only be provided by (1−221) and (1−222). The ultimate premises on which the final conclusion, not-M, rests are (1−2221), (1−221) (1−21), and (1−11). Unfortunately one of them, (1−221), is identical with the conclusion. The argument is question-begging in the sense of being circular, i.e., using the conclusion as one of the premises. The role of the crucial consideration 'If M and S, then not-V' is to motivate the justification of (1−22) and thus force the Aristotelian to use (1−222) and hence not-M as grounds.

This argument, however, is not the same as the original one, given in

Salviati's first speech, and containing only the last four lines of this reconstruction. It is only by logical analysis that the speakers can construct this deduction from the original segment; since its rhetorical uselessness follows (rather obviously) from the structure of this deduction, that shows that the rhetorical analysis is being grounded on the logical one. It follows that one cannot carry out the former in abstraction from the latter (though, to repeat, two *distinct* points are being made in the two points of view). Moreover, it is also obvious that there is no way of identifying the begging of the question by just looking at the initial segment of the argument '(1–11), ∴ (1–1); but (1–2); ∴ (1–0)'; that is, by just looking at Salviati's first speech in the Galilean text under consideration; one also needs the other parts of the passage, which provide evidence that this segment is really part of the longer one.

How much context need one provide in rhetorical analysis? After all, in Galileo's book the discussion does not end where our quotation ends. I think the answer must be that one needs as much context as is needed to justify the rhetorical point one is making. Though some passages may be too short to allow any rhetorical analysis to get started, the present passage is obviously sufficient since it provides a good illustration of a central concept of the new rhetoric, the concept of *petitio principii*. Further insight into this concept can be gained as follows.

Galileo does not ask the Aristotelian why (1–11) is supposed to be true and exactly how it implies (1–1). Let *us* pursue these questions, beginning with the latter. First the Aristotelian would probably say:

(1–12)   If W, then not-V.
(1–11)   If M, then W.
(1–1)   ∴ If M, then not-V.

The answer to 'why (1–12)?' would probably be:

(1–122)   If W, then not-S.
(1–121)   If not-S, then not-V.
(1–12)   ∴ If W, then not-V.

(1–122) would be obvious. But, why (1–121)? Probable answer:

(1–1211)   If V, then S.
(1–121)   ∴ If not-S, then not-V.

Why (1–1211)? Because:

(1–12112) If not-M, then if V then S.
(1–12111) Not-M.
(1–1211) ∴ If V, then S.

Thus the justification of (1–12) would contain not-M as a premise. This would indicate a *second* circularity in the original argument.

Now, is (1–11) true? The Aristotelian would say so because:

(1–112) If V, then if M then W.
(1–111) V.
(1–11) ∴ If M, then W.

The first premise here is obvious, but in order to justify the second he would probably give the same justification as for (1–2), thus making the original argument trebly circular.

It should be noted that both this third circularity and the first are the result of the attempt to justify V empirically, by grounding it on S. But the Aristotelian might resort to an *a priori* justification. He could argue, for example, that bodies really fall vertically because it is their nature to move in a "real" straight line toward the center of the earth. This would be circular insofar as 'actual vertical fall' here means "actual rectilinear motion toward the center of the earth". Insofar as there is a reference to the 'nature' of bodies, we would have a disguised refusal to give a justification. Hence the final conclusion would be resting on a proposition equally in need of support.

The Aristotelian could also argue that bodies actually fall vertically because it is their nature to move in a 'real' straight line toward the center of the universe, and this center coincides with the center of the earth. Then Galileo would question this second premise. The Aristotelian could support it *a priori* or *a posteriori*. The empirical argument, as Galileo shows elsewhere[11] is itself circular; the *a priori* argument would presuppose that the earth stands still and so would render circular the original objection to the motion of the earth.

The revised Aristotelian argument is the following:

(2–2122) If M and S, then not-V.
(2–2121) If not-V, then X.
(2–212) ∴ If M and S, then X.
(2–211) Not-X.
(2–21) ∴ Not-(M and S).
(2–2) ∴ If M, then not-S.
(2–1) S.
(2–0) ∴ Not-M.

The basic problem here is that (2—211) is as much in need of support as (2—0). The argument begs the question not in the sense that its conclusion is one of the ultimate premises, but in the sense that among its premises there is a proposition which is no better known than the conclusion.

A rhetorical question would immediately arise already at the stage where this new argument is given, namely, how this argument (2) is related to the original (1). From a logical point of view they are different arguments; however, rhetorically speaking, they are really two versions of the same ambiguous argument. The ambiguity in this root argument derives from the confusion of apparent vertical fall and actual vertical fall, which ambiguity is present in Aristotle's original statement of the argument and which is retained even in Salviati's statement in his first speech in the passage. So one of the things that Galileo accomplishes in this section of his book is to distinguish these two meanings of vertical fall. It also follows that what we have here is also a beautiful illustration of what Perelman regards as "another difference of paramount importance between [rhetorical] argument and formal proof. The standard logical calculi are formulated in artificial languages in which any one sign has one, and only one, meaning; in natural languages the same word often has different meanings."[12] Though Galileo is not using any formal calculus, he is using logical analysis to strip Aristotle's original 'rhetorical' argument of some of its 'rhetoric'. That is, Galileo is trying to turn the problem at hand from one where only rhetorical analysis would be relevant to one suceptible to some logical analysis. Of course, Galileo is not eliminating completely the rhetorical aspects of the problem, nor would he want to; but he is using logic so that the rhetorical problem is moved to a different point. This means that the line of separation between rhetoric and logic shifts as knowledge grows; or to be more exact, more and more arguments enter the domain of logic after crossing the borderline with rhetoric; or, again, since I believe that the two are not separate domains but distinct aspects, what happens is that there is a growth of problems of the kind where the question of adherence becomes reduced to the question of establishing truth.

Let us now consider some arguments which explore the consequences of the most important propositions which are indisputable in this context:

(3—1)   If M and S, then not-V.
(3—2)   If not-M and S, then V.
(3—3)   If M and V, then not-S.
(3—4)   If not-M and V, then S.

(3—5)   S.
(3—6)   If S and M, then not-V; from (3—1).
(3—7)   If S, then if M then not-V; from (3—6).
(3—8)   If M, then not-V; from (3—7) and (3—5).
(3—9)   If V, then not-M; from (3—8).
(3—10)   If S and not-M, then V; from (3—2).
(3—11)   If S, then if not-M then V; from (3—10).
(3—12)   If not-M, then V; from (3—11) and (3—5).
(3—13)   V if and only if not-M; from (3—12) and (3—9).
(3—14)   If M, then if S then not-V; from (3—1).
(3—15)   If M, then not-(if S then V); from (3—14).[13]
(3—16)   If (if S then V), then not-M; from (3—15).
(3—17)   If not-M, then if S then V; from (3—2).
(3—18)   (If S then V) if and only if not-M; from (3—17) and (3—16).
(3—19)   If M, then if V then not-S; from (3—3);
(3—20)   If M, then not-(if V then S); from (3—19).
(3—21)   If (if V then S), then not-M; from (3—20).
(3—22)   If not-M, then if V then S; from (3—4).
(3—23)   (If V then S) if and only if not-M; from (3—22) and (3—21).
(3—24)   Not-M if and only if (S if and only if V); from (3—23) and
           (3—18).
(3—25)   M if and only if not-(S if and only if V); from 3—24.

(3—1) is Galileo's crucial consideration, his 'thought-experiment'. (3—2),
(3—3), and (3—4) are obvious truths, which the Aristotelians themselves had
implicitly or explicitly accepted. (3—5) is an empirical fact.

(3—13) tells us that to say that falling bodies actually fall vertically is
equivalent to saying that the earth stands still. Hence it seems difficult to
avoid circular reasoning as long as one grounds the immobility of the earth
on actual vertical fall. Besides allowing Galileo to criticize the original
Aristotelian argument, (3—13) is useful to him as a step toward his ultimate
aim of justifying the motion of the earth. For he now knows that to do this
he has to deny that bodies actually fall vertically. His formulation of the
principle of inertia[14] is a step toward the justification of that denial.

(3—18) is a somewhat weaker assertion than (3—13). It may be interpreted
as saying that to conclude actual vertical fall from visible vertical fall is
equivalent to assuming the immobility of the earth. Hence, as long as one
gives that kind of empirical justification of actual vertical fall when justifying
the stability of the earth, one is arguing in a circle. It should be noted that

the step from (3—14) to (3—15) cannot be invalidated by the theory of material conditionals since this theory is irrelevant here. Or, to be more exact, one is assuming the principle that

$$\text{if (if S then not-V) then not-(if S then V)}$$

which is truth-functionally invalid, but rhetorically sound, and likely to be linguistically true of the natural language 'if-then'.[15]

(3—24) tells us that to say that the earth stands still is equivalent to regarding S and V as equivalent. Hence, justifying not-M involves a failure to distinguish S from V. And (3—25) tells us that to distinguish apparent from actual vertical fall is to assert that the earth moves. Now, the first thought that comes to mind is to use these facts to prove the motion of the earth by arguing that since it is one thing for bodies to be seen to fall vertically and another for them to actually fall vertically, it would follow that the earth moves, by (3—25). Hence Galileo could be claimed to have, with his criticism of the Aristotelian argument, not merely refuted the argument, but also its conclusion, i.e. proved the motion of the earth. The fact that Galileo did not claim so much should make us suspect that there is something wrong with this suggestion.

The problem with the suggestion must be that the distinction between S and V is not subsisting in a Platonic heaven, but is something dependent on M, an empirical fact. The consequences of this are somewhat surprising. For it would seem that the nonequivalence between S and V is a conceptual matter, accessible by pure thought. But if this were so one would be able to prove the motion of the earth in the manner indicated, which seems an invalid procedure. Hence the distinction between S and V is either not conceptual, or some conceptual distinctions are empirical at least in the sense of having an empirical origin. The last alternative seems the correct one. In other words, the use and misuse to which (3—25) could be put suggests for us an asymmetry between empirical and conceptual knowledge: empirical knowledge can generate conceptual knowledge, but not vice versa.

In conclusion we may say that the approach to the study of reasoning suggested by the new rhetoric promises to be a very fruitful one if we distinguish properly between rhetorical and logical considerations, if we keep in mind the primacy of the latter, and if proper attention is paid to the details of actual reasoning, for example by the careful study of texts such as Galileo's *Dialogue*. It has also emerged that the concept of question-begging is indeed the one with which to begin for a proper appreciation of the distinctions *and* of the relations between logical and rhetorical analysis. Finally, the logical which is primary here does not refer without qualification to orthodox formal logic, which is viewed as *a theory* of reasoning, but to the activity of reasoning itself.

NOTES

[1] See, for example, A. Tarski, *Introduction to Logic,* pp. xiii-xiv.

[2] See, for example, D. Kalish and R. Montague, *Logic.*

[3] S. Toulmin, *The Uses of Argument.*

[4] C. Perelman and L. Olbrechts-Tyteca, *The New Rhetoric*; C. Perelman, *The Idea of Justice and the Problem of Argument*; idem, 'The New Rhetoric'; idem, 'A Reply to Henry W. Johnstone, Jr.'.

[5] Perelman, 'New Rhetoric', p. 146; idem, *New Rhetoric*, pp. 110–14.

[6] Perelman, 'New Rhetoric', p. 146; idem, *New Rhetoric,* p. 113.

[7] *Ibid.*

[8] This inference from (6.1) is problematic since if the 'if-then' in these sentence schemata is interpreted as a 'material conditional' then (6.2) is not a logical consequence of (6.1). The full resolution of this problem would 'entail' writing a book like A. R. Anderson and N. D. Belnap's *Entailment,* or like E. W. Adams's *The Logic of Conditionals,* which the reader is advised to study before making superficial objections. Here the following comments will have to suffice. Partly I would want to say that I am reconstructing Galileo's reasoning, and if it is invalid, that is not the fault of my reconstruction; and if one objects that I ought to use the Principle of Charity, I can assure him that I have already done so, and I challenge him to find a more accurate reconstruction. Partly I would say that there is no good reason why the 'material' interpretation of the 'if-then' should be accepted here, that I have deliberately not symbolized it by means of the 'horseshoe' symbol but retained the more vague natural language expression. Partly I would say that perhaps the 'if-then' of these schemata should be interpreted as formal implication; then 'not-(if S then V)' would mean that S does not formally imply V, which does follow from 'S formally implies not-V', under the plausible assumption that we are dealing with consistent sentences. Partly one could try to solve the problem by adapting certain ideas from Adams's *Logic of Conditionals:* that it is more proper to conceive of conditionals as having probabilities rather than truth values (pp. 69–102); that their probability is *not* the probability of being true but the ratio of the probability of the conjunction of antecedent and consequent to the probability of the antecedent (pp. 5ff.); and that inferences involving conditionals are best evaluated in terms of the probabilistic soundness criterion that it should be impossible for the premises to be probable while the conclusion is improbable (p. 1). Partly I would say that I am advocating a practice-oriented logic or theory of reasoning, and thus it is logical principles that are to be tested by and derived from appropriately selected actual reasoning, and not the other way around. Partly I would refer the reader to the comments I make when this and similar problematic inferences arise again, e.g., at the end of this chapter, and in Sections 4, 8, and 12 of Chapter 16.

[9] For details on this, see Chapter 5 and Chapter 16, Section 10.

[10] The numbering of the propositions in the two main arguments below follows the numbering system introduced in Chapter 14. But nothing here depends on such labeling.

[11] G. Galilei, *Dialogue,* tr. Drake, pp. 34–37; cf. Chapter 16, Section 3 below.

[12] Perelman, 'New Rhetoric', p. 144.

[13] For the problematic character of this step and of the one in (3–20), see comments below and Note 8 above.

[14] Galilei, *Dialogue,* tr. Drake, p. 147.

[15] See Note 8 above.

CHAPTER 13

THE LOGIC OF SCIENCE AND THE SCIENCE OF LOGIC: TOWARD A SCIENCE OF REASONING

We have seen that neither formal logic, nor the psychology of reasoning, nor the new rhetoric, offer completely sound approaches to the science of reasoning, though each has something valuable to contribute. Formal logic is rigorous and systematic but *formal*, i.e., it is no empirical science; the psychology of reasoning is sound insofar as it is empirical, but its experimental method has intrinsic limitations and needs to be supplemented, and perhaps replaced, by an observational, historical, introspective approach; the new rhetoric is valuable for the generality of its aim but needs to be made more concrete, vis-à-vis both the materials of reasoning and logical analysis. Let us continue our critical review of available approaches.

INDUCTIVE LOGIC

Inductive logic is defined by some as the study of logical probability, or probable inference, or the degree of confirmation which the premises lend to the conclusion of an argument, with formal validity as the special case where the logical probability of the conclusion on the basis of the premises is one. Others define it as the study of those types of arguments that are most common in the empirical sciences, and hence from this point of view inductive logic is doubly relevant to our inquiry since it could offer us both a logic of science and a science of logic, that is both a theory of scientific reasoning and a scientific theory of reasoning. Because of this presumed generality, and because of this emphasis on nonmathematical reasoning, inductive logic may appear to provide the sort of thing we are looking for in our attempt to develop a science of reasoning, and hence it deserves to be examined. Moreover, the question arises of how inductive logic relates to the new rhetoric since one of the central features of the argumentation studied by the latter is, as Perelman himself states, that it is strong or weak, more or less convincing, rather than valid or invalid, that is, rather than conclusive in an all-or-none fashion, as is the case with formal proof.

Most recent work in inductive logic is too interdependent on formal logic, the mathematical theory of probability, and statistics.[1] Mostly for this reason, it is neither sufficiently empirical nor sufficiently concrete. It does not deal significantly and realistically with actual scientific reasoning, which is alleged to be its central subject matter. Rather it theorizes ad infinitum about artificial and trivial arguments, such as simple induction by enumeration, which seldom occur in the actual life of science. Nevertheless, if one examines the more elementary parts of the subject,[2] it appears to contain the seeds for remedying the insufficient generality of formal logic. Whether this appearance corresponds to reality, is the question presently to be answered.

The fundamental problem haunting elementary inductive logic is nothing less than whether there are such things as inductive arguments for it to study. For one thing, there are the Popperian objections.[3] Popper is convinced no one has succeeded in solving Hume's problem of the justification of induction, which is the question of whether one is justified in reasoning from experienced instances to other instances not yet experienced.[4] Most philosophers would probably agree that induction has not been justified.[5] Popper then argues as follows: since it is undeniable that humans learn from experience, and since because of Hume's arguments people could not learn inductively from experience, it follows that induction is seldom if ever used in reasoning; that is to say there is no such thing as inductive reasoning, and hence nothing for inductive logic (as usually conceived) to study. Coupled with this destructive criticism Popper makes the constructive suggestion that the important task is to determine how people actually learn from experience. He then adds that in science one finds the best and most explicit instances of learning from experience, so it is these that deserve careful study. Now, though Popper and his followers have practiced some of these preachings,[6] their results have been far from solid.[7] A basic flaw is that, because of their underlying empiricism, they have paid attention mostly to epistemological rather than logical matters; insofar as they have examined the latter, they have been unduly constricted by their falsificationism and fallibilism. In short, they have not paid careful enough attention to the reasoning found in actual, historical science. However, the general tenor of Popper's approach to scientific ('inductive') reasoning is acceptable: what is needed is an analysis, free of inductivist preconceptions, of the actual forms of reasoning in science. In conformity with such a program, the analysis of the arguments in a book like Galileo's *Dialogue* would seem to be very worthwhile.

Aside from this Popperian anti-inductivist criticism, unfortunately the problem of the existence of inductive arguments can be seen by remaining within the confines of Wesley Salmon's *Logic*,[8] and Salmon is certainly no foe of induction. This book consists of two main parts dealing respectively with deduction and with induction; the part on deduction is a standard elementary treatment of formal logic; the other part may be taken to deal with inductive logic. This part contains an unresolved ambiguity, namely whether inductive logic is the study of inductive arguments or the study of inductive correctness. On the one hand, he discusses a number of types of arguments which everyone would agree in regarding as instances of inductive arguments, for example, induction by enumeration, statistical syllogism, argument from authority, analogy arguments, and causal arguments;[9] however, he never formulates or makes use of a general definition of inductive argument. What looks like such a definition is really a definition of *correct* inductive argument, that is, an argument such that if all the premises are true then the conclusion (1) is probably true but not necessarily true, and (2) contains information not present even implicitly in the premises.[10] However, he never *shows* that any of the various types of admitted inductive arguments are inductively correct in this sense, specifically that they satisfy condition (1). Of course, he has intuitively plausible things to say about various faults that may be present in inductive arguments and various ways of remedying these faults, but there is no proof or even argument that when these inductive arguments are in proper form, their conclusion is rendered probable by the premises. Nor is this due to the elementary nature of the book, because advanced texts and treatises do not do this either; that is, they do not show that the probabilistic definition of inductive correctness properly describes actual arguments, but rather merely that it can be made to describe various formal structures, whose relation to actual arguments is extremely problematic. In short, if inductive logic deals with arguments definable as inductive, then we don't know that it deals with anything real because we have not even been told when an argument is so definable, whereas, if inductive logic deals with the inductive correctness of arguments, then we don't know that any real argument has the property as defined (ultimately because of Hume).

Moreover, it is a fact, admitted by Salmon himself,[11] that any 'inductive' argument can be transformed into a 'deductive' one. Salmon dismisses this problem by saying that the additional premises in such a transformation are usually problematic as regards their acceptability, and that hence the transformation does not increase the overall argument *soundness* (which is a function of the truth of premises and of the logical connection between them and the conclusion). This argument, however, should be *irrelevant* to Salmon, from his own point of view, according to which logic deals mainly with the premise/conclusion connection, and not with the truth of premises. Hence, from the latter point of view, the distinction between induction and deduction evaporates.

It is possible, however, to make a constructive elaboration of Salmon's *Logic*. We could define inductive logic to be the study of inductive arguments, and then define the latter to be arguments having an *inductive form*. An inductive form would then be characterized by a semi-ostensive definition, in terms of the various types of argument discussed by Salmon in his chapter on induction, or appropriate modifications thereof. So an inductive argument would be one having one of the following forms, or *a form like them*:

(1) Induction by enumeration: Z% of the observed members of F are G.
∴ Z% of F are G.

(2) Statistical syllogism: Z% of F are G.
x is F.
∴ *x* is G.

(3) Argument from authority: *x* asserts *p*.
∴ *p*.

(4) Argument against the man: *x* asserts *p*.
∴ Not *p*.

etc.

Now, the clause 'or a form like them' would make this definition open-ended and would invite research to discover other forms like these. Some of these are discussed by Salmon himself, namely analogy arguments and causal arguments. But the form he gives for analogy arguments is unsatisfactory, and for causal arguments he gives up the attempt to define them in terms of a form. For the last type of 'inductive' argument discussed by Salmon, hypothetico-deductive arguments in science, he provides a form, but this over-laps somewhat with the deductive form of 'affirming the consequent', so that the question would arise why he discusses the inductive logic of what he treated as a deductive type of argument in his discussion of conditional arguments. The answer would be, because he believes that hypothetico-deductive arguments (which affirm the consequent) are common in science. But for the general logician interested in all types of human reasoning, this would suggest that when a common type of argument is deductively invalid, then the logician should construct or apply different, less stringent, more realistic principles of evaluation. And this is one of the main points advocated in the 'new rhetoric'. At any rate, in 'inductive logic' as well as in the new rhetoric, it becomes apparent that it is necessary to define and study common types of actually occurring reasoning. To have at one's disposal a collection of interesting and important arguments is then of the utmost importance. Few collections possess the realism of those which constitute Galileo's *Dialogue*.

The inductive *correctness* of such arguments could be investigated in an analogous manner. Modifying properly the approach followed by Salmon, we could say that the inductive correctness of arguments having the various inductive forms depends on a number of identifiable, though not rigorously formalizable, conditions. For example, inductions by enumeration depend on whether the 'observed F's' are sufficiently numerous and sufficiently unbiased with respect to the total class of F's; statistical syllogisms depend on whether the percentage 'Z' is sufficiently high and whether the premises embody all relevant evidence; arguments from authority depend on whether x is a reliable authority concerning 'p', which in turn depends on a number of particular conditions; arguments against the man depend on whether x is a reliable anti-authority; etc. The identification of such conditions would obviously have to be carried out in a quasi-empirical, concrete, historical manner.

THE LINGUISTICS OF REASONING

At a memorable symposium[12] Bar-Hillel lamented the fact that formal logicians have devoted so little effort to the study of argumentation in natural language. Calling the situation "one of the greatest scandals of human existence",[13] he challenged "anybody here to show me a serious piece of argumentation in natural languages that has been successfully evaluated as to its validity with the help of formal logic".[14] In response to this challenge, the participants produced a bibliographical list containing 66 items.[15] An examination of these items will easily convince one of the fairness of Bar-Hillel's words with which the printed version of the symposium discussion concludes:

I asked for examples of application of formal logic to the evaluation of argumentation in natural languages. I am first of all grateful to those of you who called my attention to papers which I do not know; still all the examples – if I am not mistaken – were exclusively taken either from philosophical discourse or from mathematical discourse. Still, since all these discourses were held in natural languages, I will have to qualify somewhat my original formulations, though only very slightly so.[16]

On another occasion Bar-Hillel called "one of the major philosophical problems of our time"[17] the following problem:

Is extant formal logic, deductive and inductive . . . sufficient for the formalization of all arguments, of whatever degree of conclusiveness, that are used in ordinary language, or at least in the sciences, or at least in the natural sciences (where by formalization I mean now "those preparatory operations in applied logic, whereby sentences of ordinary language are fitted to logical forms by interpretation and paraphrase"), or is some better-developed formal logic necessary for this purpose, or is formal logic altogether insufficient for this purpose?[18]

In calling this a philosophical problem Bar-Hillel did not imply that it should concern only technical philosophers; rather he was making a rhetorical point since he was here addressing the XII International Congress of Philosophy (1958). In fact, when speaking to logicians, as he was in the symposium mentioned above, he would try to stress the fact that their neglect represented something of a betrayal of the history of their discipline;[19] whereas, when speaking to linguists, he would try to point out the importance of this problem to their work.[20]

I completely agree with Bar-Hillel about the urgency of the problem. Also I accept his historical thesis that this lamentable situation results from the fact that argumentation in natural languages developed historically into a no-man's-land between logic and linguistics.[21] I also find very plausible his analysis of the relevance of the problem to linguistics; he feels that semantics is the central part of linguistics,[22] and he claims that almost all linguists with the qualified exception of Lyons[23] and of Weinreich,[24] have failed to see "that meaning relations between linguistic entities are essentially deducibility relations and, therefore, logical relations. They missed this fact partly because a large number of the rules stating meaning relations can indeed be easily and practically formulated in the form of entries in traditional monolingual dictionaries (or lexica) .... Unfortunately the obvious efficiency of lexica has blinded almost all the linguists who gave any thought to that issue into believing that all of semantics is exhausted by lexicology, though a number of them paid some attention to the fact that other rules were needed at least for the purpose of determining how the meaning of longer linguistic entities is to be composed out of the meanings of their component smaller entities supposedly taken care of by the lexicon."[25] This has left "semantics, in an atheoretical, sometimes even anti-theoretical, bloodless, and anemic state".[26]

If Bar-Hillel's critiques are nothing less than devastating, his constructive suggestions are rather scarce. Nevertheless I believe they possess in promise what they lack in accomplishment. First, I believe he is right in suggesting that the chief obstacle to the understanding of argumentation in natural languages is "the essential dependence of communication in such languages on *linguistic co-text* (viz. the utterances, if any, that preceded the communicative act under scrutiny) and the *extra-linguistic context* (the general background in which this communicative act was performed, the motives that brought it about, the cognitive and emotional background of the participants in it, etc.)".[27] Second, "to attempt abstraction is always in order. But it is never *a priori* guaranteed that by abstracting one will be in a better position to solve a given set of problems, since for this purpose, sooner or later, the abstractive step will have to be followed by a concretive one — abstraction by *concretion*; otherwise, obviously, one will never come back to the original problems".[28] Third, "we have to face the decisive choice: to apply formal logic either to the sentences of natural languages 'as is', or to sentences in one of the constructed language systems that 'correspond' (in a sense that has yet to be clarified) to the sentences that occurred in the original argument".[29] Fourth, since "even the most developed methods known today for a direct logical treatment of natural language sentences [the theories of Montague and his associates[30]] are of limited use only ... then there remains only the 'indirect' method".[31] Next, Bar-Hillel suggests that the indirect method should be such that the original argument is reconstructed "in some *normalized, sterilized, regimented, systematized* language to the sentences of which it is then possible to apply the rules of formal logic";[32] from this, "three formidable questions arise: (a) What is the exact nature of that language system ... ? (b) What is the exact nature of that formal logic that will be applied to the sentences of that language system? (c) What is the theory that guides us in performing this fateful transition from utterance of natural languages to statements of the regimented language?"[33] According to Bar-Hillel too much work has been done in dealing with problems (a) and (b), and practically none to deal with (c). Having taken us this far, Bar-Hillel unfortunately stops.

Now, my primary concern is with (c). My suggestion is to lay the groundwork for such a theory by actually performing these 'fateful transitions' in a sufficient number of cases and with respect to natural language argumentation that has intrinsic and other interest. That is, I will be concerned first and foremost with reconstructing the arguments ín a book such as Galileo's *Dialogue*; then on the basis of such reconstructions one may construct some theory relating the original material to the reconstructions.

THE PRACTICE OF REASONING

The problems about formal logic and the science of reasoning discussed so far have all dealt with what may be called logical theory, in the sense in which this may be contrasted with logical practice. That is, it has been argued that the *theory* of reasoning needs to be empirical enough to take into account the actual reasoning processes that occur in the mind of *homo sapiens,* general enough to study rhetorical and/or inductive arguments as well as the nature of deduction, and concrete and realistic enough to take into account the fact that almost all reasoning is expressed in *natural* languages. All of these critiques are points to keep in mind when theorizing about or reflecting upon reasoning; the next one is made from the point of view of someone interested in the practical use of formal logic, that is to say, in using formal logic in one's own reasoning (about matters *other* than formal logic). I believe it is Michael Scriven who has made this criticism most forcefully.[34] He has pointed out that there is overwhelming evidence from educational psychology suggesting that the transfer of learning from one field to another is very little; and when applied to formal logic this means that to learn formal logic may make one proficient about solving problems in formal logic but has little effect on one's reasoning skill concerning everyday or other problems.[35] To this empirical argument, Scriven adds the following more conceptual one. However adequate a theory of reasoning is, by the very fact of being a theory it would consist of a number of systematically interrelated technical concepts in terms of which one would try to solve various problems about reasoning. Now arguments and reasoning problems are entities that can be found very frequently in everyday life; if one wants to solve one of these problems by the help of the theory of reasoning (formal logic), then one has first to formulate the problems in the theoretical terminology, then apply the theoretical principles and formal calculus of the theory to reach a theoretical conclusion, and finally translate this theoretical conclusion into the terms of the original problem so that it can be seen to be a solution to it as originally presented, and not to some other problem. Unfortunately, Scriven adds, as a matter of socio-historical fact, everyday reasoning problems are normally such that the theoretical encoding and decoding steps are as problematic and debatable as the original problem; hence the undeniable rigor, precision, and reliability of the intra-theoretical (formal-logic) manipulations are of no practical consequence.

The general cultural and pedagogical importance of these considerations by Scriven is undeniable and difficult to exaggerate. It is not clear, however, how his emphasis on the practice of reasoning can be made relevant to formal logic. After all, the latter is the (theoretical) study of reasoning, and Scriven himself has stressed the fact that the practical application of *any* theory does not obey the same laws as the theory itself; for example, the theory may be exact, but its practical application will necessarily be inexact; or the theory may be part of a science, but its practical application will necessarily be an art. In other words, is Scriven saying merely that if one is interested in practicing reasoning and improving reasoning skills, then the study of formal logic should have a relatively low priority, and that if one is interested in the practical application of formal logic, then one's expectation better not be too high? Or is Scriven also saying that if one is interested in the science of logic, i.e., in theorizing about reasoning, then he should pay the proper attention to the practice of reasoning? With regard to the former claims, Scriven's point is well made. But I believe he is also claiming the latter, in connection with which a problem arises, namely, how is this latter critique distinct from the earlier ones from the point of view of psychology, rhetoric, and linguistics? For, after all, the psychologists are saying that formal logic does not pay the proper attention to how humans actually reason, the adherents of the new rhetoric are saying that formal logic tends to ignore the argumentative practice of lawyers, jurists, moralists, preachers, students of the human sciences, and laymen; and Bar-Hillel is saying that formal logic ignores the practice of argumentation in natural languages. I believe Scriven's point is different, and I would describe its novelty to be the suggestion that the theory of reasoning should consist of *reasoning about reasoning*! That is, our logic books should be full of arguments about arguments, rather than full of theories, generalizations, formulations and systematizations of principles and concepts about arguments which would require a subsequent application in order to be of practical use. Certainly, that is the character of his latest book, *Reasoning*. Thus the main element of the science of reasoning would be primarily reasoning practice in the sense of analysis and evaluation of actual arguments, and the theory of reasoning would have the secondary place of formulating and systematizing principles for the critical analysis of arguments.

In summary, because the value of formal logic ultimately depends on its contributions to the improvement of actual reasoning, and because the theoretical character of formal logic makes its practical application to actual reasoning problems highly suspect, the most valuable thing for the logician to do is to engage in and produce arguments which analyze and evaluate other arguments about ordinary matters. Contributions to logic so conceived would be conclusions about arguments actually put forth as a matter of history, based on their critical analysis. The contribution envisaged here concerns the arguments in support of the earth's motion put forth by Galileo in his *Dialogue*. This inquiry will have a double logical relevance since even a superficial look at the book shows that most (though not all) of Galileo's own arguments are themselves about arguments, specifically about all the classical and contemporary objections to the earth's motion. In other words, Galileo's book consists mostly of a critical examination of all the available arguments and evidence against the earth's motion; the rest consists of the presentation of favorable arguments, partly arising from considerations made in the course of those critiques. What this means is that in the *Dialogue* Galileo primarily acts as a logician in the sense under discussion of argument analyzer or practioner of logic. But this means also that many of our own arguments will be two steps removed from the subject matter of the earth's motion, insofar as they will be arguments about Galileo's arguments about the anti-Copernican arguments; nevertheless this will introduce variety and depth in our own investigations. So, at any rate, these investigations promise to be logical in at least two ways.

Actually there is a third way in which such an investigation would be of relevance to logic. This involves considerations of the nature of science and follow a line of reasoning suggested by Scriven himself.[36] It is a fact that the desire that logic be a science is very much on at least the lips of logicians.[37] Agreeing with this goal, Scriven would argue[38] that the proper model for ('empirical') science ought not to be celestial mechanics, which is a very untypical science on account of the isolated character of the solar system. He would say that the proper (proper because realistic) model for science should be something like engineering or medicine, where progress is measured in terms of solutions to concrete practical human problems. According to this model of science, for logic to be scientific would be for it to produce results about actual reasoning problems, that is to produce justified conclusions about arguments of the past or present. Now, to add my own twist to this argument, I would say that since Galileo is regarded by almost everybody as a model scientist, and since logicians want very much to be scientific, it will be very important for them to examine his work to see to what extent it can be used to justify their own techniques of formalization and abstraction. They might discover that, or at least would be led to ponder whether, (a) their anti-empirical orientation is a version of the empty scholasticism that he fought, (b) formalization and abstraction are only sometimes appropriate, and (c) it is important to discuss methodological and philosophical problems (though one doesn't have to go to the extreme of metaphysics). If besides making such discoveries they would also discover that an essential feature of his work was argument analysis, then perhaps they would be more willing to truly scientificize logic since they could thereby also logicize science.

THE CRITICISM OF REASONING

Finally a word should be said about Stephen Toulmin's approach to logic and his critique of formal logic since his *Uses of Argument* is one of the earliest statements of the recent literature critical of formal logic. Toulmin embodies and expresses all of the critiques that I have discussed under the separate headings: he has the psychologist's concern with actual human reasoning, the interest in legal reasoning and non-compelling argumentation exhibited by the adherents to the new rhetoric, the kind of sensitivity to the natural languages in which argumentation is usually expressed that Bar-Hillel recommends, and the practical orientation that is so much at the heart of Scriven's concern. Moreover, he ends his critique by advocating a type of investigation that is very close to the one I have been groping toward in my restatement and justification of the above critiques. For example, according to Toulmin

logic conceived in this manner may have to become less of an *a priori* subject than it has recently been . . . Accepting the need to begin by collecting for study the actual forms of arguments current in any field, . . . we shall use ray-tracing techniques because they are used to make optical inferences, presumptive conclusions and 'defeasibility' as an essential feature of many legal arguments, axiomatic systems because they reflect the pattern of our arguments in geometry . . . But not only will logic have to become more empirical; it will inevitably tend to become more historical. To think up new and better methods of arguing in any field is to make a major advance, not just in logic, but in the substantive field itself: great logical innovations are part and parcel of great scientific, moral, political or legal innovations. In the natural sciences, for instance, men such as Kepler, Newton, Lavoisier, Darwin and Freud have transformed not only our beliefs, but also our way of arguing and our standards of relevance and proof.[39]

However, it should be noted that my endorsement of the various critiques has been a qualified one in most cases. My qualification and criticism of such critiques have attempted to define a sense in which logic is more important to the study of reasoning than is commonly supposed; for example, I argued that formal logic is more relevant to the kinds of experiments considered by Wason and Johnson-Laird than they suppose, and that some of their psychological models are inadequate because of an inadequate formal-logical analysis of their own evidence; I also argued that the rhetorical analysis of arguments presupposes their logical analysis, that only the actual performance of the logical analyses required in the reconstruction of natural language argumentation can provide the groundwork for a linguistics of reasoning, and that logical practice in the sense of argument analysis is fundamental in the improvement of reasoning skills. In short, I have been arguing that logic is the key to the study of reasoning, though such logic may not be formal logic as usually conceived. I think it is necessary to make a similar point in connection with Toulmin's views.

Besides advocating a logic with the features mentioned in the above quotation, Toulmin wants to make logic into a comparative study in the sense that "validity is an intra-field, not an inter-field notion. Arguments within any field can be judged by standards appropriate within that field, and some will fall short; but it must be expected that the standards will be field-dependent, and that the merits to be demanded of an argument in one field will be found to be absent (in the nature of things) from entirely meritorious arguments in another".[40] This is the result of his conviction that "all the *canons* for the criticism and assessment of arguments, I conclude, are in practice field-dependent, while all our terms of assessment are field invariant in their *force*".[41] Toulmin realizes[42] that this leads to the problem of whether the differences between the standards of different fields are irreducible, and whether a *general* logic is possible. His answer seems to be that a general account of the *structure* of arguments is possible; in fact, he provides one.[43] However, a general theory of validity presumably is not possible, which would explain his emphasis in the present book on description as opposed to prescription. I think his answer is unsatisfactory for two reasons. One is that if the layout of arguments is common to all fields (which is not to say that all arguments have the same structure, but rather that the classification of arguments into several kinds does not derive from their structure), if then the layout is general then the possibility would arise of interfield argumentation about the canons of assessment. Of course, I would agree with Toulmin that a responsible interfield criticism of standards cannot be too externalistic and reconstructionist, if it is to have a good chance of success; familiarity with and awareness of the standards internal to the field being criticized would be very important. Nevertheless, it would not seem to be in principle impossible to have general criteria, even if we accept the notion of fields of arguments.

A second objectionable feature of Toulmin's account here is that it tends to neglect the interdisciplinary nature of most creative reasoning. Toulmin's own book is a good example of this since it is an attempt to introduce jurisprudential standards of assessment in traditional logical questions. A classical example is, of course, Galileo whose supreme achievement was the introduction of standards of mathematical reasoning into physical arguments, as well as the introduction of standards of physical arguments into astronomical ones;[44] to respect the traditional integrity of the three fields, mathematics, physics, and astronomy, would have made it impossible for modern science to emerge. I do not think Toulmin would deny this; and in the above quotation where he speaks of the historical character of logic, he comes close to making the same point. Nevertheless, it is undeniable that his book contains a strong tendency toward acritical or uncritical description. Of course, to emphasize sympathetic understanding and description is certainly a good rhetorical device in opposition to the irresponsible prescriptionism which is the stock in trade of so many logicians and epistemologists. However, a general emphasis on acritical comparison would not be good for logic. The point is not to avoid criticism, but to avoid unfair criticism which fails to understand the argument being criticized.

The way to avoid unfair criticism is to take account of the context and problem situation in which the argument under consideration was given. The comparative method that Toulmin wants to inject into logic may be taken partly to contain this suggestion, and to that extent it is of course quite acceptable. The contextual method not only is a very powerful tool in the critical understanding of a given argument, but it contains an almost built-in check against abuses and excesses of the kind to which a formal logician would be inclined to succumb. For example, the propriety of generalizing the desirability of Galileo's introduction of mathematical standards into the study of motion depends on whether or not the situation is sufficiently similar to Galileo's; hence it would not be proper to say *today* that it is proper to mathematize the study of reasoning, though this may have been proper a century ago; moreover, Galileo was not merely mathematizing motion, but physicalizing mathematics by putting physical content into mathematical arguments. Analogously, it would not be proper to say that more mathematical modelling is needed in the science of psychology, since the problems facing this science are and have always been of a different nature than those of physics.[45] Similarly, just as one may find proper Toulmin's attempt to introduce standards of jurisprudence into logical matters, so one may also find appropriate Wason's attempt to reform jurisprudence by injecting a little more logical awareness into it.[46]

In conclusion, a *general* logic seems possible if we concentrate on the critical understanding of actual reasoning, which concentration carries along with it an empirical, historical, concrete or contextual, practical, and critical orientation.

NOTES

[1] See, for example, H. E. Kyburg, Jr., *Probability and Inductive Logic,* and its bibliography on pp. 199–247. Noteworthy exceptions are M. Hesse's *Structure of Scientific Inference* and L. J. Cohen's *The Probable and The Provable.*

[2] E. g., W. Salmon, *Logic*, pp. 13–17, and 81–117.

[3] K. Popper, *The Logic of Scientific Discovery*; idem, *Objective Knowledge*, pp. 1–31; and J. Agassi, 'The Role of Corroboration in Popper's Methodology'.

[4] Popper, *Objective Knowledge*, p. 4.

[5] See, for example, B. Skyrms, *Choice and Chance*, pp. 1–69.

[6] J. Agassi, *Towards an Historiography of Science*; I. Lakatos and A. Musgrave (eds.), *Criticism and the Growth of Knowledge*.

[7] See my *History of Science as Explanation* and my Essay-review of *Criticism and the Growth of Knowledge*.

[8] Pp. 1–17, and 81–117.

[9] Salmon, *Logic*, pp. 81–105.

[10] *Ibid.*, pp. 13–14.

[11] *Ibid.*, p. 17.

[12] Y. Bar-Hillel, 'The Role of Formal Logic in the Evaluation of Argumentation in Natural Languages', an intersectional symposium at the Third International Congress for Logic, Methodology, and Philosophy of Science (Amsterdam, August 26, 1967), printed in J. F. Staal (ed.), 'Formal Logic and Natural Languages'.

[13] *Ibid.*, p. 256.

[14] *Ibid.*

[15] *Ibid.*, pp. 281–83.

[16] *Ibid.*, p. 281.

[17] Y. Bar-Hillel, *Aspects of Language*, p. 108.

[18] *Ibid.*, pp. 107–108.

[19] Staal (ed.), 'Formal Logic and Natural Languages', pp. 256–57; Bar-Hillel, *Aspects of Language*, pp. 202, and 206–207.

[20] *Aspects of Language*, pp. 182–201.

[21] *Ibid.*, p. 182.

[22] *Ibid.*

[23] J. Lyons, *Structural Semantics*; cf. Bar-Hillel, *Aspect. of Language*, p. 183, n. 1.

[24] U. Weinreich, 'Explorations in Semantic Theory', in *Current Trends in Linguistics*, edited by T. A. Sebeok, pp. 395–477.

[25] *Aspects of Language*, p. 183.

[26] *Ibid.*, p. 182.

[27] *Ibid.*, pp. 207–208.

[28] *Ibid.*, pp. 208–209.

[29] *Ibid.*, p. 209.

[30] Bar-Hillel's intended reference is to Montague, whose papers have posthumously been collected in *Formal Philosophy*, edited by R.H. Thomason. The reference is anachronistic in the sense that nowadays one would have to examine the game-theoretical semantics approach stemming from J. Hintikka's work, which can be found in his *The Semantics of Questions and the Questions of Semantics*, and in E. Saarinen, ed., *Game-Theoretical Semantics*. Such an examination is beyond the scope of the present work.

[31] *Aspects of Language*, p. 212.

[32] *Ibid.*, p. 213.

[33] *Ibid.*

[34] M. Scriven, 'Definitions, Explanations, and Theories', p. 100; idem, *Reasoning*, pp. xiv-xv.

35 Scriven, *Reasoning*, p. xiv.
36 M. Scriven, 'The Frontiers of Psychology: Psychoanalysis and Parapsychology'; idem, 'A Possible Distinction Between Traditional Scientific Disciplines and the Study of Human Behavior'; idem, 'Psychology Without a Paradigm'; idem, 'Science: The Philosophy of Science', *International Encyclopedia of the Social Sciences* (1968 ed.), 14: 83–92; idem, 'A Study of Radical Behaviorism'; idem, 'Views of Human Nature'.
37 See, for example, W. V. Quine, *Methods of Logic*, p. 1.
38 Scriven, *Reasoning*, p. xv.
39 S. Toulmin, *The Uses of Argument*, pp. 257–58.
40 *Ibid.*, p. 255.
41 *Ibid.*, p. 38.
42 *Ibid.*, p. 39.
43 *Ibid.*, pp. 94–145.
44 S. Drake, *Galileo Studies*, pp. 95–112.
45 Scriven, 'Views of Human Nature'.
46 P. C. Wason and P. N. Johnson-Laird, *Psychology of Reasoning*, pp. 218–28; and P. C. Wason, 'The Drafting of Rules'.

PART III

THEORY OF REASONING

CHAPTER 14

PROPOSITIONAL STRUCTURE:
THE UNDERSTANDING OF REASONING

It is now time to formulate a number of basic principles of logical theorizing. Logic is conceived here, of course, as the theory of reasoning. In this chapter we shall be concerned with only one of its two main problems.

PROPOSITIONAL STRUCTURE

In the theory of reasoning there are two primary aims, the understanding of reasoning, and the evaluation of it. Combining the two we might say that its goal is the critical understanding of reasoning. Generally speaking, the phrase 'the understanding' of reasoning is meant to convey the same idea as the following: the explanation, the analysis of reasoning. The phrase 'the evaluation' of reasoning is meant to convey the same idea as the following: the assessment of, the criticism (favorable or unfavorable) of, judgment upon reasoning. We may say that analysis and evaluation study different aspects of reasoning. Analysis involves the study of the nature or structure of reasoning, evaluation the study of its value or worth. In analysis one is concerned with finding out what the reasoning is, in evaluation with finding out how good (i.e., valid) it is. From a third point of view, the understanding of reasoning may be said to involve the description of reasoning; while its evaluation involves prescriptions for it.

Reasoning is the activity of the human mind consisting of the giving of reasons for conclusions, or the reaching of conclusions on the basis of reasons. More exactly it is the interrelating of our thoughts in such a way as to make certain thoughts dependent on others.

Reasoning is linguistically expressed in what are called *arguments*. An argument is a basic unit of reasoning in the sense that it is a piece of reasoning sufficiently self-contained as to constitute by itself a more or less autonomous instance of reasoning. The occurrence of reasoning is normally indicated, and can always be explicitly indicated, by the use of what may be called *reasoning indicators*. These are words like the following (or phrases synonymous with such words): therefore, thus, so, hence, consequently, because, since, for. Reasoning indicator words, however, are only hints, since it is possible to express reasoning without them and since it is possible for them to have other meanings which do not indicate reasoning; but we will not consider such complications at the moment.

Reasoning indicators serve to interconnect what may be called the *propositional components* of an argument. A *propositional component of an argument* is any part of an argument which is capable of being accepted or rejected by itself. A *proposition* is any propositional component stated as a complete sentence so that it can stand by itself.

For example, consider the following passage:

All the circularly moving bodies seem to lag behind and to move with more motions than one, except the first sphere (that is, the *primum mobile*); therefore the earth, moving around its own center and being placed at the center, must be moved with two motions and must fall behind; but if this were the case, the risings and settings of the fixed stars would have to vary, which is not seen to happen; therefore the earth is not moved.[1]

The only reasoning indicator here is the word 'therefore' in both of its occurrences. The propositional components are as follows:

(PC1)   All the circularly moving bodies seem to lag behind and to move with more motions than one, except the first sphere (that is, the *primum mobile*).

(PC2)   The earth, moving around its own center and being placed at the center, must be moved with two motions and must fall behind.

(PC3)   If this were the case, the risings and settings of the fixed stars would have to vary.

(PC4)   Which is not seen to happen.

(PC5)   The earth is not moved.

The propositions would be the following:

(P1)   Same as (PC1).

(P2)   Same as (PC2).

(P3)   If the earth were moved with two motions and fell behind, the risings and settings of the fixed stars would have to vary.

(P4)   It is not seen to happen that the risings and settings of the fixed stars vary.

(P5)   Same as (PC5)

As a second example, consider the following passage:

The earth cannot move circularly, because such a motion would be a forced one and therefore not perpetual. The reason that it would be forced was that if it were natural, the earth's parts would also naturally move in rotation, which is impossible because the nature of these parts is to be moved downward in a straight line.[2]

The reasoning indicators here are as follows: 'because' in both of its occurrences, and the phrase "the reason that _____ was that _____". The propositional components are as follows:

(PC1) The earth cannot move circularly.
(PC2) Such a motion would be a forced one.
(PC3) Not perpetual.
(PC4) If it were natural, the earth's parts would also naturally move in rotation.
(PC5) Which is impossible.
(PC6) The nature of these parts is to be moved downward in a straight line.

These may be expressed as propositions:

(P1) Same as (PC1).
(P2) The earth's circular motion would be a forced one.
(P3) The earth's circular motion would not be perpetual.
(P4) If the earth's circular motion were natural, its parts would also naturally move in rotation.
(P5) It is impossible that the parts of the earth should move in rotation.
(P6) The nature of the earth's parts is to be moved downward in a straight line.

An argument may thus be conceived as a series of propositions some of which are being based on others, where the interconnections are expressed by means of reasoning indicators. The simplest possible argument contains two propositions; such a simple argument can always be expressed in either one of two standard forms which are logically equivalent: (1) A, therefore B; or (2) B because A. In both (1) and (2), B is the *conclusion* and A is the *reason* or *premise*. In other words, although both 'therefore' and 'because' are reasoning indicators, they indicate different ways to express reasoning; the proposition preceding the word 'therefore' is the reason or premise, the one following it is the conclusion; whereas the proposition preceding 'because' is the conclusion, and the one following it is the reason. The three-dots symbol '∴' is used as an abbreviation for the word 'therefore' and a convenient way of referring to simple arguments; so that 'A, ∴ B' is short for (1), or equivalently (2).

A *complex argument* is one which is made up of *at least* two subarguments combined in such a way that the conclusion of one subargument is simultaneously a reason of the other subargument. The simplest complex argument is one of the form: A because B, and B because C (or equivalently: C, therefore B; therefore A). Here B is the reason of the subargument 'A because B' (or 'B, therefore A') and also the conclusion of the subargument 'B because C' (or 'C, therefore B'). Every proposition in a complex argument falls into one and only one of the following categories: *intermediate proposition, final reason, final conclusion*. An *intermediate proposition* in a given complex argument is a proposition which serves as the conclusion of one subargument and as a reason of another subargument. A *final reason* in a given complex argument is a proposition which is a reason of some subargument but not the conclusion of any subargument. The *final conclusion* in a given complex argument is a proposition which is the conclusion of some subargument but not the reason of any subargument. In the example here, A is the final conclusion, B is the one and only intermediate proposition, and C is the one and only final reason.

The *propositional structure* of reasoning refers to the interrelationships among its various elements, i.e. among its various subarguments and among the various propositions. Such structure may be pictured in a *structure diagram* which is constructed in accordance with the following rules:[3]

(1) Label each proposition in the argument by some number or letter.

(2) Represent each proposition by enclosing its number or letter in a closed figure.

(3) The fact that a given proposition is a reason supporting another is indicated by a solid line leading up from the first to the second.

(4) The diagram will have at the top a proposition which is supported by one or more other propositions but which does not itself support any other proposition; this proposition at the top is the final conclusion of the argument being diagrammed.

(5) The diagram will have one or more propositions which support other propositions but which are not themselves supported by anything else; such propositions are the final reasons of the argument being diagrammed.

(6) Some structure diagrams may have propositions which both support, and are supported by, other propositions; such propositions are the intermediate propositions of the argument being diagrammed. That is, they are propositions which are logically placed between the final reasons and the final conclusion of the argument and which are reasons from the point of view of what they immediately support and conclusions from the point of view of what they are immediately supported by.

These rules are to be understood so that the diagrams for the two arguments considered above would be the following:

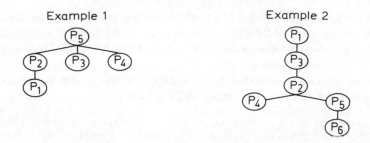

In the first example, P5 is the final conclusion; P1, P3, and P4 the final reasons; and P2 is an intermediate proposition: it is the conclusion of the subargument 'P1, ∴ P2' and one of the reasons of the subargument 'P2, P3, P4, ∴ P5'.

In the second example P1 is the final conclusion; P4 and P6 are the final reasons; and P3, P2, and P5 are intermediate propositions. P3 is the reason of 'P3, ∴ P1' and the conclusion of 'P2, ∴ P3'; P2 is the reason of the latter, and the conclusion of 'P4, P5, ∴ P2'; and P5 is one of the reasons of the latter, and the conclusion of 'P6, ∴ P5'.

There is a *standard way of labeling* propositions in structure diagrams:

(A) Label the final conclusion 'C'.

(B) Label all intermediate propositions and final reasons by an 'R' and use numerals as follows to distinguish among the various reasons:

(1) Reasons which support the final conclusion are labelled R1, R2, etc.

(2) Reasons which support the same intermediate reason, will have all digits the same, save the last one (e.g. reasons supporting R12, will be labelled R121, R122, R123, etc.).

The standard diagrams for the two arguments above would be, respectively, the following:

Of course, although some of the symbols in these two diagrams are the same, they correspond to different propositions. The point of this standard labeling is to give each proposition a label that indicates relatively easily and clearly its logical function in the argument.

These ideas may be further illustrated by considering other passages where Galileo states arguments very explicitly. For example:

> . . . since when the ship stands still the rock falls to the foot of the mast, and when the ship is in motion it falls apart from there, then conversely, from the falling of the rock at the foot it is inferred that the ship stands still, and from its falling away it may be deduced that the ship is moving. And since what happens on the ship must likewise happen on the land, from the falling of the rock at the foot of the tower one necessarily infers the immobility of the terrestrial globe.[4]

The logical structure of this argument may be pictured as follows:

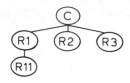

The various propositions are:

C: The terrestrial globe is immobile.
Rl: From the falling of the rock at the foot of the mast it is inferred that the ship is standing still, and from its falling away it may be deduced that the ship is moving.
R2:   What happens on the ship must likewise happen on the land.
R3:   On the land the rock falls at the foot of the tower.
R11:   When the ship stands still the rock falls to the foot of the mast, and when the ship is in motion it falls apart from there.

A fourth example is part of the argument that natural straight motion is impossible:

> . . . straight motion being by nature infinite (because a straight line is infinite and indeterminate), it is imposible that anything should have by nature the principle of moving in a straight line; or, in other words, toward a place where it is impossible to arrive, there being no finite end. For nature, as Aristotle well says himself, never undertakes to do that which cannot be done, nor endeavors to move whither it is impossible to arrive.[5]

The structure is:

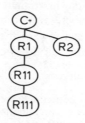

The propositions are:

C: It is impossible that anything should have by nature the principle
         of moving in a straight line.
R1:   Straight motion is motion toward a place where it is impossible to
         arrive.
R2:   Nature never undertakes to do that which cannot be done, nor
         endeavors to move whither it is impossible to arrive.
R11:   Straight motion is by nature infinite.
R111:   A straight line is infinite and indeterminate.

The fifth example is:

Generation and corruption occur only where there are contraries; contraries exist only among simple natural bodies, movable in contrary motions; contrary motions include only those made in straight lines between opposite ends; of these there are but two, namely, from the middle and toward the middle; and such motions belong to no natural bodies except earth, fire, and the two other elements; therefore generation and corruption exist only among the elements.[6]

Its structure is:

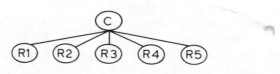

The propositions are:

C: Generation and corruption exist only among the elements.
R1:   Generation and corruption occur only when there are contraries.
R2:   Contraries exist only among simple natural bodies, movable in
         contrary motions.
R3:   Contrary motions include only those made in straight lines
         between opposite ends.

R4:   Of motions made in straight lines between opposite ends, there
        are but two, namely, from the middle and toward the middle.

R5:   Motions from and toward the middle belong to no natural bodies
        except earth, fire, and the two other elements.

The sixth and last example here is:

. . . because the third simple motion, namely, the circular, about the middle, has no
contrary (because the other two are contraries, and one thing has but one contrary)
therefore that natural body to which such motion belongs lacks a contrary and, having
no contrary, is ingenerable and incorruptible, etc., because where there are no contraries
there is no generation, corruption, etc. But such motion belongs to celestial bodies alone;
therefore only these are ingenerable, incorruptible, etc.[7]

Its structure is:

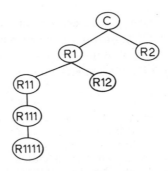

C: Only celestial bodies are ingenerable, incorruptible, etc.

R1:   That natural body to which circular motion belongs is ingenerable
        and incorruptible, etc.

R2:   Circular motion belongs to celestial bodies alone.

R11:   That natural body to which circular motion belongs lacks a
        contrary.

R12:   Where there are no contraries there is no generation, corruption,
        etc.

R111:   The third simple motion, namely, the circular, about the middle,
        has no contrary.

R1111: The other two simple motions are contraries, and one thing has
        but one contrary.

THE RECONSTRUCTION OF ARGUMENTS

Some arguments are such that what they need is not so much analysis, but rather synthesis, as it were. For example, in Galileo's *Dialogue* some arguments occur in the course of long, often poetical, and usually dramatically intense passages, and they need to be *reconstructed*. In general, a reconstruction of an argument is a restatement of it such that no logically extraneous propositions are included and such that all logical interconnections among the stated propositions are explicitly and clearly indicated, by means of reasoning indicators.

For example, Galileo's critique of the ship experiment, which takes about four pages of text,[8] may be reconstructed as follows: It is not true that when the ship is in motion the rock falls away from the foot of the mast because once the rock is dropped it will tend to move in two directions: (1) vertical, or toward the center of the earth, and (2) horizontal, or around the center of the earth along with the ship. The rock will tend to move in a horizontal direction with the speed of the ship because the nature of bodies is such that, when left to themselves, they continue in a state of rest, if they are at rest, or in a state of uniform motion, if they are in motion. That the latter is probably true may be seen as follows: balls sliding down a smooth inclined plane move down the incline with a continuously increasing speed; and balls climbing up a smooth inclined plane move with a continuously decreasing speed; therefore, balls moving along a plane going neither up nor down move with a speed that neither increases nor decreases, that is, a ball moving in a horizontal direction will move in that direction with constant speed.

Galileo's discussion of vertical fall[9] has the following two elements.

(a) Reconstruction of the Aristotelian argument from vertical fall: The earth does not move because bodies thrown vertically upward fall vertically back to the same place from which they were thrown and this could not happen if the earth were moving; for if the earth were moving, then the place of ejection would move along with it while the projectile was going up and down, and if that place moved then the body would fall some distance away from it.

(b) Reconstruction of Galileo's criticism: The Aristotelian argument is a 'begging of the question' fallacy because the way the Aristotelians would support their premise that bodies fall vertically would be to say "because they are *seen* to fall vertically". Now, to say this assumes that apparent vertical fall implies actual vertical fall. And to show the latter you have to assume that the earth is at rest because if the earth moves, then apparent vertical fall does *not* imply actual vertical fall (it would imply actually slanted fall).

Another example is the Analytical Summary presented above (Chapter 2), which is a reconstruction of the whole main argument of the book.

THE INTERDEPENDENCE OF REASONS

Let us now examine a passage from the *Dialogue* which is noteworthy: (1) because it is easily overlooked on account of its literary structure, which has the character of a parenthetical discussion; (2) because it contains an intrinsically interesting argument, indeed a distinct Aristotelian argument for celestial inalterability, in addition to the one based on contrariety of motions and to the one based on the alleged lack of observed changes; (3) and because it contains arguments which can serve as good examples for some additional ideas in logical analysis. The passage[10] contains a discussion of what may be called a teleological argument for celestial inalterability. To be more exact, it can be reconstructed as a series of three things: (A) an Aristotelian teleological argument for celestial inalterability, (B) a Galilean critical counterargument, and (C) an Aristotelian defensive reply to the counterargument. The reconstructions are as follows.

(A) *Reconstruction of the Aristotelian teleological argument for celestial inalterability*. There are two ways to show that there cannot be celestial mutations. First, mutations would make celestial bodies imperfect since celestial mutations would be superfluous. They would be superfluous because celestial bodies need only light and motion to fulfill their purpose. And this is so because their purpose is to serve man.

Second, celestial mutations would be vain and useless, and nature does nothing in vain. They would be vain and useless because they would not occur for the benefit of man, and they would not occur for the benefit of celestial inhabitants (since there are no such creatures).

(B) *Reconstruction of the Galilean counterargument*. This argument is deficient for several reasons. First it assumes that change makes things imperfect, and this assumption is false because change seems to make things more perfect. This is so because it is usually the less perfect things that are devoid of change (a desert is less perfect than a garden, a dead animal is less perfect than a living one); on the other hand, when things are both perfect and devoid of change (for example, precious metals), they are precious because they are scarce.

Second, it seems unlikely that the purpose of celestial bodies is to serve the earth since this would mean that divine and eternal things would be serving a base and transitory body.

Third, to say that there are no men on the celestial bodies is probably correct, but from this it doesn't necessarily follow that there are no changes there, since (from this same proposition) it does not necessarily follow that there are no creatures there; in fact, to think that it does follow would be like arguing by someone who had lived all his life in a forest, that there could not be things like ships or fish. Now, if there could be creatures there, then there would be changes for their sake.

Fourth, the conclusion seems wrong since some terrestrial changes derive in part from celestial bodies and to cause a change without being changed seems unlikely, so that celestial bodies too must be alterable.

Finally, to say that the purpose of celestial bodies is to serve man is unjustified because the only argument you could give is the following: "For the purpose of everything else is to serve man. For example, horses, herbs, cereals, fruits, birds, beasts, and fishes exist for the comfort and nourishment of men." Now, the problem with this argument is that at best it shows only that the purpose of all *terrestrial* things is to serve man; and from this we cannot conclude that the purpose of celestial bodies is also to serve man (especially for an Aristotelian who distinguishes sharply between the two domains and believes that different principles apply).

(C) *Reconstruction of the Aristotelian reply to the counterargument.* This criticism is invalid for two reasons. First, it is not true that the original (Aristotelian) argument assumes that change makes things imperfect. In fact, the claim that change makes things imperfect is nowhere explicitly stated in the argument; nor is this claim implicitly employed as the reason why change would make celestial bodies imperfect since the reason is given explicitly and it is that celestial mutations would be superfluous. (In other words, the assumption in question is not being made in the original argument because there is no explicit or implicit step in it saying "mutations would make celestial bodies imperfect because mutations make things imperfect"; instead the step reads "mutations would make celestial bodies imperfect because celestial mutations would be superfluous".)

Second, though various parts of the earth are transitory and subject to change, the earth as a whole is not transitory but as eternal as the celestial bodies; thus if celestial bodies serve the earth, this means that certain eternal bodies are serving another eternal body, and there is no absurdity in that.

Several comments should be noted about these reconstructions. First, the Aristotelian argument has two apparently distinct versions in the text, and it would be inaccurate to ignore either one; so I have kept both versions in my reconstruction. Second, the first point in the Galilean counterargument is indeed incorrect, as the first point in the Aristotelian reply correctly points out. Third, the second Galilean criticism also has a cogent objection in the second Aristotelian reply. Fourth, the third and last Galilean points are plausible ways of integrating the text discussions pertaining to those topics, respectively; however, the reconstructed last point involves more interpretation and more addition of ideas not present in the text. Fifth, the only way to integrate the discussion about precious metals in the Galilean critique seems to be as in the second part of the first point; it has *prima facie* plausibility, but merely *prima facie*, for Simplicio could add the following third point to his reply: "Third, to say that precious metals are precious because they are scarce is true only in the sense *'primarily* because they are scarce', since their relative scarcity does contribute somewhat to their preciousness. At any rate, the proposition is irrelevant; the relevant one would be: precious metals are perfect because they are relatively scarce; but this latter proposition is not true since their scarcity has nothing to do with their *perfection*." But there is no textual justification for including this in the reconstruction.

These reconstructed arguments provide good examples for what I shall call the interdependence and the independence of reasons. Two reasons which immediately support the same proposition (and which would thus have the same number of digits as subscripts in a diagram) are *interdependent* when each depends on the other in order to support that proposition and each alone is insufficient or incomplete to provide that support. When reasons are interdependent the rules for standard labeling apply without change. When two reasons are not interdependent, they are called *independent*. So two reasons are independent of each other when each does not depend on the other to support the conclusion based on them (whether or not each alone is sufficient to provide that support, i.e., whether or not each is interdependent with *some other* reason). In other words, two reasons may be independent of

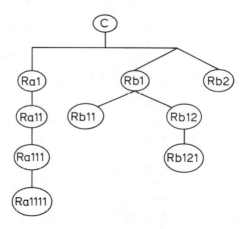

each other and still be interdependent with other reasons. The standard labeling of independent reasons is as follows: the small letters 'a', 'b', 'c', etc. should be placed after the label for the proposition they support, to distinguish one *set* of reasons from another independent set; and then these small letters should be carried for each proposition, as in the above diagram.

Applying these ideas, the diagram for the reconstructed argument A above would be the one just drawn. The various propositions are:

C: There cannot be celestial mutations.
Ra1:   Mutations would make celestial bodies imperfect.
Ra11:   Celestial mutations would be superfluous.
Ra111:   Celestial bodies need only light and motion to fulfill their purpose.
Ra1111:   The purpose of celestial bodies is to serve man.
Rb1:   Celestial mutations would be vain and useless.
Rb2:   Nature does nothing in vain.
Rb11:   Celestial mutations would not occur for the benefit of man.
Rb12:   Celestial mutations would not occur for the benefit of celestial inhabitants.
Rb121:   There are no celestial inhabitants.

Here, Ra1 and Rb1 are independent, whereas Rb1 and Rb2 are interdependent, and so are Rb11 and Rb12.

The structure diagram for B would be:

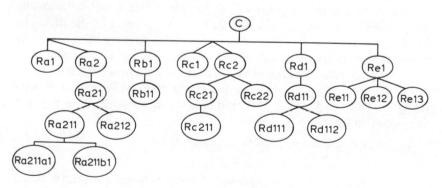

Some of the more important propositions are:

C: Argument (A) is deficient.
Ra1:   Argument (A) assumes that change makes things imperfect.

Ra2:   The assumption that change makes things imperfect, is false.

Ra211a1:   A desert is less perfect than a garden.

Ra211b1:   A dead animal is less perfect than a living one.

Ra212:   When things are both perfect and devoid of changes (for example, precious metals), they are precious because they are scarce.

Rc1:   To say that there are no men on the celestial bodies is probably correct.

Rc2:   From (Rc1) it does not necessarily follow that there are no changes on celestial bodies.

Rc21:   From the same (Rc1) it does not necessarily follow that there are no creatures on celestial bodies.

Rc211:   To think that from (Rc1) it does follow that there are no creatures on celestial bodies, would be like arguing by someone who had lived all his life in a forest, that there could not be things like ships or fish.

Rd1:   The conclusion of argument (A) seems wrong.

Rd11:   Celestial bodies too must be alterable.

Re1:   To say that the purpose of celestial bodies is to serve man is unjustified.

Re11:   The only argument you could give is the following: "For the purpose . . . . nourishment of men".

Note that the giving of examples to support a generalization (which is happening in 'Ra211a1, Ra211b1, ∴ Ra211') normally generates independent reasons. Note also that in a list of critical remarks, each critique is usually independent of the others; here the Ra's are independent of the Rb's, etc. Finally note that the Ra subargument of B deals with an assumption made presumably in connection with Ra1 of A; that the Rb subargument of B relates to proposition Ra1111 of A; that the Rc subargument of B relates to proposition Rb121 of A; that the Rd subargument of B tries to directly refute the final conclusion C of A; and that the Re subargument of B relates to proposition Ra1111 of A.

The propositions and diagram for argument C are:

C: The criticism of argument B is invalid.

Ra1:   It is not true that argument Λ assumes that change makes things imperfect.

Ra12:   The claim that change makes things imperfect is not implicitly employed (in argument A) as the reason why change would make celestial bodies imperfect.

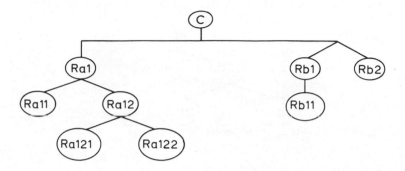

Ra122:   The reason is that celestial mutations would be superfluous.
Rb1   If celestial bodies serve the earth, this means that certain eternal
             bodies are serving another eternal body.
Rb2   There is no absurdity in 'that'.

One advantage of this idea of the independence of reasons is that it allows
us to integrate into a unified whole separate arguments, if they have the same
final conclusion. For example, consider the following passage:

This principle being established then, it may be immediately concluded that if all integral
bodies in the world are by nature movable, it is impossible that their motions should be
straight, or anything else but circular; and the reason is very plain and obvious. For
whatever moves straight changes place and, continuing to move, goes ever farther from
its starting point and from every place through which it successively passes. If that were
the motion which naturally suited it, then at the beginning it was not in its proper place.
So then the parts of the world were not disposed in perfect order. But we are assuming
them to be perfectly in order; and in that case, it is impossible that it should be their
nature to change place, and consequently to move in a straight line.

Besides, straight motion being by nature infinite (because a straight line is infinite
and indeterminate), it is impossible that anything should have by nature the principle
of moving in a straight line; or, in other words, toward a place where it is impossible to
arrive, there being no finite end. For nature, as Aristotle well says himself, never undertakes to do that which cannot be done, nor endeavors to move whither it is impossible
to arrive.[11]

The second paragraph contains one argument whose structure was analyzed
above (section on 'Propositional Structure'). The first paragraph contains the
following argument:

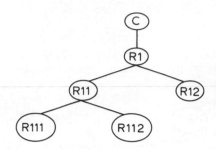

The propositions are:

C: It is impossible that it should be the nature of integral bodies to
            move in a straight line.
R1:   It is impossible that it should be their nature to change place.
R11:   If the motion which naturally suited bodies were to go ever
            farther from their starting point and from every place through
            which they successively pass, then the parts of the world would
            not have been disposed in perfect order.
R12:   The parts of the world are in perfect order.
R111:   Whatever moves straight changes place and, continuing to move,
            goes ever farther from its starting point and from every place
            through which it necessarily passes.
R112:   If the motion which naturally suited a body was to go ever
            from its starting point and from every place through which it
            successively passes, then since the beginning it was not in its
            proper place.

Combining the two arguments we get:

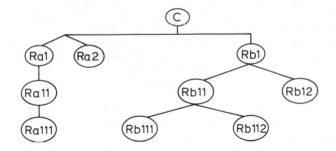

LATENT STRUCTURE

The Aristotelian teleological argument is also a good example to illustrate certain ideas about what I shall call latent propositional structure. The application of these ideas to that argument allow us to give substance to the intuitive judgment that the two versions of the Aristotelian argument have something in common, in spite of appearances. As I mentioned above, the appearances are that we have two distinct versions, the only common proposition being the conclusion. Probing their latent structure allows one to discern other points of contact.

The *latent propositional structure* (or more simply, the *latent structure*) of an argument consists of those propositions which are not explicitly stated in the argument but are implicitly assumed or taken for granted by the giver of the argument. In one case, a proposition is part of the latent structure, or more simply a proposition is latent, when for a particular step of the argument it is needed, in addition to the other explicit propositions involved in that step, to fully justify that step. In another case, a proposition is latent when it is one of the reasons being implicitly used to justify one of the *final reasons* in the explicit argument. Because of their position in structure diagrams, in the first case the propositions may be called *horizontally latent*, in the second case *vertically latent*. In a structure diagram all latent structure is drawn in dotted lines.

For example, consider the first version of the teleological argument (first paragraph of reconstructed argument A above). Its explicit structure is Diagram A. (Call *explicit structure* the interrelationship of the propositions which are explicitly stated). If we include the latent structure, we would have Diagram B.

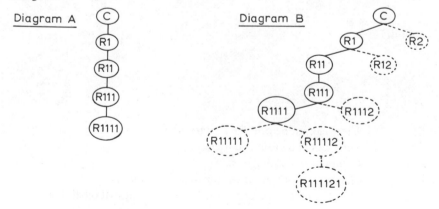

Diagram A   Diagram B

C: There cannot be celestial mutations.
R1:   Mutations would make celestial bodies imperfect.
R2:   Celestial bodies are perfect.
R11:   Celestial mutations would be superfluous.
R12:   Anything in which something superfluous happens is imperfect.
R111:   Celestial bodies need only light and motion to fulfil their purpose.
R1111:   Their purpose is to serve man.
R1112:   The only way in which celestial bodies can serve man is to move
             and send light to the earth.
R11111:   The purpose of all terrestrial things is to serve man.
R11112:   The purpose of celestial bodies is the same as the purpose of
             terrestrial things.
R111121:   The principle of anthropocentrism (e.g., "man is lord of the
             universe, not just the earth"; or "man is the measure of all
             things".)

For a second example consider the second version of the teleological argument (second paragraph of reconstructed argument A above). Its explicit
structure is Diagram C; if we include the latent structure, we get Diagram D.

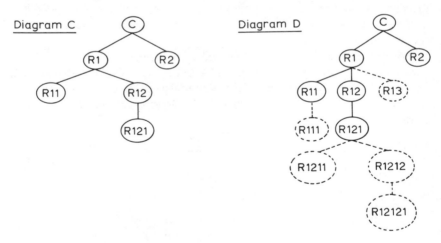

C: There cannot be celestial mutations.
R1:   They would be vain and useless.
R2:   Nature does nothing in vain.
R11:   They would not occur for the benefit of man.
R12:   They would not occur for the benefit of celestial inhabitants.

R13:   Only things beneficial to man or other intelligent creatures are
            not vain and useless.

R111:   The only way in which celestial bodies can serve man is to move
            and send light to the earth.

R121:   There are no such creatures.

R1211:   There are no men there.

R1212:   Man is the only inhabitant of the universe.

R12121:   The principle of anthropocentrism (e.g. "man is lord of the
            universe, not just the earth", or "man is the measure of all
            things".)

Note than in Diagram B, R2, R12, and R1112 are horizontally latent,
whereas R11111, R11112, and R111121 are vertically latent. In Diagram D,
R13 is horizontally latent, whereas R111, R1211, R1212, and R12121 are
vertically latent. Note also that R1112 of Diagram B is the same proposition
as R111 of Diagram D, and that R111121 of Diagram B is the same proposition as R12121 of Diagram D. What does this mean? These identities are the
"points of contact" that one's logical judgment perceived earlier.

Let us apply the notion of latent structure to the argument from vertical
fall[12] which was examined in detail in other contexts earlier.[13] The original
Aristotelian argument is the following:[14]

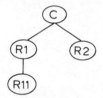

C: Not-M.
R1: If M, then not-V.
R2: V.
R11: If M, then W.

Now, it we include the latent structure below R2, we would get:

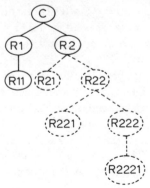

R21:   S.
R22:   If S, then V.
R221: Not-M.
R222: If not-M, then if S then V.
R2221: If not-M and S, then V.

The circularity here derives from the identity, C = R221. If we include the latent structure under R1, we would get:

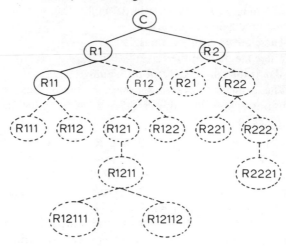

The propositions would be:

R12:   If W, then not-V.
R121:   If not-S, then not-V.
R122:   If W, then not-S.
R1211:   If V, then S.
R12111:   Not-M.
R12112:   If not-M, then if V then S.
R111:   V.
R112:   If V, then if M then W.

The second circularity derives from the identity, C = R12111. And since R111 = R2, if the latent structure under R2 were repeated for R111, C would then reappear a third time as one of the propositions latent vertically down from R111.

To summarize, we have formulated a number of basic principles for the logical analysis of reasoning, by using as illustrative examples various arguments and passages from Galileo's *Dialogue*. The main result has been that for the understanding of reasoning, which is a pre-condition of its evaluation, one needs to identify its propositional structure, which refers to the inferential interrelationships among its propositions. The concept of proposition here developed is one derivative from that of reasoning, so that the identity of individual propositions is dependent on the role they play in the various steps of an argument. The principles will be indispensable for carrying out the analyses of Chapter 16.

NOTES

[1] G. Galilei, *Dialogue*, tr. Drake, p. 137.
[2] *Ibid.*, p. 133.
[3] These rules are adapted from R. B. Angell, *Reasoning and Logic*, pp. 369–93. The same is true for most of the ideas in this chapter.
[4] Galilei, *Dialogue*, tr. Drake, p. 144.
[5] *Ibid.*, p. 19.
[6] *Ibid.*, p. 39.
[7] *Ibid.*
[8] *Ibid.*, pp. 144–48. For more details, see Chapter 16.
[9] *Ibid.*, pp. 139–41. For more details, see Chapters 8 and 12.
[10] *Ibid.*, pp. 58–62. See also, Chapter 16.
[11] *Ibid.*, p. 19.
[12] *Ibid.*, pp. 139–41.
[13] In Chapters 8 and 12, besides earlier in this chapter.
[14] The abbreviations here are those of Chapter 12.

CHAPTER 15

ACTIVE INVOLVEMENT:
THE EVALUATION OF REASONING

Besides possessing propositional structure, reasoning can be evaluated as being good or bad. The favorable evaluation of reasoning seems to arouse less interest among both practical reasoners and logical theorists. It is not clear why this is so: perhaps it is because the truth shines more brilliantly than error and hence needs less display to prevail and flourish than error does in order to be avoided. I am inclined to think that a fundamental reason is the fact that contrasts are more instructive and significant, for unfavorable evaluation is much more likely to involve considerations about good *and* bad reasoning if for no other reason than that one's own unfavorable evaluation, in order to be adequate and well-grounded, must be itself an instance of good reasoning. In short, in unfavorable evaluation one is (presumably) being exposed to both good and bad reasoning, the bad reasoning being unfavorably evaluated, and the good reasoning justifying the unfavorable evaluation; hence, one gets a better intuitive feeling and theoretical understanding of good and of bad reasoning, and of the difference between the two. Thus, the criticism of reasoning, as we may call the unfavorable evaluation of reasoning, is a topic very much worth studying.

FALLACIES

The only explicit attempt by the great majority of contemporary logicians to deal with the criticism of reasoning is the accounts of fallacies found in elementary logic textbooks. Before we proceed further, it will be good to examine them.[1] These accounts usually contain four elements: a general definition of the concept of fallacy, a description of various practices which are categorized as fallacies, a classification of fallacies into various groups, and an illustration of the descriptions of fallacies with examples.

As concerns the concept of a fallacy, there is a tendency to regard any logically incorrect argument as a fallacy. But then, either explicitly by giving a definition of a more specific notion, or implicitly by limiting the discussion to only certain kinds of logically incorrect arguments, the concept of a fallacy effectively becomes that of a type of common but logically incorrect argument.

The things which are categorized as fallacies include such practices as appeals to force, to pity, to authority, *ad hominem* arguments, begging the question, *dicto simpliciter*, converse accident, equivocation, amphiboly, composition, division, *post hoc ergo propter hoc*, hasty induction, affirming the consequent and denying the antecedent.

The classification of fallacies usually divides them into four major groups. One group is made up of fallacies variously called linguistic, verbal, semilogical, or of ambiguity; a second group of ones variously called psychological or of irrelevance; a third group is made up of so-called deductive, logical, or formal fallacies; a fourth group of so-called inductive or material ones.

The fourth element of such accounts, the examples of the various types of fallacies, is usually rather meager. It consists mostly, if not exclusively, of more or less artificially constructed examples for the purpose of illustrating the various descriptions of fallacies. Examples of fallacies actually occuring in the history of thought or in contemporary investigations and controversies are rare.

What is wrong with such accounts of fallacies? One problem concerns the paucity of actual examples, just mentioned. It is in fact puzzling that logic textbooks shouldn't be able to come up with more examples of fallacies actually committed given that fallacies are supposed to be *common* errors in reasoning. One gets the suspicion that logically incorrect arguments are not that common in practice; that their existence may be largely restricted to logic textbook examples and exercises.

If someone doubts the fact being referred to let him consult some textbooks.[2] Let him consult Wesley C. Salmon's *Logic* and discover as I, out of methodological duty, have just done once again before writing the next phrase, that no actual example is given for any of the fallacies mentioned above. Let him consult R. J. Kreyche's *Logic for Undergraduates* and fail to find a single actual example of the above mentioned allegedly common fallacies. Let him consult Cohen and Nagel's *Introduction to Logic and Scientific Method* and discover that, though their account of what they call 'abuses of scientific method' may be interpreted as containing some actual examples, their account of formal, verbal, and material fallacies does not. Let him consult Fearnside and Holther's *Fallacy, The Counterfeit of Argument* and discover that, though the examples are more numerous, better, and less artificial than in more standard logic texts, it is not clear how many of them would remain fallacious when put in the mind or mouth of actual persons in an actual situation.

Let him consult Monroe Beardsley's *Thinking Straight*, which is more practically oriented than most books. He will find that, though the exercises often concern actual examples, these are usually either prejudicially edited or inadequate illustrations of the various fallacies and argument forms. If these exercise examples had been good illustrations, one would find the author giving more actual examples in his textual discussion. Whereas one finds *only one* actual example of the fallacies mentioned above. On p. 217 (3rd edition, 1966) one finds a passage which may have been somewhat edited, since no reference of the source is given. Even so, the passage is not a very good example of an *ad hominem* fallacy, since for this purpose the author is forced to misinterpret the argument being advanced, which is actually an argument from analogy. Be that as it may, the point here is that even in such a practically oriented textbook as this one, one finds *only one* alleged example of a fallacy actually committed.

Finally let him consult what is perhaps the most ambitious, popular, and widely acclaimed introductory logic textbook, Copi's *Introduction to Logic*. He will find there in the author's account of fallacies only three actual examples. The historical question that Stalin is reported by Harry Hopkins to have asked Churchill at Yalta, "And how many divisions did you say the Pope had available for combat duty?", is given as an example of the fallacy of appeal to force (p. 74). The second actual example, which is supposed to illustrate the fallacy of appeal to pity, is part of the plea to the jury made by the famous lawyer Clarence Darrow at the trial of Thomas I. Kidd (p. 78). The third one, and in Copi's own words a "considerably more subtle example" of the appeal to pity fallacy, is Socrates' refusal in Plato's *Apology* to make an appeal to pity. This example is perhaps too subtle. And rather subtle one will find practically all of the eighteen or so quotations included in the exercises among a larger number of less subtle but more artificial examples.

The conclusion I wish to draw from such "consultations" is not that errors in reasoning are probably not common in real life, but that there probably are no common errors in reasoning. That is, logically incorrect arguments may be common, but common types of logically incorrect arguments probably are not.

The problem I wish to raise here is, do people actually commit fallacies as usually understood? That is, do fallacies exist in practice? Or do they exist only in the mind of the interpreter who is claiming that a fallacy is being committed?

The next problem I wish to raise concerns the classification of fallacies. It does not concern the question of the various groups into which fallacies should be subdivided, which is a problem often mentioned in textbooks and which actually turns out to be just a question of what names to give to the various groups. The problem is that a given alleged fallacy cannot be understood to be a fallacy unless it is classified as belonging to a certain group. In fact, the various groups derive from the various reasons given practices are fallacies. In other words, the inadequacy of the classification of fallacies can only derive from the inadequacy of the justification of their fallaciousness. Hence, since the arbitrariness of the classification is usually admitted, the arbitrariness of the justification should also be admitted.

In other words, the real problem here is whether any given alleged fallacy is really a fallacy and why, and not what the various kinds of fallacies are. In fact, if a fallacy is defined as a type of common but logically incorrect argument, the various types would have to be the following: (1) arguments claiming to be deductively valid but which are not; (2) arguments claiming to be inductively strong but which are actually inductively weak; (3) arguments claiming to have *some* inductive strength but which have none.

The problem then, with this element of logic textbooks accounts of fallacies, is the failure to recognize that the problem is not one of classifying a disputed practice which can be shown to be a fallacy on other grounds, but one of showing that it is a fallacy. For there is no way for an argument to be a fallacy without falling into one of the three above mentioned classes. This makes one suspect that many of the disputed practices usually regarded as fallacies may be either not fallacies or not always fallacies.

To investigate this in more detail, let us examine the second element of textbook accounts of fallacies, the description of various devices which I wish to call by the neutral term of "disputed practices". One problem with these descriptions is that they are usually prejudicial in the sense that their fallaciousness is built right into the description. There would be nothing wrong with this were it not for the fact that they then become logician's fictions or at best practices seldom found in reality (actual life). There is a pattern in these biased descriptions, and it is the following. If the disputed practice is a type of inductive argument, namely one claiming that the conclusion is only strongly, but not conclusively, supported by the premises, then the practice will be described as a type of deductive argument, namely one claiming that the conclusion is conclusively supported by the premises. If the disputed practice is a type of what might be called a partial argument, namely one claiming that the conclusion is only partly, but not too strongly supported by the premises, then the practice will be described as a type of allegedly inductive strong argument. One might think that the pattern runs out of material here, but it can be extended as follows: if the disputed practice is a type of non-argument, namely not an attempt to support one proposition with others, then it will be described as an argument claiming that certain propositions provide at least some support for another (the conclusion). Finally, if the disputed practice is an argument having as conclusion a special type of proposition, then it will be described as an argument having another conclusion; the pattern (or shall I say the fallacy?) is that of exaggerating the strength of the connection claimed between various assertions or of creating one where none is claimed.

The pattern can be illustrated by considering some of the disputed practices which textbook writers find most abhorrent. One of these is the so-called fallacy of affirming the consequent. I need not remind the reader that this fallacy is defined as that committed when a proposition is inferred from a conditional of which it is the antecedent and from the consequent of the conditional. But though it may be that all the textbook examples of arguments having the form "If P, then Q; Q; $\therefore P$" are indeed fallacies, that does not mean that actual arguments having this form are *normally* fallacious.

In order to declare such an argument fallacious the logician must interpret it to be a deductive argument, namely an argument that claims to be formally valid. In other words, when the argument given says "If P, then Q; Q; $\therefore P$" the logician must interpret this to mean "$P \supset Q$; Q; therefore, as a necessary consequence of them alone, P".

But the argument could also mean: "Q; the fact that P would explain the fact that Q; therefore, no other explanation of Q being available, we may presume that P". Under the second interpretation there is nothing logically wrong with the argument. Hence it is not arguments having the form of affirming the consequent that are fallacies, but deductive arguments having that form. That is, to show that the actual argument is a fallacy, the logician has to argue that it is deductive. This will usually be a difficult, if not insurmountable, task, since most such arguments are inductive; and evidence for this is the fact that the textbook writer usually does not even attempt to show that the argument does claim to be formally valid.

Another alleged fallacy beloved of textbook writers is the *post hoc ergo propter hoc* manner of reasoning. This is described in W. C. Salmon's *Logic* as "concluding that B was caused by A just because B followed A" (pp. 101–2) and in Copi's *Introduction to Logic* as "the inference that one event is the cause of another from the bare fact that the first occurs earlier than the second" (p. 82). No justification is given why these interpretations are preferable to the following: "concluding that B was caused by A *partly because B followed A*" or "the inference that one event is the cause of another from the fact, *among others*, that the first occurs earlier than the second". These latter interpretations should be preferred because they are more accurate in the sense that they correspond more closely to a type of reasoning in which people actually engage.

Included in a third group of fallacies are usually appeal to force and appeal to pity. A typical description of the first is "appealing to force or the threat of force to cause acceptance of a conclusion" and of the second "appealing to pity for the sake of getting a conclusion accepted" (Copi). These descriptions are prejudicial in their reference to a conclusion, since a conclusion is by definition a proposition which is part of an argument and which is being supported by other parts of the argument called premises. Hence appealing to force or pity to cause acceptance of a *conclusion* means giving an argument in which the conclusion is supported not by appealing to evidence but by appealing to force or pity. Of course *these* arguments are fallacies of irrelevance, but irrelevant are also those notions of appeal to force and to pity to actual appeals to force and pity. These could non-prejudicially but along the same lines be described as "appealing to force or to pity to cause acceptance of a certain proposition or to cause certain action". When so described, they can be seen to be methods, among others, of which giving an argument is one, in order to cause acceptance of a certain proposition. And the appropriateness of the method depends on the context. What does not depend on the context is the truth of the claim that it is a category mistake to regard typical actual appeals to force and to pity as fallacies; being non-arguments they cannot be logically incorrect arguments.

The conclusion to be drawn from the above discussion is that the concept of a fallacy as a type of common but logically incorrect argument is a chimera since the various disputed practices usually referred to as fallacies are either not common or not logically incorrect or not arguments. The chimera is probably the result of either hasty generalization or over-simplification or formalistic prejudice. That is to say, it is the conclusion of arguments having the form of a hasty generalization, or of ones whose premises neglect too many relevant considerations, or ones based on a formalistic prejudice. By the latter I mean valuing too highly the form that a given argument may have and thinking that the only useful notion of form is one such that arguments having a given form are either all correct or all incorrect; this then leads to regarding the disputed practices as always logically incorrect, whereas sometimes they are and sometimes they aren't.

My discussion may be interpreted to support and to be supported by two traditional philosophical doctrines, that evil is unreal and that the real is rational.

To establish a connection with the doctrine of the unreality of evil it suffices to regard fallacies as logical sins, or erroneous reasoning as logical evil. This in turn involves the idea that truth, logical correctness, validity, and rationality are values like goodness, beauty, and utility; the most recent historical appearances of this idea were perhaps in the philosophy of Benedetto Croce and of John Dewey. My discussion would then be giving a meaning to the unreality of logical evil and would be indicating the sense and extent to which it is true: actually occurring logically incorrect arguments are not very common because, to find one, the logician usually has to exaggerate the strength of the logical connection between premises and conclusion being alleged by the argument giver.

The real which in this context the rationalist would claim to be rational is the actual reasoning practiced by people in their various activities. The rationalist is *a priori* skeptical about whether fallacies are as common as introductory logic textbooks would want to make him believe. The realist on the other hand, that is to say the one who has a sense of reality, feels (*a posteriori*) that people are simply not as irrational as those textbooks would lead one to believe. The rationalist will try to find ways of interpreting actual arguments so that they are logically correct; the realist will try to find as accurate a reconstruction of actual arguments as possible. In so doing the realist discovers that actual arguments do not tend to be logically incorrect as much as textbooks lead one to believe; that is to say he discovers that the rationalist is right. Realism and rationalism in this case coincide.

Of course, a thoroughgoing rationalist may be inclined to go to the absurd extreme of claiming that no actual argument is ever fallacious. Absurd if for no other reason than that would mean that the usual logic textbook accounts supporting their concept of a fallacy are logically correct. To be sure the rationalist might in this quest try to find evidence that those accounts are not arguments, and hence not logically incorrect for categorial reasons. He may find rationality in them by categorizing them differently. I personally don't know what this category would be, but I doubt very much that the rationality involved would be pedagogic rationality. The realist in me prevails here and parts company with the rationalist.

UNDERSTANDING VERSUS EVALUATION

If reasoning cannot be criticized as easily as logic textbooks would have it, that is not to say that it cannot be criticized at all. What one can do is to study concrete instances of argument criticism and then derive whatever theoretical lessons one can; this corresponds to the approach advocated earlier (Chapter 13). What I plan to do here is to study Galileo's critiques of various Aristotelian arguments as found in the *Dialogue*. The basic principles of analysis will be the framework of propositional structure discussed in the last chapter; moreover, each of my discussions will be based on a specific indicated passage from Galileo's book and will have three parts: a reconstruction of an argument widely accepted in Galileo's time, which will be referred to either as the object argument or as the Aristotelian argument; a reconstruction of Galileo's critique as an argument about the object argument, which will be referred to either as the Galilean criticism or as the meta-argument; a logical commentary on my part containing discussions of concepts presupposed in Galileo's critique and evaluations of Galileo's meta-arguments. Most of these evaluations of mine will be implicit in my reconstructions of Galileo's critiques, whose relatively obvious plausibility usually will not require any explicit discussion.

One central lesson that will be emerging is that it is both possible and effective to evaluate arguments 'actively' in the sense that the inferential interrelationships among the propositions involved are tested by reasoning at the level of, and largely in terms of, the object argument and by checking whether what follows from asserted premises is the conclusions drawn in the object argument or other propositions. This lesson is supported both by the procedure of Galileo, who usually reaches conclusions different from the Aristotelian ones, and by my own procedure in which I will usually reach the same conclusions as Galileo; the active character of my procedure lies primarily in the intricacy of the reconstructions on which my agreement with Galileo is grounded; moreover, *accuracy* for all of these reconstructions is claimed.

To appreciate the theoretical significance of this last claim it should be stressed that one of the most fundamental problems faced by the serious theorist and practitioner of reasoning is that of understanding *vs.* evaluation. The problem derives from the fact that, besides being characterizable as something possessing propositional structure, as explained in the last chapter, reasoning is something susceptible of being good or bad, namely of being evaluated favorably or unfavorably. Moreover, the following principle would seem uncontroversial: the unfavorable evaluation of a given argument is worthless if it involves a misunderstanding of the argument; that is, demonstrated understanding is a necessary condition for responsible negative criticism. Now, to demonstrate understanding, normally an *accurate reconstruction* of the argument is needed; then the question arises as to how elaborate, charitable, and explicit one should be in reconstructing the argument, since assumptions implicit in the original must be stated explicity in the reconstructed argument.

Scriven's solution to this problem is what he appropriately calls the Principle of Charity[3], which he formulates as follows:

the assumptions you identify mustn't be too strong or they will be an unfair reconstruction of the argument, since they will be fairly easily refuted even though the argument might still be perfectly sound. On the other hand, the assumptions mustn't be too weak, or they won't connect the stated premises to the conclusion. (Or they won't support an independent part of the conclusion that has so far received no support.) The assumption mustn't be a triviality of definition or fact − since it then isn't worth mentioning. Nor can it be a mere assertion of the fact that the arguer thinks this is a sound argument, since that's not worth mentioning. It must be something new but on the other hand, it must still be true and relevant.[4]

As Scriven goes on to add, the Principle of Charity makes pretty good sense in the abstract; the difficulty is, however, to put it into practice in a judicious manner.[5] The way to learn this is not by explicit formal elaborations and qualifications of the principle, but rather by applying it to concrete instances; or at least, such concrete applications are a pre-condition for an eventual more formal or exact statement of the principle.[6]

Another aspect of the problem is that an asymmetry seems to exist in the relations between understanding and favorable criticism on the one hand and between understanding and unfavorable criticism on the other. That is, the requirement of understanding plays a less important role, if at all, in favorable evaluation: if the original argument has been reconstructed in such a way that the reconstructed argument is sound, valid, or has significant merit, then such a reconstructed argument acquires for the logician an importance and value which it retains even if it is an inaccurate reconstruction. The point is as simple as it is subtle, and in recent literature it has not escaped the explicit mention of a philosopher such as Richard Montague.[7]

Why should this asymmetry exist? I believe the only way of explaining it is to accept the kind of 'practical' approach to logic being advocated here. For to admit the asymmetry is like saying that a good reconstructed argument may be regarded as a logical achievement (relatively independently of its accuracy vis-à-vis the original argument), whereas a bad reconstructed argument will not be counted as a logical achievement unless it is an accurate reconstruction. That is, the good argument has intrinsic logical value, presumably because or in the sense that logic is itself the practice of reasoning, and hence a good instance of reasoning is a logical contribution. But where is the logical value of the bad but accurately reconstructed argument? Presumably it lies in the argument that goes from the original to the reconstructed argument; the accuracy of the reconstruction is another way of talking about the goodness of this high level argument. The logical accomplishment may thus again be regarded as an instance of logical practice, or reasoning, as the "practical" approach to logic advocates.

Practical logic so conceived also explains the desirability of something like the Principle of Charity: adequate understanding is necessary because otherwise our reconstruction will be inaccurate, and an inaccurate reconstruction is a case of a bad or unsound higher level argument going from the original to the reconstructed one.

Thus the Principle of Charity is being used everywhere in my analysis. But this is not to say that I shall inquire directly or to any appreciable extent into whether Galileo himself is using such a principle, i.e., whether the arguments criticized by him are accurate reconstructions. This would be beyond the scope of my investigation. However, it is important to note that, in view of what I have called above his 'active' method of evaluation, Galileo is implicitly conforming to the Principle of Charity. In short, Galileo's meta-arguments have logical interest primarily because they constitute good criticism of the object arguments (as I have reconstructed them), whereas my reconstructions (of object arguments and of meta-arguments) have logical interest partly (indirectly) because of the logical value of Galileo's meta-arguments, and partly (directly) insofar as my reconstructions are validly grounded, i.e., are accurate. Moreover of course, in my commentary below I will be articulating general concepts for the understanding and evaluation of reasoning, in accordance with the approach to logic being advocated here.

NOTES

[1] Such accounts have been criticized by C. L. Hamblin, *Fallacies*, pp. 9–49. His criticism overlaps with mine and has had some effect. For example, I. M. Copi's fourth edition of his *Introduction to Logic* was the first re-edition of this book to appear after Hamblin's; in it Copi says that "thanks largely to the useful criticisms by Professor C. L. Hamblin in his book *Fallacies*, some corrections have been made in Chapter 3" (p. viii). When we look at this chapter dealing with 'Informal Fallacies' (pp. 72–107), we find, however, that the 'corrections' really amount to nothing more than cosmetic changes; see, for example, pp. 73, 78, and 82.

[2] The books mentioned in my text are, I believe, a representative cross-section of books

that have been traditionally well-known, widely-used, and influential. More recently (in the seventies) a trend has emerged toward more realistically oriented textbooks. The best such books, however, de-emphasize the fallacy approach: for example, Michael Scriven's *Reasoning*, which has probably the best discussion of the evaluation of actual, everyday arguments, avoids the concept of fallacy altogether; Robert J. Fogelin's *Understanding Argument: An Introduction to Informal Logic*, which has probably the best collection of actual, classic selections of arguments, emphasizes analysis and understanding, rather than criticism. On the other hand, the fallacy-oriented textbooks tend to define so many types of fallacies that most of what they call 'fallacies' involve faults and problems other than errors of *reasoning*. The result is an unacceptable amount of violence to the fundamental distinction between an argument and a non-argument; this is bound not only to cause confusion to the student, but it affects the authors' own judgment of what is an argument and what isn't. For example, Howard Kahane, on p. 28 of *Logic and Contemporary Rhetoric*, misinterprets an insult as an error of reasoning when he gives as an example of "the fallacy of *ad hominem* argument" Spiro Agnew's famous remark that "a spirit of national masochism prevails, encouraged by an effete corps of impudent snobs who characterize themselves as intellectuals". (For an analysis of some actual and non-fallacious *ad hominem* arguments, see my "The Concept of *Ad Hominem* Argument in Galileo and Locke".) Another example is the invitation by S. Morris Engel, on pp. 63–64 of *With Good Reason: An Introduction to Informal Fallacies*, to treat as a 'fallacy of ambiguity' an explanatory remark found in John Maynard Keynes's *Treatise on Money*: "It is enterprise which builds and improves the world's possessions. If enterprise is afoot, wealth accumulates whatever may be happening to Thrift; and if enterprise is asleep, wealth decays whatever Thrift may be doing". (For a significant and fallacious argument exploiting ambiguity, see my 'Galileo's Space-Proportionality Argument: A Role for Logic in Historiography'; for a significant but nonfallacious use of ambiguity in reasoning, see my 'Logic and Rhetoric in Lavoisier's Sealed Note: Toward a Rhetoric of Science".) It is obvious that the cavalier way in which these admittedly realistically fallacy-oriented authors dispose of highly controversial issues can only encourage superficiality.

[3] M. Scriven, *Reasoning*, pp. 71–73, 76–77, and 75.

[4] *Ibid.*, pp. 173–174.

[5] *Ibid.*, p. 174.

[6] This fits very well with Scriven's conception of logic as reasoning about reasoning, or at least with my interpretation of Scriven's idea. See above, Chapter 13.

[7] See J. F. Staal (ed.), 'Formal Logical and Natural Languages (A Symposium)'; Montague's remark can be found on p. 275.

CHAPTER 16

GALILEO AS A LOGICIAN:
A MODEL AND A DATA BASIS

For several reasons it is now time to engage in a minute critical analysis of the arguments put forth in Galileo's book. In part, it is time to evaluate the arguments which in Chapter 2 were merely described; in part it is useful to reanalyze the *Dialogue* in accordance with the results of Chapter 7, which suggest the idea of Galileo as being first and foremost a methodologist of reasoning; in part it is important to give content to the program outlined in Chapters 11, 12, and 13; in part it is important to put to work the technical framework developed in Chapter 14; finally one needs examples of Galileo's critiques of arguments, in accordance with the suggestions in the last chapter.

It turns out that almost all passages in the *Dialogue* can be analyzed as being explicitly arguments about arguments. The higher level arguments will be called *meta-arguments*, while the lower level ones will be called *object arguments*. Object arguments are usually arguments about natural phenomena advanced either by Aristotle, or by his followers, or by others whom Galileo is criticizing. Meta-arguments are arguments advanced by Galileo in criticism of various features of the object arguments and are thus only *secondarily* about natural phenomena. (However, some of Galileo's criticism is substantive and pertains to the factual truth of a premise or presupposition of an object argument, so that some meta-arguments, or parts thereof, are *directly* about natural phenomena.)

It is the extent of the context under consideration and the explanatory power of the reconstruction which determine whether or not one of Galileo's physical arguments is regarded primarily as part of a meta-argument: the governing consideration is whether or not a particular physical argument can be reconstructed as having the function of serving to criticize some object argument contained in the passage. For example, Galileo's own arguments about natural motion (F43—57),[1] about the ship experiment (F170—5), and about centrifugal motion (F217—24, 238—44) are primarily meta-arguments about corresponding Aristotelian or geostatic arguments. Conversely, some of Galileo's meta-arguments occur in a context such that they are primarily designed to support some of his physical theses by criticizing counter-arguments; such are his arguments criticizing the geostatic explanation of tides (Fourth Day) and of the motion of sunspots (F372—83), those criticizing the

Peripatetic explanation of the telescopic appearance of the moon (F95–112) and of the moon's secondary light (F112–24), and the one criticizing the Aristotelian notion of the heliocentrism of planetary motions (F346–68).

The analysis of *all* meta-arguments in the *Dialogue* would have been unnecessarily long and excessively tedious; hence the following criteria of selection have been used. First, I have neglected meta-arguments which are critiques of counter-arguments and whose immediate context makes them part of a physical argument which is not itself a meta-argument; examples are the second group of arguments in the preceding paragraph. Second, I have neglected passages which are primarily expository or narrative in character: e.g., the discussion of the earth-moon similarities (F87–95) and differences (F124–32); the discussion of Aristotle's authority (F132–9); the presentation of the probable arguments for the earth's diurnal motion (F139–50); and the preliminary presentation of the geostatic arguments (F150–9). Third, I have neglected physical arguments whose immediate context does not make them part of a meta-argument, such as those concerning conservation and composition of motion (F180–93), and concerning retrograde motion (F368–93). The fourth criterion I have used is that of manageable length, so that by starting at the beginning of the book I stopped only when I thought I had gone through a sufficiently large number of analyses, mindful of quality and variety, besides quantity. This led me to stop with the centrifugal motion argument (F214–44) which is approximately at the middle of the book (pagewise). This place is a very proper one since the rest of the Second Day and the second half of the Third Day (F383–442) have already been explicitly analyzed as meta-arguments in the analytical summary of my discussion of the book's logical structure (Chapter 2); similarly, Galileo's discussion of the 1572 nova at the beginning of the Third Day (F299–346) is obviously, and was explicitly analyzed as, a meta-argument about Chiaramonti's argument that the nova was sublunary.

Finally, it will be seen below that Galileo's meta-arguments are very suggestive and instructive logically and lend themselves very readily to the formulation of principles and clarification of concepts at a first level of logical theorizing; this I have done in the subsections labeled 'Comments'. It should also be noted that the discussion in each section below has two aspects, one of logical theory and one of logical practice; this is reflected in the title of each section which includes both the name of an actual argument and one or more notions useful in the theory of reasoning. This two-sidedness is analogous to the concreteness of the methodological content elaborated earlier (Chapter 5), where methodological ideas were being illustrated with scientific examples. In the present discussion ideas in the theory of reasoning are illustrated by actual reasoning.

Some clarifications of a technical nature are needed before we proceed further. Each analysis below consists of three parts: an object argument, a meta-argument, and comments. The object arguments and meta-arguments are reconstructions of the corresponding passage, and all their propositions have been numbered basically in accordance with the standard labeling discussed earlier (Chapter 14): in order to distinguish the propositions of the object argument and of the meta-argument in a given section, the letter 'A' precedes the numbers of all propositions of the former (as a reminder that Aristotelian arguments are usually meant), and the letter 'G' precedes all the numbers of all propositions of the latter (as a reminder that Galilean arguments are being referred to). The object argument will usually have only one final conclusion, to which the single digit '1' has been assigned, so that all its supporting propositions can then be numbered in the standard way with numerals whose first digit will be '1'. The meta-argument will usually have more than one independent subargument, and hence the distinct final conclusions of these have been assigned the labels G1, G2, G3, G4, etc., (the 'G' being a reminder that Galileo's own arguments are being referred to); all supporting propositions for each such final conclusion, say Gn, are then numbered in the standard way by adding digits to the Gn in accordance with the rules. Such notation achieves two purposes simultaneously: it describes unambiguously the logical-propositional structure of all object arguments and meta-arguments to be discussed, and it labels uniquely each Aristotelian and Galilean proposition in the passages under discussion.

Given such numbering the construction of the corresponding structure diagrams is a merely mechanical task, and hence such diagrams have been drawn only in those few cases which seemed particularly interesting. For added convenience, the various labels in a diagram have not been enclosed in a circle or closed figure.

On some occasions in my comments, reconstructed arguments are stated for which it is relatively more debatable whether they are contained in the text, though it is clear that parts are. In such case the label 'FG' is used as a prefix before the standard numbering by numerals.

In stating the various object arguments and meta-arguments, the labels of their constituent propositions are written in parenthesis just before the propositions themselves. On the other hand, the statement of a meta-argument will usually also contain labels of object argument propositions used to express various claims about those propositions; in such cases no parentheses have been used around the labels of these object argument propositions, so as not to create visual confusion with the labels of the meta-argument propositions. Analogously, in my comments, I refer to propositions of both object arguments and meta-arguments by using their labels without parentheses.

All references are to Volume VII of Galileo's *Opere*. Readers interested in logical theory and the theory of reasoning may simply ignore these references; others are thereby provided with the means of engaging in that type of logical practice which is the reconstruction of arguments from original texts.

Finally, the order of the sixteen sections that follow is merely that in which the arguments appear in Galileo's book. Since from the point of view of the theory of reasoning this order is neither one of ascending nor one of descending significance, it is advised that theorists of reasoning *do not* read them in this order, and that they read them in an order best suited to their special interests. For example, those interested primarily in formal logic should read Sections 4, 8, 12, and 11 before any others. Those primarily interested in informal logic should begin with Sections 15, 3, 9, and 6. Those primarily interested in the logic of science should begin with Sections 2, 14, 16, and 10. Others can use the various section titles, which attempt to be descriptive; a list of these titles may be found at the very beginning of the next chapter.

1. CONCLUSIONS VS REASONS VS CAUSES, AND THE THREE-DIMENSIONALITY ARGUMENT (F33–38.24)[2]

Perhaps it is a mere coincidence, but the fact is that the *Dialogue* begins with an implicit discussion of the fundamental logical distinctions between a proposition and its justification and between a proposition and its implications. The argument under discussion is the following:

(A1) The world is perfect because (A11) it has the three dimensions of length, width, and depth and (A12) these are all the dimensions that exist; and (A11) the world has three dimensions because (A111) three is a very special number (in that three is (A1111) the number of parts that everything has, namely beginning, middle, and end; (A1112) the number used in sacrifices to the Gods; and (A1113) the least number of things required before the word 'all' can be applied to refer to them collectively.)

The relevant proposition here is the three-dimensionality of the world, A11. Galileo is here accepting this proposition, but neither its alleged implication (A11, ∴ A1), nor its alleged justification (A1111, A1112, A1113; ∴

A111; ∴ A11). In short, Galileo is agreeing that the world is perfect and that it has all three dimensions, but denying that there is a connection between the two propositions such as to ground perfection on three-dimensionality; and he is also denying that three-dimensionality is to be grounded on the special properties of the number three. He justifies his denial of (A11, ∴ A1) by suggesting an alternative justification of the perfection of the world, namely that it is the work of God; and he justifies his denial of (A111, ∴ A11) by arguing that the world is three-dimensional because exactly three mutually perpendicular lines can be drawn through any given point.

If we call the Aristotelian argument here criticized by Galileo, the three-dimensionality argument, then we may say that Galileo's own argument is one about the three-dimensionality argument. Galileo is arguing that (G1) the three-dimensionality argument is doubly invalid, because (G11) perfection is improperly grounded on three-dimensionality, and because (G12) the latter is improperly grounded on the special properties of the number three; and he justifies proposition G11 by saying that (G111) the cause of the world's perfection is God, and he justifies proposition G12 by saying that (G121) the cause of the world's three-dimensionality is the fact that three and only three mutually perpendicular lines can be drawn through a given point.

How correct is Galileo's own argument? I would agree with Galileo's final conclusion, G1, and with the two reasons he gives, G11 and G12. That is, I accept G1, Galileo's justification (G11, G12, ∴ G1), and also propositions G11 and G12. Why do I accept G11? For Galileo's reason or for some other reason? Galileo's justification of G11 could be elaborated as follows:

(G112) The world is perfect.
(G111) The cause of this perfection is God.

When so elaborated I would accept the justification, but not the premises as such. My difficulty would be with accepting G112 and consequently with accepting G111. Do I have a justification of G11? To be sure, I would say that I don't see that there is any connection between perfection and three-dimensionality; but this would be merely another way of expressing G11. However, I believe I could use my rejection of G112 as my reason for G11; that is, I would say that the perfection of the world is improperly grounded on its three-dimensionality because the world is not perfect.

What of G12? Here I would accept both Galileo's premise G121, and his justification (G121, ∴ G12), which one may want to elaborate by adding as a latent proposition (G122) that the world is three-dimensional.

Thus my own argument (F) about Galileo's argument (G) would be the following: All the inferential steps of Galileo's argument (G11, G12, ∴ G1; G111, G112, ∴ G11; and G121, G122, ∴ G12) are acceptable; moreover, all of its propositions are acceptable except for G112 and G111; these two propositions together amount to saying that the world is perfect because God created it.

Let us summarize as follows. Some of the important propositions being discussed are:

(a) The world is perfect.

(b) The world is three-dimensional.

(c) The number three has special properties.

(d) The world is perfect because God created it.

(e) The world is three-dimensional because three is the exact number of mutually perpendicular lines that can be drawn through a given point.

The Aristotelian argument is:

(a)
|
(b)
|
(c).

Galileo's critique is:

My conclusions are:

(1) If a and d, then not-$(b, ∴ a)$.

(2) If b and e, then not-$(c, ∴ b)$.

(3) b.

(4) e.

(5) ∴ not-$(c, ∴ b)$.

(6) Not-a.

(7) Not-d.

(8) Not-a, ∴ not-$(b, ∴ a)$.

Formalism aside, Galileo's technique here is first of all one of partial agreement, with respect to (a) and (b). Then he gives causal explanations of these, namely (d) and (e). Then he uses these explanations together with the facts being explained to refute the alleged logical connection between (a) and (b), and the logical ground for (b). The net result is the logical dissociation of propositions (a) and (b) and their causal association, respectively, with God and with three-fold mutual perpendicularity.

Another logical lesson we may derive is that it is very important to distinguish our conclusions from our reasons, and reasons from causes. Conclusions and reasons are distinct because one may accept a conclusion, e.g. (b), but not the alleged reason, e.g. (c); or one may accept a reason, e.g. (b), but not the conclusion drawn from it, e.g. (a). Causes and reasons are distinct in the sense that a cause is given in a context where some fact is agreed upon but what produces it is being disputed, e.g. when Galileo says that the world is perfect because God created .it, or that the world is three-dimensional because only three mutually perpendicular lines are possible; whereas a reason is being given in a context where some conclusion is disputed and accepted facts are appealed to in order to settle the dispute; for example, for Galileo what is in question is neither the world's perfection nor its three-dimensionality but whether three-dimensionality implies perfection and is implied by the special properties of the number three.

2. LOGICAL EVALUATION, AND THE NATURAL MOTION ARGUMENT (F38.25–57.4)

The Aristotelian argument being criticized here is (F38.25–43.11):

(A1231) There are two kinds of simple motions, straight and circular.

(A123) ∴There are two kinds of natural motions, straight and circular.

(A122) But, straight natural motion is toward or away from the center of the universe.

(A121) And, circular natural motion is around the center of the universe.

(A12) ∴Natural motions are toward, away from, and around the center of the universe.

(A11) But, for each natural motion there is a corresponding natural element.

(A1) ∴Earth and fire are the elements whose natural motions are, respectively, toward and away from the center of the universe; and the heavenly bodies are the bodies whose natural motion is circular around the center of the universe.

Galileo's critique (F38.25–43.11) is as follows: (G1) The basic premise of the argument, A1231, may be accepted since (G11) simple motions are those along simple lines, (G12) simple lines are those whose parts are congruent to one another, and (G13) the example of the cylindrical helix may be dismissed. However, (G2) none of the conclusions drawn from this basic premise follow from it: (G21) proposition A123 does not follow because (G211) what follows from A1231, i.e. from a consideration of only such simple motions, is that there is only one type of natural motion, the circular, and that straight motion is merely the simplest means of acquiring the natural state of rest at the proper place or of circular motion; (G22) proposition A12 does not follow because (G2211) straight natural motion would be natural motion along any straight line, and (G2212) circular natural motion would be natural motion alone any circular path, and hence (G221) both A122 and A121 are false; (G23) proposition A1 would not follow from A11 and A12 because (G231) all that would follow from them is the *existence* of *some* elements whose natural motion is straight and of one whose natural motion is circular, and (G232) some independent justification would be needed to equate these elements with earth, fire, and the heavenly bodies.

It is important to notice that Galileo seems to be concentrating on criticizing the validity of the Aristotelian argument, that is, the various inferential steps of the argument supporting A1, which is very appropriate since this argument was supposed to be an *a priori* one. The criticism has the very interesting form of claiming that one proposition does not follow from another because what really follows from it is some other proposition different or inconsistent with the first. This type of logical evaluation might be labeled 'active', for the critic engages in reasoning at the same (object) level as the argument being evaluated, rather than reasoning only at the metalevel, so that he produces a counterargument. The evaluation is nevertheless logical (and not merely scientific) since it centers around the interrelationships of propositions. It should also be noted that proposition G211 is not being justified within this argument, but rather that it will be justified in the next one about natural circular motion; it may be regarded as one further conclusion that may be drawn from the final one in the next argument. It is this logical connection which gives the natural circular motion argument of the next section scientific value in the sense of the science of logic, independently of its scientific merits in the sense of substantive natural science; that is, the argument to be reconstructed in the next section has without doubt the logical-evaluative function of showing that the Aristotelian natural motion argument is invalid, in particular that proposition A123 does not follow from A1231.

Galileo's natural circular motion argument (F43.12–57.4)

(G122)   Straight motion cannot be the natural state of integral bodies.

(G121)   Straight motion cannot naturally be perpetual.

(G12)   ∴Straight motion cannot be natural.

(G112)   But, circular motion can be the natural state of integral bodies.

(G111)   And, circular motion can naturally be perpetual.

(G11)   ∴Circular motion can be natural.

(G1)   ∴The only type of natural motion is circular. (I.e., all natural motion is circular.)

And

(G23)   Straight motion is a good means of restoring order out of disorder.

(G22)   Straight motion is a possible means of creating order.

(G21)   Straight motion is a natural way of acquiring circular motion.

(G2)   ∴Straight motion is the simplest means of acquiring the natural state of rest at the proper place or of circular motion.

Now, (G122) straight motion cannot be the natural state of integral bodies first because (G122a1) if it were then such bodies would be changing place, and (G122a2) if integral bodies change place that means that the universe is not in perfect order, and second because (G122b1) if it were then such bodies would have a tendency to move through an infinite distance, since (G122b11) a straight line is in principle infinite.

And, (G121) straight motion cannot naturally be perpetual because (G1211) when straight motion is violent it obviously cannot be perpetual, and (G1212) when straight motion is nonviolent (i.e., spontaneous or 'natural') it can be shown that it cannot be perpetual. This is so because (G12121) nonviolent straight motion must be accelerated, and (G12122) accelerated nonviolent straight motion cannot be perpetual. The reason for the latter is that (G121221) when a body moves with accelerated nonviolent straight motion it must be approaching a place toward which it has a natural inclination (for (G1212211) otherwise it would not be accelerating), and (G121222) when it has reached this place the acceleration must stop. And the reason for the former is that (G121211) nonviolent straight motion is the simplest way for a body to move from one place to another toward which it has a natural inclination, (G121212) in moving from one place to another toward which it has a natural inclination a body will acquire speed continuously and gradually (since (G1212121) there would be no reason to acquire one degree of speed rather than another), and (G121213) if a body acquires speed continuously and gradually then its motion is accelerated.

Also, (G112) circular motion can be the natural state of integral bodies because (G1121) a natural state of circular motion preserves order.

Finally:

(G1111113)   acceleration occurs naturally when a body approaches the point toward which it is inclined to move;

(G1111112)   retardation occurs naturally when a body moves further away from the point toward which it is inclined to move;

(G1111111)   circular motion around a point toward which a body is inclined can be interpreted as motion where the body is simultaneously approaching and moving further away from that point;

(G111111) ∴ circular motion around a point toward which a body has natural inclination would be subject to simultaneous acceleration and retardation;

(G11111) ∴ such circular motion would be uniform;

(G1111) ∴ it could be perpetual;

(G111) ∴ circular motion can naturally be perpetual.

Comments. The accuracy of this reconstruction may be justified as follows. The argument in support of G122 can be found on F43.12–43.30; the argument in support of G121 can be found partly on F56.20–56.25 (the part containing proposition G12122), and partly on F44.17–45.10, for the part containing G12121. The argument supporting G111 can be found on F56.9– 56.20. The argument supporting G112 can be found on F55.3–55.4 and F56.1–56.9. Proposition G23 can be found on F43.30–44.8; proposition G22 on F44.8–45.10 and F53.13–55.1; and proposition G21 on F45.11–53.6.

Though the structures are obvious from the numbering, it is useful to have them visually available in the following figures:

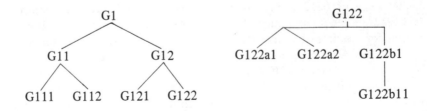

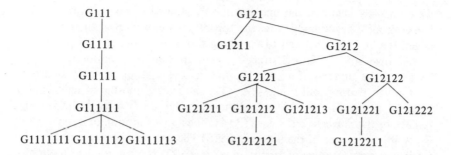

Finally, it is clear that Galileo's natural circular motion argument involves a concept of natural motion different from that of the Aristotelian argument. For the latter, natural motion is identified with simple motion; for the former, natural motion is simple motion which can be a perpetual state. There is another concept of natural motion lurking behind both arguments, namely that of natural as nonviolent or spontaneous; this is not the controversial concept in this context since such a concept would imply that only observation could tell us which types of motion were natural, whereas the present discussion is attempting an a priori determination.

Is Galileo's counter-argument sound? All the inferential steps seem to me rather plausible, but beyond saying this there is no way of answering the question, other than by comparison with Aristotle's argument: A1231, ∴ A123. Whereas this inference seems rather arbitrary, Galileo's counter-justification makes use of several Aristotelian ideas, and it is in this sense that what really follows from A1213 is G1 and G2 instead.

It should be noted that G1 is not saying that all circular motion is natural, nor that all motion is natural in the sense that all movable bodies have only circular motion. But is G1 telling us anything about actual motion? It is telling us that if an actual motion is circular, it may be natural, and if it is noncircular, it definitely is not. Primarily G1 tells us something about what can and cannot be rather than what is or is not. It will help Galileo to prove that the earth *can* move, that it is *not impossible* for it to rotate and revolve. But this is not to say that the earth *does* move. Similarly, straight motion is merely unnatural, rather than unreal.

3. EQUIVOCATION, COMPOSITION, CIRCULARITY, AND THE UNIVERSE-CENTER ARGUMENT (F57.5–62.27)

The object argument. (A15) The parts of the element earth and of the element

water, namely heavy bodies, move naturally straight downwards, for (A151)
if one drops a heavy body from the top of a tower with straight and vertical
edges, the body can be seen to fall along those edges and to land at the foot
of the tower. (A14) Natural straight downward motion is motion toward the
center of the universe, for (A1411) the natural motion of heavy bodies is
contrary to that of light bodies, (A1412) the natural motion of light bodies
is straight-upwards, (A1413) straight-upward motion is toward the circumference of the universe, and hence (A141) the natural motion of heavy bodies
is toward the center of the universe. (A13) The parts of the elements fire and
air can be seen to move naturally upwards. (A12) Natural upward motion is
straight motion toward the inside of the lunar orb. (A11) Whatever applies to
the parts applies to the whole. Therefore, (A1) the natural motion of the
whole Earth is straight toward the center and of Fire straight away from the
center.

   The meta-argument. (G1) The argument is unsound in that its last step is
really an equivocation, for (G11111) the sense of 'natural' in which the
premises are true is that of 'spontaneous' or 'nonviolent', and hence (G1111)
the sense in which A1 follows is that the *spontaneous* motion of the whole
earth would be straight toward the center (if it had been violently removed
from it), but (G1112) the sense in which A1 is being *asserted* is that the potentially perpetual state of the whole Earth is straight motion, for (G11121)
this is the sense in which A1 conflicts with a claim of the argument just
presented (Galileo's natural circular motion argument), namely its proposition G122, so that (G111*) if 'natural' in A1 means 'spontaneous' then the
inference is valid but irrelevant, and (G111**) if 'natural' in A1 refers to a
potentially perpetual state, then A1 is relevant but obviously cannot be
inferred from the premises in the sense in which they are being asserted to be
true; it follows that (G11) the argument is either valid but irrelevant or
relevant but invalid.

   Morever, (G2) the first premise, A15, is questionable insofar as it claims
that the downward motion is straight, for (G211) the actual straightness of
free fall cannot be inferred from its apparent (visible) straightness, and hence
(G21) proposition A15 does not follow from A151.

   Third, (G3) premise A11 is absurd, for (G31) if premise A15 is interpreted
in a sense in which it is true, namely to mean that the spontaneous motion of
heavy bodies is toward the center of the earth, then these two premises would
imply that the whole earth would spontaneously move toward its own center
if it were violently removed from whatever place it is in, and (G32) this is
intrinsically impossible.

Next, (G4) the second premise, A14, is questionable insofar as it refers to the center of the universe, and the supporting argument begs the question in two ways. (G41) First, (G4111) proposition A1413 is not obvious since (G41111) it speaks of the universe circumference; therefore, (G411) one may ask for a justification; (G412) the justification would have to be on the basis of the proposition that straight-upward motion is seen to be away from the surface of the earth; (G413) if so, then the argument would have to be: "(A141321) Straight-upward motion is seen to be away from the earth's surface; ∴ (A14132) straight-upward motion is toward any circumference greater than and concentric with the earth's; but (A14131) the universe circumference is concentric with the earth's; ∴ (A1413) straight-upward motion is toward the universe circumference"; but (G414) proposition A14131 which is here being assumed is essentially identical to the one it is helping to prove, namely proposition A14. (G42) Second, (G4211) proposition A141 does not follow from its stated reasons, A1411, A1412, A1413, since (G421111) in a spherical finite universe any straight motion whatever is toward its circumference, but only those motions toward the circumference which are along diameters of the universe have contrary motions which are toward its center, and hence (G42111) the natural motion of heavy bodies could be contrary to 'toward the circumference' and yet not toward the center of the universe; therefore, (G421) if proposition A141 is to follow, proposition A1413 must be interpreted to mean that upward motion is toward the circumference and along a diameter of the universe; but (G422) if this is the meaning of A1413 then it needs a justification (since (G4221) the only thing we know is that upward motion is along diameters of the earth), and (G423) if it needs a justification the only one that could be given would be based on the proposition that the centers of the earth and of the universe coincide, and (G424) if A1413 is justified by assuming such an identity then the argument supporting A14 would beg the question (since (G4241) to assert such an identity is essentially the same as to assert proposition A14).

Comments. The accuracy of these reconstructions may be justified as follows. 'A151, ∴ A15' can be found on F58.28–58.35. The subargument supporting A14 can be found on F58.38–59.7. The immediate reasons supporting A1 can be found on F57.10–57.19. Galileo's G1 argument is on F57.20–57.29 and F56.25–57.10. The G2 critique is on F57.29–57.34 and F59.16–59.18. The G3 argument is on F57.34–58.4. The G41 subargument is on F60.10–60.24. The G42 argument is on F58.5–58.8 and F60.25–61.15.

The last step of the object argument as stated gives rise to a good example of what modern logicians call the fallacy of composition, which is an argument where it is concluded that the whole has a given property just because its parts have that property. Of course, in the argument as stated one of the premises, A11, explicitly asserts this; hence, technically speaking, the argument avoids the fallacy, In practice, however, it makes little difference whether A11 is explicitly stated or not, for the problem of whether it holds in any given case must be solved on a case-by-case basis. Galileo's critique G3 may be taken to be a proof of the formal invalidity of arguments of this type; Galileo does this by calling attention to a property which it is logically impossible for a whole to have, though it is in fact true that the parts do have it.

It is interesting to note that Galileo's critique of the validity of the last step of the Aristotelian argument, namely his G1 argument, is essentially of the form: the conclusion does not follow from the premises because what follows from them is some other proposition, and this is so because if we take the premises to be true then we could validly infer a proposition about the *spontaneous* (or 'natural' in the sense of spontaneous) motion of the whole terrestrial globe, and this proposition is different from A1 because A1 as asserted must be speaking of natural motion in the sense of a potentially perpetual state of motion. Such an active evaluation of reasoning conforms to Galileo's previous critiques.

Galileo does not in this context prove G211, but he does so later on F164–9. One way of stating that later criticism is that all that follows from A151, namely from apparent straight fall, is that *if* the earth does *not* rotate, *then* free fall is actually straight; the crucial consideration in the proof is that *if* the earth rotates *then* free fall is actually slanted to the earth's surface.

The argument supporting G4211 seems to criticize the validity of 'A1411, A1412, A1413, ∴ A141' by the counterexample method of showing that it is possible for the premises to be true while the conclusion is false.

Finally, it may be useful to summarize the two ways in which the object argument begs the question. They both involve the subargument supporting A14: if one of its supporting reasons is to be acceptable it needs to be justified on the basis of a proposition essentially identical to A14, and if one of its supporting inferential steps is to be correct one needs to assume the same essentially identical proposition. We may also say that the argument begs the question in the sense that when its latent structure is constructed, the argument becomes doubly circular by having A14 reappear twice as an underlying reason for itself.

4. CONTRADICTION, AD HOMINEM, EQUIVOCATION, AND THE CONTRARIETY ARGUMENT (F62.28–71.33)

The object argument. (A12) Change does not occur unless there is contrariety; (A11) there is no contrariety among heavenly bodies, since (A111) heavenly bodies have circular motion, (A112) no motion is contrary to circular motion, and (A113) contraries have contrary motions; therefore, (A1) heavenly bodies are unchangeable. Now, to show that no motion is contrary to circular motion, consider that (A1121) there are three types of simple motions, straight-toward, straight-away from, and around the center, that (A1122) the two straight motions are contrary to each other, and that (A1123) one thing can only have one thing as contrary. [F62.31–63.17]

The meta-argument. (G1) The Aristotelian argument supporting A1 is invalid since (G11) if argument A1[3] were otherwise correct the valid conclusion to be inferred would be that either the earth as well as the heavenly bodies is unchangeable, or that the heavenly bodies as well as the earth are changeable, or that there is no connection between change and the straight/circular motion distinction. [F63.18–64.11]

Second, (G21) the argument's main intended consequence is that the earth stands still, but (G22) its main premise, A12, is more difficult to ascertain than this consequence. For (G211) the argument's final conclusion, A1, would be used as follows: "since heavenly bodies are unchangeable, and since the earth is changeable, the earth is not a heavenly body, and therefore the earth does not have the annual motion". And (G221) the earth is a very large and accessible body, but (G222) the connection between change and contrariety is impossible to find in most phenomena, e.g., in the apparently spontaneous generation of some insects, in the different rates of change in most living things, and in changes resulting from the transposition of parts. [F64.11–65.24]

Third, (G3) the contrariety argument is problematic since (G31) if one accepts it then one would have to accept the following self-contradictory argument: "(A42) heavenly bodies have contraries, since (A421) heavenly bodies are unchangeable, (A422) terrestrial bodies are changeable, and (A423) changeability and unchangeability are contraries; but (A41) all bodies which have contraries are changeable; therefore, (A4) heavenly bodies are changeable". One would have to accept this argument since (G311) propositions A422 and A423 are uncontroversial; (G312) proposition A421 is the conclusion of the contrariety argument, A1; (G313) proposition A42 follows from its immediate reasons; and (G314) if one accepts the contrariety argument then proposition A41 could be justified as follows: "(A41) all bodies which have contraries are changeable, since (A411) change does not occur unless there is contrariety and (A412) this means either that (a) change does not occur to a body unless it has a contrary, or that (b) change does not occur in a region unless there is contrariety within that region, and (A413) the latter cannot be; for (A4131) proposition (b) would imply that there is no change within the region of the elements earth and water and no change within the region of the elements fire and air, and (A4132) if so then terrestrial bodies would be unchangeable". [F65.25—67.18]

Fourth, (G4) proposition A11 is questionable since (G41) Aristotle would regard heavenly bodies as the denser parts of the heavens, (G42) if heavenly bodies are regarded as the denser parts of the heavens then differences of rarity and density exist in the heavens, and (G43) if differences of rarity and density exist in the heavens then a change-producing contrariety exists in the heavens; for (G4311) differences of rarity and density give rise to the light/ heavy contrariety of terrestrial bodies, (G4312) this contrariety gives rise to the upward and downward spontaneous motions, and (G4313) these motions are allegedly the source of terrestrial changes, and hence (G431) differences of rarity and density may be regarded as the cause of terrestrial changes; moreover, (G4321) the cause of terrestrial as well as celestial differences of rarity and density is the quantitative difference of more or less matter in a given space, (G4322) the cause of terrestrial differences of rarity and density is not the qualitative difference of heat and cold (since (G43221) the density of solid substances changes little when their degree of heat changes significantly), and hence (G432) the cause of terrestrial differences of rarity and density is the same as the cause of celestial differences of rarity and density. [F67.19—69.24]

Fifth, (G5311) premise A12 would be supported on the basis of the changes and of the up and down motions of terrestrial bodies; but (G5312) the term 'bodies' in the phrase 'terrestrial bodies' refers to the parts of the terrestrial globe; therefore, (G531) the meaning of the term 'bodies' relevant to premise A12 is that of 'parts of integral (whole) bodies'. Whereas, (G521) the meaning of the term 'bodies' in premise A11 is that of 'integral (whole) bodies' since (G5211) A11 is supported on the basis of claims about integral (whole) bodies. But, (G51) if the conclusion in A1 is to follow from the premises then the term 'bodies' must be used with the same meaning in both premises; (G52) if 'bodies' means 'parts of integral bodies' then premise A11 is groundless; (G53) if 'bodies' means 'whole bodies' then premise A12 is groundless; therefore, (G5) if the conclusion follows then the argument is groundless (i.e., either the conclusion does not follow or it is based on groundless premises). [F69.25–71.9]

Comments. (1) Galileo gives no explicit argument to justify G11. However, the following argument is implicit in the text in F63.18–64.11, and so I will still use the label 'G' and a numbering to refer the various propositions to its final conclusion G11:

(G111) There are really two sides to the contrariety argument, A1, which may be reconstructed as follows: "(A21) change does not occur unless there are contraries; (A22) there are no contraries unless there are contrary natural motions; (A23) contrary natural motions are only those straight motions made in contrary directions; (A24) there are only two natural straight motions made in contrary directions, namely toward and away from the universe center; (A25) natural straight motion toward the universe center belongs to the elements earth and water, and motion away from the center belongs to the elements fire and air; therefore, (A2) change does not occur except among terrestrial bodies". And "(A322) bodies whose natural motion is circular have no contraries, since (A32211) the other two types of simple motion are contrary to each other, and (A32212) one thing can only have one contrary, and hence (A3221) there is no circular motion contrary to natural circular motion; but, (A321) where there is no contrariety there is no change; therefore, (A32) bodies whose natural motion is circular are unchangeable; but, (A31) natural circular motion belongs only to heavenly bodies; therefore, (A3) only heavenly bodies are unchangeable".

Now, (G112) the 'natural' motions being mentioned in these arguments ought to be conceived either as nonviolent, spontaneous motions or as potentially perpetual motions, since (G1121) the Aristotelian concept of natural motions as simple has already been criticized.

(G1131) If 'natural notion' is conceived as potentially perpetual motion then there are no contrary natural motions, since (A11311) it has already been argued that in this sense of 'natural' there is only one type of natural motion, namely the circular; but (G1132) if there are no contrary natural motions then proposition A31 ought to be replaced by the proposition that natural circular motion belongs to all integral bodies, to the earth as well as to the heavenly bodies; (G1133) if the latter proposition replaces A31 then the conclusion in argument A3 would be that the earth as well as the heavenly bodies are unchangeable; therefore, (G113) if 'natural motion' is conceived as potentially perpetual motion then the valid conclusion is that the earth as well as the heavenly bodies are unchangeable.

(G1141) If 'natural motion' is conceived as nonviolent, spontaneous motion then straight motions toward and away from the center of their wholes belong, respectively, to the denser and rarer parts of all integral bodies, heavenly as well as terrestrial; (G1142) if this last clause is true then it ought to replace proposition A25 in the argument supporting A2; (G1143) if A25 is so replaced then the valid conclusion is that heavenly bodies as well as the earth are changeable; therefore, (G114) if natural motion is conceived as nonviolent, spontaneous motion then the valid conclusion is that the heavenly bodies as well as the earth are changeable.

Finally, (G115) if 'natural motion' is conceived in both ways then it follows that there is no connection between change and the straight/circular distinction; for, (G1151) propositions G113 and G114 imply that if 'natural motion' is conceived both ways then the earth and heavenly bodies are both changeable and unchangeable, and (G1152) the source of this contradiction is the proposition that change does not occur unless there are contrary motions, and (G1153) if we deny this proposition we get that it is not true that change does not occur unless there are contrary motions, and (G1154) this denial means that it is not true that change occurs if there are straight motions and does not if there are circular motions, and (G1155) this means that change is unconnected with the difference between straight and circular motions; now, proposition G1152 is true because (G11521) in the following argument the contradiction is derived from the proposition in question together with other acceptable propositions, and the proposition in question is derived from propositions explicitly used in the arguments A1 and A2: "(113/123–2) change does not occur unless there are contraries; (113/123–1) there are no contraries unless there are contrary (natural) motions; therefore, (113/123) change does not occur unless there are contrary (natural) motions; but (122) natural circular motions are not contrary, and (121) the natural motion of all integral bodies — earth as well as heavenly bodies — is circular; therefore, (12) the earth as well as heavenly bodies are unchangeable; moreover, (112) nonviolent motions towards and away from the center of a whole are contrary, and (111) nonviolent motions toward and away from the center of their whole are the natural tendencies of all the parts of all integral bodies; therefore, (11) the heavenly bodies as well as the earth are changeable; therefore, (1) the heavenly bodies and the earth are both changeable and unchangeable".

(2) Galileo's second criticism does not use any term to refer to the fault being claimed there for the contrariety argument. The fault may be described as grounding a conclusion on a premise which is more difficult to ascertain than the proposition which the conclusion is designed to support. An appropriate term might be 'useless'. To be more exact, we might say that an argument is

useless when it is contextually a subargument of a bigger argument such that the final conclusion of the first argument appears as an intermediate proposition of the bigger argument, and a final reason of the first argument is more questionable than the final conclusion of the bigger argument. For example, consider two arguments with the following structures:

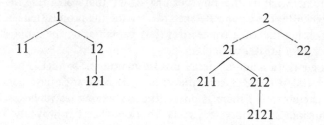

Argument 1 is useless if in the context argument 2 is being presented; if 1 = 21, 11 = 211, 12 = 212, 121 = 2121; and if either 2121 or 211 is more questionable than 2.

Is uselessness a logical or rhetorical fault? The fault clearly does not refer to the *connections* among the propositions in the various subarguments of either argument involved; hence the fault is not logical. The fault is that in the bigger argument the final conclusion is easier to ascertain than one of its final reasons; this means that there is at least one argument which has for final conclusion either the final conclusion of the bigger argument *or its negation*, and which is much better than any argument supporting the final reason in question; if so, the so-called bigger argument (#2 above) should not be given, but if it should not be given, and if in fact it was not given (since only the smaller argument was given), then the smaller argument is faulty insofar as it suggests the bigger one; so what is wrong with the original (smaller) argument is this suggestion, and this seems to be primarily a rhetorical failure.

(3) In stating the contraricty argument, supporting Λ1, Galileo uses the reasoning indicator 'if therefore' (*se dunque*, F35.62). I am not talking about the hypothetical indicator 'if . . . then _____' which connects two simpler propositions into a single, but more complex conditional proposition; rather I mean a connective which relates propositions into an argument, and which is more fully expressed as 'if therefore . . . then' with a proposition before and one after the 'then'. Moreover, the 'if therefore' is always preceded by another proposition. These purely syntactical considerations suffice to distinguish 'if therefore' from 'if-then'.

The next thing to consider is whether 'if therefore' is equivalent to 'therefore if'. The latter indicates that a conditional proposition is being stated as a conclusion from what precedes, whereas the former indicates that the (not necessarily conditional) proposition following the 'therefore' is a conclusion from what precedes. But in order to understand this fully we need to determine what is the full meaning of 'if therefore'. Let us consider Galileo's own text as a concrete example, stating the propositions in the same order and using the same connectives (but paraphrasing the propositions as I already have for greater clarity):

(1) all generation and all decay involve contraries, so that (2) change does not occur unless there is contrariety; but (3) contraries have contrary motions; if therefore (4) there is no contrariety among heavenly bodies, since (5) no motion is contrary to circular motion, then (6) heavenly bodies are unchangeable. Here, the reasoning indicators, in the order in which they occur, are: so that, if therefore, since, and then. 'So that' is unambiguous and indicates the subargument '(1), ∴ (2)'; 'since' is also unambiguous, and it indicates '(5), ∴ (4)'. Keeping in mind the latent proposition that (7) heavenly bodies have circular motions, semantical intuition tells us that (4) is being grounded on (3), (5), and (7) together. Semantical considerations also tell us that (6) is being grounded on (4) and (2). This gives the argument:

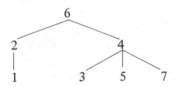

This structure corresponds to the argument as previously reconstructed, if we let 6 = A1, 2 = A12, 4 = A11, 7 = A111, 5 = A112, 3 = A113, and if we add 1 = A121 to the previous argument where it had been neglected for the sake of incisiveness.

For the present discussion proposition 5 is an unnecessary complication, since it might have been written in place of (3), leaving the latter latent. Thus the meaning of 'p; if therefore q, then r' is 'p; ∴ q; ∴ r'. This makes 'if therefore' a triadic connective, preceded by at least one clause and followed by at least two, indicating the statement of two subarguments, with the clause immediately following having the role of an intermediate proposition, the last clause the role of their final conclusion, and the preceding clause the role of their final reason.

Finally, it should be noted that 'p; if therefore q, then r' may also be expressed as 'p; if then q, therefore r' and as 'p; if then q, r'.

(4) Galileo's third critical conclusion, G3, has been stated as attributing a 'problematic' character to the contrariety argument. It would have been somewhat inaccurate and premature to use any other less general concept since the text merely asks, without answering, what this criticism proves, and since it is not immediately apparent what the criticism in fact amounts to. Let us then look at the matter more closely.

There are three things to keep in mind, the original argument A1, the critical counterargument A4, and the proposition that heavenly bodies are unchangeable, A1 = A421. Let us use the following abbreviations:

T_A: *t*errestrial bodies are changeable (*a*lterable);
T_C: *t*errestrial bodies have *c*ontraries;
H_A: *h*eavenly bodies are changeable (*a*lterable);
H_C: *h*eavenly bodies have *c*ontraries;
X: only contrariety yields change.

Three acceptable claims implicit in Galileo's critical argument G3 are:

1. T_A;
2. if not-H_A and T_A, then H_C; and
3. if H_C and X, then H_A.

On the basis of these, we could make the following deductions, where the source and justification of each line is noted in parenthesis after each line number:

4(3).	If H_C, then (if X then H_A);
5(2–3).	if not-H_A and T_A, then (if X then H_A);
6(5).	if not-H_A and T_A and X, then H_A;
7(6).	if T_A and X, then (if not-H_A then H_A);
8(–).	if (if not-H_A then H_A) then H_A;
9(7–8).	if T_A and X, then H_A;
10(1–9).	if X then H_A;
11(10).	not-(X and not-H_A);
12(11).	either not-X, or H_A.

Line 8 is obvious upon reflection; it is, in fact, an instance of the so-called Law of Clavius. Line 10 says that if one accepts the connection between change and contrariety, he ought to accept heavenly change; hence, the Aristotelians ought to accept heavenly change. In other words, they are

inconsistent in accepting the connection together with heavenly incorrupti-
bility (line 11), and they should reject one or the other (line 12). It should be
noted that the contradictory propositions in Galileo's counter-argument G3,
namely propositions A4 = H_A and A421 = not-H_A, do not render it invalid;
rather they invalidate, i.e., render false, the conjunction of two (Aristotelian)
propositions, A41 and A421. In other words, Galileo's argument is a partial
(but valid) reductio ad absurdum of heavenly unchangeability, partial
because the reductio depends on the assertion of the connection between
change and contrariety; from the point of view of the *conjunction* of these
two ideas, it is a full reductio ad absurdum. Thus heavenly unchangeability is
not refuted in the abstract, but only within the Aristotelian framework.

Let us now see what follows about the original contrariety argument A1.
The argument corresponds to proposition:

13.   if not-H_C and X, then not-H_A.

Putting this together with previous results we get:

14(10—13).   if not-H_C and X, then H_A and not-H_A; [4] and
15(14).   not-(X and not-H_C).

Line 15 tells us that the antecedent clauses of line 13 are inconsistent. This,
of course, does not affect the *formal* validity of the corresponding argument;
all it means is that its premises are contradictory. So there is a contradiction
hidden within the premises of the contrariety argument. However, because of
this contradiction, the argument is practically worthless and rhetorically
ineffective, for the premises could never be all true, and hence the situation
could not arise where one could say that since all the premises are true,
therefore so is the conclusion. In other words, from the point of view of the
acceptance of the conclusion (heavenly unchangeability), one cannot be
properly persuaded by being told that he ought to accept not-H_A because he
accepts or ought to accept X and not-H_C, for in view of line 15, he ought not
to accept both of these, and hence if he does, he should be shown *this* inconsistency, rather than being invited to reason in accordance with line 13.
We might thus label the contrariety argument 'self-contradictory' and note
that this is more of a rhetorical than logical impropriety.

Next, the following proposition follows directly from line 10, since if H_A
follows from X alone, it will follow from X plus other propositions:

16(10).   if not-H_C and X, then H_A.

When we compare this with the contrariety argument expressed in line 13, we can see the arbitrariness of this argument. Once more, the contrariety argument is not *formally* invalidated by line 16, but rhetorically speaking, and from the point of view of practical reasoning, one might ask why one should *draw* conclusion not-H_A when its opposite could just as legitimately be drawn. In the light of line 16, the contrariety argument might be labeled arbitrary, and for this reason rhetorically ineffective.

Moreover, in the present situation 'if-then' has perhaps an inferential meaning, rather than being a material conditional.[5] If the 'if-then' is interpreted inferentially, then line 16 means that not-H_C and X imply H_A, and hence that 'not-(if not-H_C and X, then not-H_A)'; this would then invalidate the contrariety argument. In other words, if, by line 16, one can *really* infer H_A from not-H_C and X, rather than merely hypothetically and abstractly do so, as one does in formal logic, then one cannot also really infer not-H_A. Thus, in the sense of actual reasoning and active evaluation previously noted, one might say that the conclusion of the contrariety argument does *not* follow because what follows is instead the denial of it.

Next, by the same technique as before, from line 10 one may also deduce

17(10).   if not-H_C and X and T_C, then H_A.

The interesting thing that this tells us is that the contrariety argument is incomplete, another rhetorical, alogical fault. In the contrariety argument one infers not-H_A from two premises, not-H_C and X. Line 17 tells us that if the Aristotelians had also considered another proposition acceptable to them, namely T_C, as a premise in the context of the same argument, they would have arrived at a different (opposite) conclusion.

Let us now consider the following deduction:

18(9–13).   if T_A and X and not-H_C, then H_A and not-H_A;
19(18).   not-(T_A and X and not-H_C);
20(19).   either not-T_A, or not-X, or not-(not-H_C).

The last line does not depend on the assertion of terrestrial changeability T_A. It says that the Aristotelians ought to reject either terrestrial change, or the connection between change and contrariety, or heavenly lack of contrariety. This is analogous to the previously justified conclusion of Galileo's first criticism. The main difference is that the alternatives in line 20 include heavenly lack of contrariety rather than heavenly unchangeability.

Finally, a formal way of investigating the formal validity of the contrariety argument is by putting line 13 in the antecedent of a conditional and seeing what follows. Let us begin with the following obvious assertion:

21(−).   if line 13 and not-H_C and X, then not-H_A;

then:

22(10−21). if line 13 and not-H_C and X, then H_A and not-H_A;
23(22).   not-(line 13 and not-H_C and X);
24(23).   either not-line 13, or H_C, or not X;
25(24).   if not-H_C and X, then not-line 13.

Line 24 says that one ought to reject either the contrariety argument, or heavenly unchangeability, or the connection between change and contrariety. Line 25 says that if one accepts the latter two ideas, then one should reject the contrariety *argument*. Another interpretation is that if one claims consistency for not-H_C and X, then one has to reject the validity of the argument. This corresponds to our previous remark that its validity derives from the contradiction in its premises. Since the Aristotelians accept heavenly lack of contrariety and the connection, it seems that they should reject the argument; hence the argument should be invalid for them. Does it follow from this that the argument is invalid *per se*? Of course, it should not be forgotten that line 25, like everything else in this discussion depends on lines 1, 2, and 3. The Aristotelians might, of course, reject some of these, though it is not clear on what grounds. Of the three, line 3 might seem most vulnerable since it corresponds to an argument suggested by the contrariety argument itself, or to be more exact, they seem to be both instances of the same manner of reasoning. In fact, they are what I have previously called two *sides* of the same argument; however the two sides are different enough so that it is possible to accept line 3 and reject line 13; for, disregarding X, they are the converse of each other; and if the connection between change and contrariety is held in only one direction, we would accept that whenever there is contrariety there is change, but not that whenever there is no contrariety there is no change. This would receive additional support from empirical considerations. In this manner the contrariety argument may actually be invalidated. Of course, it could still be written in a formally valid manner, with an unjustified premise, i.e., with the connection between change and contrariety stated more strongly that is warranted; in this case its problem would be this premise, and I feel inclined to say that this would not change the 'logic' of the situation, meaning the substantive issue. But if one accepts this manner of speaking, one is no longer equating logic with formal logic.

(5) Simplicio objects to argument A4 used in Galileo's criticism G3 by comparing it to the liar's paradox:

The Cretan said, "I am lying".
Either the Cretan was lying, or he was telling the truth.
If he was lying, then "I am lying" was a lie.
If "I am lying" was a lie, then he was telling the truth.
∴ If the Cretan was lying, then he was telling the truth.
But if he was telling the truth, then "I am lying" was true.
And, if "I am lying" was true, then he was lying.
∴ If the Cretan was telling the truth, then he was lying.

In this paradox, from each one of the alternative assumptions one can deduce its opposite. However, in the argument in question, though one can deduce heavenly changeability from heavenly unchangeability, one cannot deduce unchangeability from changeability. Hence, whereas the liar's paradox gives rise to an infinite series of deductions, the present argument terminates once heavenly changeability has been derived. The following makes this clear:

(A52)
(A5112) ⟩ Change occurs if and only if there are contraries.
(A5122)
(A5121) The elements earth and water are the contraries of the elements fire and air (since (A51211) the natural motion of the former is contrary to that of the latter).
(A512) ∴ Elemental bodies are changeable.
(A5111) But heavenly bodies have no contraries (since (A51111) their natural motion is circular).
(A511) ∴ Heavenly bodies are unchangeable.
(A51) ∴ Heavenly bodies are the contraries of elemental bodies (since the former are changeable and the latter are unchangeable).
(A5) ∴ Heavenly bodies are changeable.

The structure of this argument is:

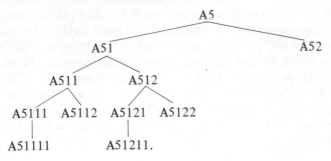

(6) Galileo's critical argument G4 is interesting because it is basically *ad hominem* in the (seventeenth century) sense that it derives a conclusion, G4, previously unaccepted by an opponent from propositions accepted by him but not necessarily by the arguer; in the present case the main such proposition is that heavenly bodies are the denser parts of the heavens. In saying that argument G4 is 'basically' *ad hominem* I am implicityly calling attention to the fact that, in the way in which it is expressed, the argument is not *completely ad hominem*; for the immediate premises for the conclusion (propositions G41, G42, and G43) are such that Galileo would accept them, the first because of its character of (historical) report, the last two because of their conditional structure; this subargument could have been stated in such a way as to involve directly the disputed proposition and then would have been explicitly *ad hominem*: "proposition A11 is questionable since heavenly bodies are the denser parts of the heavens, and hence the contrariety of rarity and density exists in the heavens". This is the way that subargument 'G431, G432, ∴ G43' is stated.

Rhetorically speaking, that is, from the point of view of persuasion, *ad hominem* arguments are very effective since the arguer thereby provides his opponent with reasons for accepting the conclusion; though the *ad hominem* argument would not constitute a reason for someone who did not accept the problematic premise(s), the opponent in the case does accept it (them), and hence if the argument is otherwise correct, he is induced to accept the previously unaccepted conclusion.

Let us now examine in more detail the above suggestion that conditional propositions provide a logically impeccable means of expressing *ad hominem* arguments. In G4 Galileo is saying essentially that: "(G41') if one believes that heavenly bodies are the denser parts of the (mostly extremely rarified) heavens, then he ought to believe that there is a change-producing heavenly contrariety; therefore, (G4') Peripatetics ought to believe in heavenly contrariety, since (G42') they believe that heavenly bodies are the denser parts of the heavens". So far the argument is logically correct, indeed formally valid. The rest of the argument may be reconstructed as follows: "(G411') if one believes that heavenly bodies are the denser parts of the heavens, then he believes that the terrestrial contrariety of upward and downward motions causes terrestrial changes (since (G4111') both beliefs are part of the Aristotelian system); (G4121') but if the contrariety of upward and downward motions causes terrestrial changes, then differences of rarity and density are the ultimate cause of these changes (since (G41211') differences of rarity and density cause the contrariety of upward and downward motions of terrestrial bodies), and (G4122′) if differences of rarity and density are the ultimate cause of terrestrial changes, then differences of rarity and density among heavenly bodies cause heavenly changes (since (G41221′) terrestrial and celestial differences of rarity and density have the same nature, for (G412211′) they derive both from the quantitative difference of more or less matter in a given volume); therefore, (G412′) if the contrariety of upward and downward motions causes terrestrial changes, then differences of rarity and density among heavenly bodies cause heavenly changes; but (G413′) if differences of rarity and density among heavenly bodies cause heavenly changes, then there is a change-producing heavenly contrariety; therefore (G41′) if one believes that heavenly bodies are the denser parts of the heavens, then he ought to believe that there is a change-producing heavenly contrariety."

This argument has the following structure:

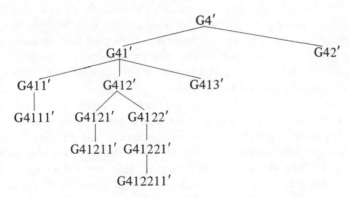

All propositions here are conditional, except for G4′, G42′, G4111′, G41211′, G41221′, G412211′; I believe they are all acceptable and that their inferential connections are correct. This may be compared with G4, whose structure is:

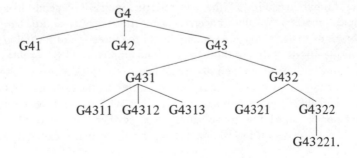

Let us now see what can be said by way of generalization. Suppose we have the *ad hominem* argument 'R1 since R11; but R2; ∴ C', where the final reasons R11 and R2 are accepted by the opponent but not by the arguer. The latter, however, does accept that *if* R11 then R1 and that if R1 and R2 then C. If the *ad hominem* argument is inferentially correct, then these conditions will be true; putting the two together we get 'if R11 and R2 then C'. Since the 'opponent' accepts R11 and R2 but not C, this shows his beliefs are inconsistent, i.e., that he should either reject 'not-C' or one of R11 or R2. This shows that the central value of *ad hominem* reasoning is logical.

(7) What does criticism G5 amount to? That is, what further conclusion could be drawn from G5? One reformulation would be to say that either the conclusion does not follow, or else one of the premises is groundless; that is, that the argument is either logically invalid or based on an unjustified premise. We could also say that if the premises are well grounded then the conclusion does not follow; this is not exactly the same as saying that if the premises are true then the conclusion is false, and hence that the argument is formally invalid; it is however to say that even if the premises are accepted the conclusion need not be, and hence that the argument is rhetorically invalid, i.e., ineffective from the point of view of persuading someone about the conclusion on the basis of the premises.

The critique is also a good illustration of the interplay between logic and rhetoric. On the one hand, it might seem that the formal validity of the contrariety argument depends on the nonambiguous use of a particular term, which is not a formal-logical characteristic of the term; for one might have considered G531, G521, and G51 together and drawn the conclusion that the argument is logically invalid, and then logical invalidity would have been a consequence of a non-formal fact. On the other hand, it should be noted that the ambiguity of the term 'bodies' is established by logical considerations in the sense of considerations pertaining to the understanding of the reasoning involved in arriving at the two premises A11 and A12. Thus the formal invalidity of the contrariety argument depends ultimately on logical matters broadly conceived, and this suggests a broadening of the notion of logic. Is this broadening one in the direction of including rhetorical considerations? I do not think so since the ambiguity of the problematic term is being grounded on nonrhetorical features of the contextual arguments supporting the two problematic premises, that is, the argument 'A111, A112, A113, ∴ A11' and the implicit argument for A12; each of these subarguments is being assumed not to be ambiguous. Rather the broadening involves a greater emphasis on logical context, in the sense that the logical validity of a given argument may depend on that of a bigger argument which includes the subarguments needed in the context to support the premises of the given argument; the fuller argument may be such that if we take its final reasons to be true, the final conclusion does not follow because those final reasons lead to intermediate propositions in which a term is used ambiguously; or the full argument may be such that if we take the final conclusion to be true, then at least one of the final reasons from which it follows is obviously groundless.

Criticism G5 may be taken to charge the contrariety argument with the fallacy of equivocation, and it helps us to conceive this fallacy as an argument where a term is used ambiguously in such a way as to conceal the fact that if the term is given one meaning then the conclusion obviously does not follow, and if it is given the other meaning then one of its premises is obviously groundless.

(8) The arguments in this section can serve to illustrate and justify some claims about 'unless' (*se non*) and 'only' (*solo*). Many logic textbooks advise their students to translate 'p unless q' as the inclusive disjunction '$p \vee q$' or equivalently as the material conditional '$-q \longrightarrow p$'.[6] This advice reflects both linguistic insensitivity and inadequate understanding of actual reasoning. It seems more accurate[7] to interpret 'p unless q' to mean 'p if and only if not-q' and 'only A's are B' as 'all A's are B and all non-A's are non-B'. When 'unless' and 'only' are so understood, then the intuitively valid inferences in argument A2 and *in part* in argument A3 become also theoretically and reflectively valid, whereas the orthodox logician's interpretations would render them invalid. In fact, these arguments should be formalized as follows, where 'iff' is an abbreviation for 'if and only if':

(A21) No change iff no contraries;

(A22) no contraries iff no contrary motions;

(A23) contrary motions iff contrary directions;

(A24) contrary directions iff toward and away from center;

(A25) toward and away from center iff terrestrial bodies;

(A2) ∴ no change iff no terrestrial bodies.

And

(A32) All bodies with natural circular motion are unchangeable;

(A31) all heavenly bodies have natural circular motion, and all non-heavenly bodies lack natural circular motion;

(A3) ∴ all heavenly bodies are unchangeable, and all nonheavenly bodies are changeable.

There is no problem with this version of argument A2, where it should be noted that Galileo thinks that A25 ought to be replaced by: (A25′) toward and away iff denser and rarer parts of integral bodies. However, in argument A3, the second clause of proposition A3 does not follow, and the second clause of A31 is superfluous. Galileo does not mention this problem, which he must have regarded as trivial. It could be argued that since 'nonheavenly' in this context means 'terrestrial', since the second clause of A3 does follow from A2, and since in the text the second argument, A3, is stated immediately after the first, therefore the second clause of A3 is being grounded on proposition A2, and therefore both parts of A3 follow. However, this would not explain the presence of 'only' in the statement of A31, i.e., the problem of the superfluousness of the second clause of our A31 would remain; in fact, this presence of 'only' indicates that the 'only' of A3 is being grounded thereupon. Hence argument A3 would seem to be inescapably invalid (in part). At any rate, Galileo's critical modification of argument A3 would not have this fault since he wants to replace A31 by the proposition that (A31′) all integral bodies have natural circular motion, and thereby conclude merely that all integral bodies are unchangeable.

5. THE COUNTER-EXAMPLE METHOD, AND THE A POSTERIORI ARGUMENT (F71.34–83.11)

The object argument. (A11) No changes have ever been observed to occur in the heavenly bodies; (A12) changes are observed routinely on the earth; therefore, (A1) the heavenly bodies are unchangeable [F71.34–72.30]

The meta-argument. (G11) If there were changes in the heavenly bodies, most of them could not be observed from the earth, since (G111) the distances from the heavenly bodies to the earth are very great, and (G112) on the earth changes can be observed only when they are relatively close to the observer. And, (G121) if there were changes in the heavenly bodies large enough to be observable from the earth, these large changes would not be observed unless careful, systematic, exact, and continual observations were made; but, (G122) no such observations have been made (by the argument's proponents); therefore, (G12) if there were changes in the heavenly bodies large enough to be observable from the earth, they might not have been observed. Therefore, (G1) the a posteriori argument (i.e., the object argument) is logically invalid. [F72.31–74.11]

Second, (G2) the argument is invalid because (G21) no terrestrial changes would be noticeable to an observer on the moon before some particular very large terrestrial change had occurred, and yet (G22) terrestrial bodies are obviously changeable and would have been so even before that occurrence. [F74.12–74.23]

Third, (G3) the argument as stated is a stronger version of the one that may correspond more closely to Aristotle's letter, namely: "(A11′) no one has ever observed any generation or decay of heavenly bodies in the heavenly region; therefore, (A1′) the heavenly region is unchangeable". For (G31) this argument would have the additional fault of sanctioning the following clearly invalid argument: "(A11″) no one has ever observed any generation or decay of terrestrial globes in the terrestrial region; therefore, (A1″) the terrestrial region is unchangeable". (G32) What is wrong with the literal version of the Aristotelian argument is that it implicitly contrasts heavenly bodies to such terrestrial bodies as cities rather than to terrestrial globes. [F74.23–75.8]

Fourth, (G4) premise A11 is false since (G41) changes have now been observed both within heavenly bodies and in the heavens. For example, (G411) comets constitute changes in the heavens, regardless of whether they originate there or in the terrestrial region [F76.17, F77.1–77.12]; (G412) the new stars of 1572 and of 1604 are observed generation and decay in the heavens [F76.16–76.19, F82.13–82.36]; and (G413) sunspots are changes on the body of the sun since (G4131) they appear and disappear at random in the solar disc, (G4132) their apparent motion across the solar disc is slower at the edges and faster in the middle, and (G4133) their apparent size and shape is narrower at the edges and wider in the middle [F79.1–79.25]. On the other hand, (G414) the arguments against these recent observations are worthless; for example, (G4141) the author of the *Anti-Tycho* uncritically assumes that comets are phenomena to which parallax theory is applicable [F77.7–77.8], (G4142) he uncritically rejects contrary observations not confirming his thesis [F77.9–77.12], and (G4143) he is inconsistent in his attitude toward comets and toward new stars (since (G41431) he attempts to locate the former in the terrestrial region but regards the latter as irrelevant on the grounds that they do not represent changes within well-established heavenly bodies) [F77.13–77.18, F82.13–82.36]; and (G4144) the interpretations of sunspots as planets circling the sun and obscuring parts of it conflict with the evidence of their random appearance and disappearance, and of their changing apparent speed and shape [77.19–80.9].

Finally, (G5) the argument's conclusion A1 is false; for the fact that (G511) changes have now been observed in the heavens shows that (G51) heavenly bodies are changeable [F75.9–76.11, F80.16–82.7]

Comments. (1) It is interesting to note the presence of the second premise,

A12, in the a posteriori argument. It strengthens the argument by providing evidence that heavenly changes should be observable. Of course, Galileo's critique G1 invalidates this by pointing out some relevant differences between heavenly and terrestrial changes. Though invalid, the a posteriori argument does attempt to appeal to the right kind of evidence; the problem is merely that it does not go far enough. In other words, the argument does not depend on the assumption that everything that is not observable does not exist, as it would if proposition A11 were the only premise; rather it assumes some proposition to the effect that if something is observable when occurring on the earth then it will be observable when occurring elsewhere. The problem with this is that it makes a difference where the observer is located.

Galileo's critique seems to interpret the a posteriori argument as an explanatory one, in the sense that it presents the conclusion about heavenly unchangeability as the explanation for the difference of observation mentioned in the premises. Galileo then suggests two other ways of explaining the difference: it may be due to the great distance between the earth and the heavenly bodies and/or to the lack of sufficiently careful observations of the heavenly bodies. This does not invalidate the Aristotelian *explanation*, in the sense of showing that it, i.e. conclusion A1, is false; rather, the alternative explanations invalidate the *argument*, i.e., the inferential link between premises and conclusion, since there is no reason to prefer the Aristotelian explanation to the one suggested in Galileo's critique. To conceive Galileo's critique in this manner also makes it clear that he is not accepting the premises as such, but that he is merely saying that even if we accept them, that does not force us to accept the conclusion. In fact, he will soon question premise A11.

(2) In his second criticism Galileo is using the method of counterexample. He envisages a similar situation involving the observation of the earth from the moon; in such a situation the corresponding premises of an a posteriori argument for the unchangeability of the earth would be true, and yet we know that the conclusion is false. This shows that it is possible for premises A11 and A12 to be true and conclusion A1 false, and hence that A1 does not follow from A11 and A12. The method of counterexample is, of course, the technique favored by formal logicians to prove invalidity, and it is a relatively 'active' method of criticizing reasoning insofar as it involves the creation of a counterexample.

It is interesting to compare the counterexample method to the one mentioned earlier in which one argues that a conclusion does not follow because what follows is some other proposition different from it. In the present argument, Galileo might have said that the most that would follow from the premises is that there have been no changes in the heavenly bodies *so far* (i.e., during the period of human observation). It seems clear that the method of alternative conclusion, as the latter technique might be called, is active insofar as one has to think of the conclusion that does follow; it is also more constructive than the counterexample method since, besides showing what is wrong with the given argument, it shows what could be right.

Let us look at this in more detail. Let a be the argument 'R1, R2, $\therefore$ C'. The counterexample method begins by formulating another argument a' having the same logical form as a, but with obviously true premises and obviously false conclusion. Then one argues as follows: since a and a' have the same logical form, if a is valid then a' is also valid; but a' is invalid, because it has true premises and false conclusion; therefore, a is also invalid. In the method of alternative conclusion one argues that if R1 and R2 are true then C' would also be true (perhaps on the basis of other premises R3, R4, etc.), where C' is such that if C' is true then C is not; if follows that if R1 and R2 are true then C is false; that is, R1 and R2 imply not-C; given the consistency of R1 and R2, which is usually obvious or unproblematic, it follows that R1 and R2 do not imply C, i.e., that a is invalid. Though the method of alternative conclusion depends on the consistency of the original premises and the truth of the other ones used − R3, R4, etc. − this does not make it less effective than the counterexample method, since the latter depends on the identity of logical form between a and a', which is almost always problematic. Moreover, the rhetorical effectiveness of the latter depends on the counterexample having some substantive and topical similarity to the original argument; an abstract counterexample will be unconvincing, and this will normally be an indication of the unsoundness of the justification that the original argument has the proposed form. In using the other method this problem does not arise, even though the additional premises R3, R4, etc., may be unfamiliar, since they are necessarily relevant from the point of view of substance and topicality.

(3) Galileo's criticism G3 constitutes a negative evaluation of argument A1 in spite of the fact that it claims that A1 is better than the more literal version of A1. For here Galileo strengthens his previous two criticisms by pointing out that there he has criticized a stronger version of the argument and not a straw man. To be sure, the strengthening is operative only at the rhetorical level, but it is rather effective at this level.

Regarding the literal version of A1, Galileo is somewhat explicitly using the method of counterexample. However, here we have a good example of the limitations of this method, for we see that the counterexample introduced in

G31 tells us that something is wrong with the literal version, but it does not tell us exactly what is wrong, or why it is wrong, or why it is wrong in the sense of making us *understand* what went wrong with it, so that we might be able to avoid such an error in future reasoning. Thus a limitation of the counterexample method is perhaps its relative ineffectiveness in improving the practice of reasoning.

It is G32 that begins to diagnose the trouble, which turns out to be part of an equivocation problem. For what G32 means is that the latent proposition in the argument 'A11', ∴ A1' ' is A12': we constantly observe generation and decay of terrestrial bodies in the terrestrial region; now, this proposition is needed in the argument since it together with A11' creates a contrast which is then accounted for by A1'; but the contrast is spurious since 'bodies' in A11' means integral bodies, whereas 'bodies' in A12' means *parts* of integral bodies; hence there is nothing to explain; hence, there is no reason to conclude A1'; in short, A1' does not help to explain the alleged facts mentioned in the premises A11' and A12' because there is nothing to explain.

(4) The main interest of Galileo's critique G4 is that, though it pertains to (one of the premises of) the a posteriori argument, it is itself a positive argument in support of a thesis, G41. Here we should note that this proposition is supported not only directly with positive evidence (G411, G412, and G413), but also indirectly by criticism of available counterevidence and counterarguments, namely by the argument supporting G414. In other words, argument G4 is not only a meta-argument but also an argument in support of heavenly changes; nevertheless, in the course of this object-level reasoning, Galileo finds it useful and necessary to engage in some evaluation of counterarguments. This supports a claim about the evaluative nature of reasoning *per se*, whereas the other passages so far support a claim about the ratiocinative nature of the *evaluation* of reasoning.

(5) Galileo's criticism G5 indicates that in the present case the falsehood of a premise falsifies the conclusion. This is obviously not true in general, but in this case one contributing factor is the fact that both premise A11 and conclusion A1 are denials of existence claims; hence the denials of these denials are positive and very informative propositions.

It would be wrong to fault Galileo's argument 'G511, ∴ G51' by saying that this manner of reasoning is inconsistent with his criticism of the validity of the a posteriori argument, given in G1 and G2, and that this reasoning is only correct from the point of view of the empiricism of the Aristotelians. For in G1 and G2 Galileo did not criticize the presupposed empiricism of the a posteriori argument, but rather interpreted the latter as an explanatory argument, drawing an arbitrary explanation for conclusion. Hence Galileo could defend his G5 reasoning by criticizing explanations of G511 other than G51.

6. CHARITY, AND THE TELEOLOGICAL ARGUMENT
(F83.12–87.14)

The object argument. (A1) Changes among terrestrial bodies enhance the perfection of the earth; for example, (A2) living organisms are more perfect than dead ones, and (A3) gardens more than deserts. But, (A4) heavenly changes would render heavenly bodies imperfect, since (A5) heavenly changes would be of no use or benefit to man, and hence (A6) they would be superfluous; therefore, (A7) unchangeability would enhance the perfection of heavenly bodies. Therefore, (A8) heavenly bodies are unchangeable. This is also shown by the fact that, since (A6) heavenly changes would be superfluous, and since (A9) nature does nothing in vain, (A10) there cannot be any heavenly changes.

The meta-argument. (G11) The inference from A5 to A6 is incorrect since (G111) man should not be regarded as the sole creature for whose sake and benefit the whole universe exists. Moreover, (G12) the inference from A6 and A9 to A10 is incorrect since (G121) even if we accept A9 it cannot be interpreted to mean that nature brings nothing into existence which is superfluous from the point of view of human needs. Hence, (G1) the teleological argument is unsound.

Comments. The propositional structure of the teleological argument is:

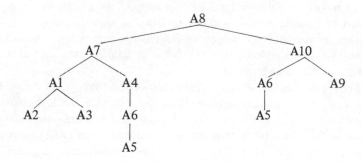

Since A5 and A6 appear in two distinct subarguments, we could also have the following diagram:

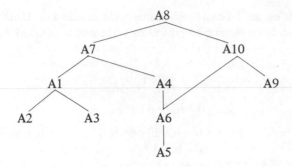

It is interesting to note that the alleged superfluousness of heavenly changes, A6, supports heavenly unchangeability, A8, in two ways: first by means of perfection considerations (A4, A1, A7), and second by means of natural teleology, A9. It is this double use of proposition A6 that would make standard labeling problematic in this case; in fact, it would be impossible for the second diagram above, and misleading for the first insofar as the double appearance of propositions A5 and A6 would be missed by the different labels they would receive.

It should be pointed out that the subargument from A2, A3, and A5 to A8 is a way of integrating into the object argument what would otherwise appear as incorrect criticism in the evaluative meta-argument; for this subargument makes it obvious that the Aristotelians are not assuming the excessively broad generalization that change makes all things imperfect. To attribute this excess would be a violation of what may be called the Principle of Charity,[8] and there is certainly textual evidence that Sagredo (i.e., one part or one mood of Galileo) commits this violation.[9] At the same time, since it has been possible to reconstruct the teleological argument as above, and since this reconstruction is still subject to Galileo's main criticism, it would be a violation of that principle on our part (vis-à-vis Galileo) not to reconstruct the teleological argument as we have.

(2) One might wonder whether the inference from A7 to A8 should be faulted for affirming the consequent. Here the reasoning would be that a fuller reconstruction of this step would be: (A81) if heavenly bodies were unchangeable then their perfection would be enhanced; but (A82) heavenly bodies are (or ought to be regarded) as perfect as possible, since, e.g., (A821) they are the work of God, or whatever; therefore, (A8) heavenly bodies are unchangeable. Here A81 would correspond to A7. However, there is no reason to prefer this reconstruction to the following: (A811) if heavenly

bodies were prefect then they would be perfect by reason of unchangeability; but, (A812) heavenly bodies are perfect (since (A8121) . . .); therefore, (A81) heavenly bodies are perfect by reason of unchangeability; therefore, (A8) heavenly bodies are unchangeable. Here there is one extra step, the original A7 has been reconstructed differently, and the problematic step has been replaced by a valid instance of affirming the antecedent. The best way of choosing between these two reconstructions would be to look at the subargument from A1 and A4 to A7. A1 seems to mean: one of the effects of terrestrial changes is greater terrestrial perfection. A4: one of the effects of heavenly changes would be greater heavenly imperfection, i.e., if heavenly bodies were changeable then they would be less perfect. Since this last proposition implies A811, the second one of the above reconstructions is to be preferred; nor does this make A1 superfluous since it is its presence that suggests this interpretation of A4.

7. EQUIVOCATION, AND THE VIOLENT MOTION ARGUMENT
(F159.29–162.14)

The object argument. (A1) The earth cannot move circularly, because (A111) such motion would be violent, and hence (A11) it would not be perpetual; it would be violent because (A1111) if the earth's circular motion were natural then its parts would also move circularly by nature, but (A1112) it is impossible for the earth parts to move circularly by nature since (A11121) the natural motion of the earth parts is straight downwards.

The meta-argument. (G1) The violent motion argument is unsound since (G11) the clause "the parts of the earth would also move circularly by nature" can mean either that these parts would move around their own centers or that they would move around the earth's center. (G12) In the first case proposition A1111 would be obviously groundless, and the step from A11121 to A1112 obviously invalid. (G13) In the second case proposition A1112 would be groundless since (G131) the reason supporting it, A11121, would be groundless: for (G1311) the concept of natural motion used in A11121 is either the one suggested elsewhere in the argument, namely in the step from A111 to A11, or it is that of actual motion under undisturbed conditions; (G1312) if we use the concept of motion suggested in the step from A111 to A11 then no straight motion could be natural, since (G13121) no straight motion can be perpetual, and (G13122) the step from A111 to A11 suggests equating the natural with the perpetual; (G1313) if natural motion is actual motion under undisturbed conditions then there is no reason to believe the natural motion of the earth parts is straight downwards, since

(G13131) the conditions under which they move straight downwards are contrived artificially and seldom occur in nature.

Comments. Galileo's critique shows that either A1111 is groundless and A1112 does not follow from A11121, or that A1112 is groundless. Since the groundlessness of A1111 can be conceived as invalidity of the step from A1112 to A111, the first alternative means that A1112 neither is implied by A11121 nor imples A111. Moreover, the groundlessness of A1112 derives partly from the validity of the step from A111 to A11; this means that if the step from A111 to A11 is valid then A1112 is false; that is, if A1112 is true then the step from A111 to A11 is invalid; hence, the second alternative is also a type of invalidity. Thus the violent motion argument is invalid in any case.

It might be objected that, though the step from A111 to A11 may *suggest* equating the natural with the perpetual, it presupposes merely that non-perpetuity is a necessary conditon for nonnatural (violent) motion, hence that *if* motion is perpetual *then* it is natural, and hence that G13122 and G1312 are questionable. In short, an Aristotelian can consistently accep̃t both propositions A11121 and the step from A111 to A11; and hence both this step and proposition A1112; so that Galileo would be violating the Principle of Charity in attributing to Aristotle the equation of the natural with the perpetual.

It is true that to make perpetuity a sufficient condition for natural motion formally validates the step from A111 to A11. However, the statement of such a sufficient condition is a mere restatement of the validity of the inference; hence it would be entirely appropriate to ask for a justification of such a sufficient condition. The equation of the natural with the perpetual, or to be more exact, with the potentially perpetual, would provide one, and it is in this sense that the step from A111 to A11 may be taken to presuppose the above-mentioned equation. This, of course, would accord with Galileo's argument in F53–57.

8. IMPLICATION VS SUPPORT, AND THE TWO MOTIONS ARGUMENT (F162.15–164.25)

The object argument. (A121) All bodies known to move circularly, except the *primum mobile*, have two motions; hence, (A12) if the earth moves circularly then it has two motions. But, (A112) if the earth has two motions then there would be a variation in the rising and setting points of the fixed stars, and (A111) no such variation is observed; therefore, (A11) the earth does not have two motions. Therefore (A1) the earth does not move circularly.

The meta-argument. (G1) The circular motion mentioned in the conclusion can refer either to axial rotation around one's own center or to orbital revolution around some point outside oneself.

(G24) If circular motion means axial rotation then the second motion mentioned in the argument is an axial rotation in the opposite direction causing the primary rotation to lag behind. (G23) If so, then the argument can be reconstructed as follows: "(A12') one terrestrial rotation would imply a second since (A121') all rotating bodies except the *primum mobile* have two opposite rotations; but, (A11') a second terrestrial rotation is impossible since (A112') if it occurred there would be variations in stellar risings and settings, (A111') which there are not; therefore, (A1') the earth cannot rotate." Now, (G2111212) if A121' is true, then if the earth were the *primum mobile* the first terrestrial rotation would not imply the second, since (G21112121) A121' means that one rotation implies a second if and only if the rotation does not belong to the *primum mobile*; but, (G2111211) if the earth rotates then *it* is the *primum mobile*, since (G21112111) to conceive the earth as having axial rotation is to conceive it as the *primum mobile*; therefore, (G211121) if A121' is true, then if the earth rotates its first rotation does not imply a second terrestrial rotation; therefore, (G21112) if A121' is true, then the first terrestrial rotation would not imply a second one; but, (G21111) if A12' follows from A121', then if A121' is true then the first terrestrial rotation would imply a second; therefore, (G2111) if A12' follows from A121', then A121' implies a contradiction; therefore, (G211) if A12' follows from A121', then A121' is false; therefore, (G21) either A12' does not follow from A121' or A121' is false. Moreover, (G22111111) the two motions argument may be interpreted as supporting the impossibility of a (first) terrestrial rotation, A1', by the impossibility of a second terrestrial rotation, A11'; therefore, (G2211111) A11' would be pointless if a first rotation were impossible independently of the second; therefore, (G221111) A11' would be pointless if a first rotation were impossible without the second being also impossible; i.e., (G22111) A11' would be pointless if it were true that the first rotation is impossible and the second is not impossible; therefore, (G2211) it is not true that the first rotation is impossible and the second not; therefore, (G221) if the first rotation is impossible then so is the second; therefore, (G22) if the second rotation were not impossible, the first would not be impossible either.

(G33) If circular motion refers to orbital revolution then the second circular motion is the daily axial rotation. (G32) If so the argument can be reconstructed as follows: "(A1″) the earth cannot have the (annual) orbital revolution because (A12″) if it did then it would also have the (daily) axial rotation, and (A112″) if it had orbital revolution and axial rotation then there would be variations in stellar risings and settings, (A111″) which there are not, so that (A11″) it can't have both revolution and rotation". But, (G31) A112″ is false, since (G311) it is not stellar variations in risings and settings that follow from the earth's joint rotation and revolution, (G3111) as it will be shown later. Therefore, (G3) if circular motion refers to orbital revolution then premise A112 of the original argument is false.

Comments. It should be noticed that I have stated neither Galileo's final conclusion (from G1, G21, G22, G23, G24, and G3) nor the conclusion, G2, to be drawn from G21, G22, G23, and G24.

The first problem to solve is that of the relationship between Galileo's G21 critique and the following one that readily comes to mind. The step from A121′ to A12′ is formally invalid if one does not assume that (A122′) the earth is not the *primum mobile*; next, one would ask the Aristotelians to justify this latent proposition; now, it is obvious that the question of whether or not the earth is the *primum mobile* is equivalent to the question of whether or not the earth rotates on its axis, which is the question the argument is trying to settle; hence the argument begs the question.

Let us use the following abbreviations:

R = the earth *r*otates;
S = the earth has a *s*econd rotation;
P = the earth is the *primum mobile*.

The two motions argument (i.e., one version of it) is then:

(A121′)   (if R then S) iff not-P;
(A12′)   ∴ if R then S;
(A11′)   not-S, since (A111′);
(A1′)   ∴ not-R.

The step from A121′ to A12′ assumes:

(A122′)   not-P.
But: R iff P.

Hence to support not-P one needs the following argument:

(A1222′)   not-R;

(A1221′)   R iff P;
(A122′) ∴ not-P.

Since A1222′ = A1′, the full argument is circular:

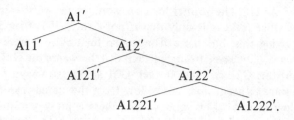

Of course, an argument cannot be circular if it is formally invalid; so this argument must be valid. Hence this critique seems to be inconsistent with Galileo's, which suggests invalidity, since G211 implies that if A121′ is true then A12′ does not follow from A121′, which is to say simply that A12′ does not follow from A121′.

Let us see whether this impression is correct. G2111212 is saying that:

(F1)   if [(if R then S) iff not-P] then [if P then not-(if R then S)],

which is right. G2111211 says that:

(F2)   if R then P.

G211121 is equivalent to:

(F3)   if [(if R then S) iff not-P] then [if R then not-(if R then S)],

which is a formal consequence of F1 and F2. G21112 says:

(F4)   if [(if R then S) iff not-P] then not-(if R then S),

which follows formally from F3 only if one adds:

(F3.1)   if [if R then not-(if R then S)] then not-(if R then S),

which is not formally valid, since it is refuted by assigning the truth-value falsehood to R. However, this formal invalidity depends on a truth-functional interpretation of 'if-then', which there is no reason to accept here. We may interpret 'not-(if R then S)' as: one cannot conclude S from R.[10] Then (F3.1) means that if given R one cannot conclude S from R, then one cannot conclude S from R. Hence, F3.1 is correct. Similarly for the argument going from G211 to the proposition that A12′ does not follow from A121′.

The difference between the two critiques is that the circularity critique is saying that the 'full' argument, i.e., the two motions argument as stated plus its latent structure, is circular and (hence formally) valid; whereas Galileo's G21 critique implies that the argument as stated is invalid insofar as A12′ does not follow from A121′. The formal logician would agree with this, giving as a reason that what follows is only that if not-P then (if R then S). Galileo is implicitly saying this, but more. The reason for Galileo's more complicated criticism stems, I believe, from the fact that he wants his G21 criticism to connect with his G22 critique. In fact, G21 amounts to saying that A12′ is groundless since either it does not follow from the stated reason A121′ or if it does follow then A121′ is false, so that there is no way of justifying A12′ here. But A12′ amounts to saying that A11′ implies A1′, hence to say that A12′ is groundless is to say that there is no reason to think that A11′ implies A1′. But if so then an obvious possibility to consider is whether or not A11′ supports A1′, in some sense different from implication. I believe that the G22 critique is an attempt to answer this question. When Galileo says that "A11′ would be pointless if . . . " he may be interpreted as saying that "A11′ would not support A1′ if . . .". The crucial consideration G2211111 amounts to interpreting the step from A11′ to A1′ as an explanatory argument, so that the G22 critique may be interpreted as follows: if A11′ supports A1′ then the fact that A1′ has to be the alleged explanation for the fact that A11′; if so then A1′ should imply A11′; therefore, if A11′ supports A1′ then A1′ implies A11′. In drawing conclusion G2211 Galileo is assuming that A11′ is not pointless, i.e., that it supports A1′, so that in G22 he concludes that not-A11′ should imply not-A1′, i.e., that A1′ should imply A11′. However, if this conclusion is to fit with the rest of Galileo's criticism, his meaning must be that the two motions argument must be assuming that A1′ implies A11′, so that, since no reason is given for this implication, no reason is being given why A11′ supports A1′. Putting this together with my interpretation of the G21 criticism, we get that: (G2) there is no reason to think that A11′ either implies or supports A1′.

I have not included this proposition in my reconstruction because there is another way in which he may be interpreted to be combining what I have called the two lines of criticism, G21 and G22. This other interpretation emphasizes Galileo's explicit assertion (F164.6–164.8) that the two motions argument (in its first version) is self-contradictory, and it reflects more accurately the tone of the text at the point where Galileo seems to combine his two lines of criticism; moreover, as it will be seen below the interpretation has great logical elegance. However, what it gains in accuracy and elegance it loses in psychological plausibility, since several nontrivial steps are omitted by Galileo; also it has a difficulty which I will discuss shortly. The interpretation would correspond to the following deduction, using earlier abbreviations.

(1) If R then S. (This is A12'.)
(2) A121'.
(3) If S then R. (This is G22.)
(4) If A121', then [if P then not-(if R then S)]. (This is G2111212.)
(5) If R then P. (This is G2111211.)
(6) If S then P. (From 3 and 5.)
(7) If P then not-(if R then S). (From 2 and 4.)
(8) If S then not-(if R then S). (From 6 and 7.)
(9) If (if R then S) then not-S. (From 8.)
(10) If [if (if R then S) then not-S] then [if (if R then S) then not-R]. (Formally valid.)
(11) If (if R then S) then not-R. (From 9 and 10.)
(12) If R then not-(if R then S). (From 11.)
(13) If [if R then not-(if R then S)] then not-(if R then S). (Inferentially though not truth-functionally valid.)
(14) Not-(if R then S). (From 12 and 13.)
(15) (If R then S) and not-(if R then S). (From 1 and 14.)

The gap is the text is due to the fact that we find only (3) on F163.25–163.27; '(2), (4), ∴ (7)' on F163.29–163.36; (5) on F163.36–164.2; and (15) on F164.3ff.

The difficulty with this interpretation stems from (3). Here the G21 line of criticism is made to depend on the G22 line, which is the opposite of the first interpretation where the possibility that not-S might *support* not-R, through (3), is raised after it has been shown that there is no reason to believe that not-S *implies* not R. Moreover, the present interpretation depends on the *assertion* of (3), which in the first interpretation is merely considered in order to conclude that the two motions argument should prove (3). Here, therefore, the question arises whether (3) is justified.

Galileo begins in G22111111 by saying that not-S is a reason for not-R and ends concluding in G22 that if S then R. This seems a curious way of reasoning, tantamount to deriving the converse from a conditional proposition. The crucial consideration is G2211111, for G22 plausibly follows from it (on the assumption, which Galileo presumably *ought not* to be granting, that the reason is not pointless). Galileo seems to be saying that if not-R is asserted because of not-S then in the absence of not-S (i.e., given S) one would not be prepared to assert not-R. In other words, though it is not true that if the falsity of p implies the falsity of q then the truth of p implies the truth of q, it is plausible that if the rejection of p leads to the rejection of q then the nonrejection of p leads to the nonrejection of q. In short, Galileo's point is perhaps more rhetorical than logical.

But now we get the following problem. If this type of rhetorical point is correct here, perhaps it could always be made whenever the argument has the *modus ponens* form (if p then q; p; $\therefore$ q), for if one accepts q because of p, then if one did not accept p one would not accept q. This generalization may be accepted, by declaring it harmless, and then the problem evaporates. However, there remains the problem that Galileo would be assuming that not-S supports not-R. This problem may be solved by saying that Galileo would then have shown that *if* not-S supports not-R *then* the argument is self-contradictory, and that therefore if one denies that it is self-contradictory one is committed to saying that the conclusion A1' is not supported by A11'. But one reason for this lack of support might be the presence of an implication, hence the criticism of the implication 'if A11' then A' ' would seem to be required.

Finally regarding criticism G3, Galileo distinguishes it sharply from the rest, calling it a case of 'false consequence' by contrast to G2 which he calls a case of 'paralogism' and 'fallacy'. This seems to amount to saying that whereas G2 charges an error of reasoning, G3 does not. How is this possible, in view of the fact that what is false according to G3 is a 'consequence', specifically A112", which is a conditional proposition and hence false because its consequent clause does not follow from its antecedent clause? The main difference is that, whereas the two motions argument gives a reason for A12, it does not for A112; there can be no mistake of reasoning unless there is reasoning, and no reasoning supports A112, which the argument takes as obvious. Presumably it is true and relatively obvious if it means A112'; it is only when it is contrued as A112" that the stellar variations do not follow. Even so one must be careful here, since certain (annual) stellar variations would indeed follow, as Galileo himself shows later (F406–416). So it would have been better for him to say here, as he says later, that the main problem about these stellar variations caused by the earth's revolution is that of determining *exactly* what the nature of these variations is, exactly enough so as to put them to experimental test. Since in the present passage Galileo does show appreciation of the two motions argument construed as supporting A1", we may take this to be the needed qualification to his otherwise simplistic-sounding claim about the falsity of A112".

9. GROUNDLESSNESS, AND THE MIXED MOTION ARGUMENT
(F166.30–169.21)

The object argument. (A12) If the earth rotated then a rock dropped from the top of a tower would not land at its foot since (A12a1) on a moving ship a rock dropped from the top of its mast lands away (stern) from its foot, and since (A12b1) if the earth rotated and the rock landed at the foot of the tower then the rock would have a mixture of two (natural) motions, toward and around the center, (A12b2) which mixture is impossible because (A12b21) on a rotating earth the air would be incapable of carrying around falling bodies like rocks (as shown by the fact that (A12b211) strong winds are observably incapable of impressing motion on heavy rocks). Since (A11) the rock falls at the foot of the tower, it follows that (A1) the earth stands still. [F166.30–167.34, and F168.6–169.5]

The meta-argument. (G1) Proposition A12a1 does not support A12 because (G11) there is great disparity between the case of a moving ship and that of a rotating earth: first, (G1111) the ship's motion is not natural whereas the earth's rotation would be, hence (G111) on a moving ship the falling rock has no inclination to follow its motion, whereas on a rotating earth all of its parts would have the tendency to rotate; second, (G1121) on a rotating earth the lower part of the earth's atmosphere would rotate along (because (G11211) this lower atmosphere is trapped between mountains, and because (G11212) it contains many earth vapors and exhalations), whereas (G1122) on a moving ship the air between the top and bottom of the mast does not follow its motion, so that (G112) the medium through which the rock falls is moving on a rotating earth but is not for the case of a moving ship. [F167.35–168.35]

(G2) Proposition A12b211 does not support A12b21 because (G21) A12b211 exemplifies the case of the air causing a body to start moving after being at rest, whereas A12b21 exemplifies the case of the air maintaining or not impeding a body's motion, and (G22) there is great disparity between the two cases, and (G23) a situation properly analogous to that of A12b21 would be a rock dropped from the paws of an eagle while carried by the wind. [F169.6–169.21]

(G3) Proposition A12b21 does not support A12b2 because (G311) on a rotating earth falling bodies would have circular motion around the center, by nature, and hence (G31) they would not need to be carried around by the air. [F168.11–168.18]

Comments. The structure of the object argument may be visualized as:

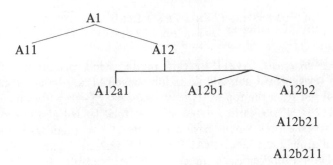

Here A12a1 has been interpreted as an immediate reason for A12 because Galileo's criticism G1 amounts to a criticism of the analogy presupposed in arguing directly from A12a1 to A12. Similarly G2 criticizes the analogy presupposed in the step from A12b211 to A12b21. It is interesting to note that Galileo strengthens his G2 criticism by mentioning explicitly, in proposition G23, a correct analogue for the situation mentioned in A12b21; in short, his G2 criticism is that the presupposed analogy does not hold partly because it is obvious that it doesn't if we look at the alleged analogues from a certain point of view, and partly because the analogy that does hold is some other specific one.

It would be possible, however, to reformulate the object argument by structuring it as follows (using the same labels as before):

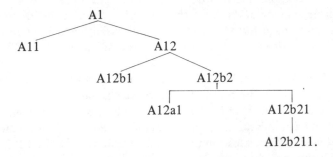

Here the step from A12a1 to A12b2 would be an explanatory argument, in the sense that the impossibility of mixed (down and around) motion would be presented as the explanation for the alleged facts of the ship experiment. Evidence for the textual accuracy of this interpretation of the falling bodies argument is a statement of it given by Galileo in a passage where he states in a preliminary way all arguments against the earth's rotation (F151–2).

Galileo's criticism could then be reformulated as alternative-explanation criticism, the alternative being that the rock might fall behind because of the nonnatural character of the motion imparted to it by the ship and/or because of air resistance. Therefore, from the evaluative point of view, this reformulation of the object argument is no better than the other. If the reformulated argument has an advantage, it is from the point of view of structure, which is such that all the steps above A12b2 are correct, so that one may conceive of the reformulated argument as one from the impossibility of mixed motion, A12b2, and then evaluate it as groundless since the subarguments supporting this impossibility are wrong. It should be noted that this groundlessness remains even if one emphasizes the fact that the impossibility is supposed to pertain to natural motions, for in that case Galileo's critique of the natural motion argument would become operative; the effect of that critique here would be that the two 'natural' motions being mixed are natural in different senses, the straight-downwards in the sense of spontaneous inclination, the circular rotation in the sense of potentially perpetual state.

10. ANALOGY, AND THE SHIP EXPERIMENT ARGUMENT
(F169.22–180.18)

The object argument. (A121) When a ship stands still a rock dropped from the top of the mast falls at the foot of the mast; but (A131) when a ship is moving a rock dropped from the top of the mast falls away from its foot; therefore, (A13) if the rock falls at the foot then the ship is standing still, and (A12) if the rock falls away from the foot then the ship is moving. But (A11) what is true of the ship must be true of the earth. Therefore, (A1) if a rock dropped from the top of a tower falls at its foot then the earth is standing still.

The meta-argument. (G1) The argument's main premise, A131, is false. For (G114113) motion is conserved on noninclined surfaces whenever accidental and external disturbances are removed, since (G1141131) on inclined surfaces downward motion is accelerated and upward motion is retarded [F171–3]; and (G114112) when a ship is moving its motion takes place on a noninclined surface, if accidental and external disturbances are removed, since (G1141121) the surface of a calm sea is horizontal; and (G114111) when a ship is moving the rock before being dropped has the same motion as the ship; therefore, (G11411) the horizontal motion the rock has before being dropped is conserved, if accidental and external disturbances are removed; therefore, (G1141) the rock will continue to move horizontally with the ship after it is dropped, if accidental and external disturbances are removed; therefore, (G114) the rock will fall at the foot of the mast on a moving ship, if accidental and external disturbances are removed. (G113) There are two relevant disturbances, the air and the rock's vertical fall; but (G112) the disturbance of the air may be ignored, since (G1121) its effect is negligible; and (G111) the rock's motion of fall does not disturb its horizontal motion, since (G1111) these two motions are not contrary, and since (G1112) they derive from distinct causes; (G11121) the fall derives from gravity, while (G11122) the horizontal motion derives from the 'virtue' impressed on the rock by the ship before the drop. The latter can be shown as follows: (G111221) the rock's horizontal motion after the drop may be regarded as projectile motion; now, (G111222) the cause of projectile motion is either the impressed 'virtue' or the medium (air) through which the motion takes place. But (G111223) the cause cannot be the medium: (G111223a11) moving air pushes light substances more easily than heavier ones, so that (G111223a1) if the air carried the projectile then a ball of cotton could be thrown further than a rock; (G111223b1) lead pendulums oscillate much longer than cotton ones; (G111223c1) if the force cannot be impressed by the thrower directly to the projectile, then it cannot be impressed by the thrower to the air; (G111223d1) arrows can be shot against the wind; and (G111223e1) arrows travel much less when shot sideways as compared to normal. It follows that (G11) the rock falls at the foot of the mast even when the ship is moving.

Comments. The argument from G1141131 to G114113 is on F171–3; the rest of the argument up to G114 is on F174; the rest up to G111222 is on F175; from there on the argument is on F175–180.18.

Galileo's critique amounts to a counter-argument designed to support a positive claim, G11, about the ship experiment; since this claim is a denial of a premise, A131, of the ship analogy argument, the latter is thereby criticized. The logical interest of Galileo's denial is twofold. First, the denial is not grounded on a direct experimental test, but rather on relatively easily available empirical facts (G1141131, G111223a11/b1/c1/d1/e1), and then these are inferentially connected with phenomena on the ship; the creation of such interrelationships is logical practice. Second, because of the structure of the object argument, the effect of G11 is that nothing follows about the earth's motion from the ship experiment: for if we replace A131 by G11 in the ship analogy argument, then A13 no longer follows, since though A13 is the contrapositive of A131 it is not the contrapositive of G11; moreover, though A12 would still follow from A121, now it would also follow (from

G11) that if the rock falls away from the foot, then the ship is standing still; this together with A12 would yield that if the rock were to fall away from the foot of the mast, then the ship is both moving and standing still, which is to say that the rock cannot fall away from the mast's foot, which merely reinforces the conjunction of A121 and G11. From A121 and G11, adding A11, it would follow that if the earth stands still the rock falls at the tower's foot, and that if the earth rotates the rock also falls at the tower's foot, from which, since the rock does fall at the tower's foot, it doesn't follow either that the earth stands still or that it rotates.

Since these considerations are essentially in Galileo's text, they strengthen my reconstruction of the ship analogy argument given above, the main problematic spot being the inferring of A12 and A13 from A121 and A131, which in the text Galileo justifies by 'the principle of the converse' (*per il converso*, F170.2). It would be superficial to object that in the statement of the argument Galileo is committing the fallacy of affirming the consequent, since (according to the commonly accepted meaning of 'converse') the converse of 'if *p* then *q*' is 'if *q* then *p*', so that he would be inferring A12 from A131 and A13 from A121. It would be no less superficial to say that, since Galileo does not accept the ship analogy argument, though he cannot be faulted for its statement, which is accepted by the Aristotelian spokesman, he could be faulted for not calling attention to this error in the statement of the argument. Such criticism would be groundless or circular since there is no more reason to equate Galileo's 'converse' with the one in contemporary logic textbooks than with the 'contrapositive' found therein; second, if Galileo's 'converse' is interpreted in the logic textbook sense, this would make nonsense of his argument that the ship experiment proves nothing about the motion of the earth, for from G11 and A121 he would have concluded that if the rock falls at the foot then the ship, and hence the earth, is both moving and standing still, which is to say that the ship experiment proves too much, rather than nothing.

Finally, it should be mentioned that, as reconstructed, the subargument to G111223 is not as explicit as it could be. The analysis of its latent structure is left as an exercise to the reader.

11. DEDUCTION, INDUCTION, RHETORIC, AND THE EAST-WEST GUNSHOT ARGUMENT (F194.12–197.1)

The object argument. (A11) If the earth rotated, gunshots would range further toward the west than toward the east; for (A1111) on a rotating earth the projectile's motion toward an eastward target would be in the same direction as that of the gun due to the earth's motion, whereas (A1112) the projectile's motion toward a westward target would be in the opposite direction as that of the gun, and hence (A111) on a rotating earth the distance between the gun and the point hit will be greater in a westward than in an eastward shot by an amount equal to the total distance traversed by the gun during the time of flight of both balls. Since (A12) gunshots toward the west and toward the east are observed to range equally, it follows that (A1) the earth does not rotate.

The meta-argument. (G1131222) If the earth's rotation implies unequal ranges for east-west gunshots, then arrow shots which range, e. g., 300 yards from a motionless cart would range 200 yards in the direction of the cart and 400 yards in the opposite direction, from a cart moving at a speed equivalent to 100 yards per time of each shot [F194.27–195.13]; (G1131221) if so, then from a moving cart both shots would range 300 yards when those in the direction of the cart are shot with a force equivalent to a range of 400-yards-from-a-motionless-cart, and when those in the opposite direction are shot with a force equivalent to a range of 200-yards-from-a-motionless-cart [F195.13–195.30]; therefore, (G113122) if the earth's rotation implies unequal ranges for east-west gunshots, then from a moving cart both shots would range 300 yards when those in the direction of the cart are shot with a force equivalent to a range of 400-yards-from-a-motionless-cart, and when those in the opposite direction are shot with a force equivalent to a range of 200-yards-from-a-motionless-cart. But (G113121) the greater and lesser amounts of shooting force give rise to correspondingly greater and lesser speeds for the arrows. Therefore, (G11312) if the earth's rotation implies unequal ranges for east-west gunshots, then from a moving cart both shots would range 300 yards when those in the same direction move with 4 degrees of speed, those in the opposite direction move with two degrees of speed, and the cart moves with one degree of speed [F195.31–196.12]. But (G11311) if from a cart moving with one degree of speed arrows are shot with equal strength in opposite directions (the strength being equivalent to three degrees of speed from a motionless cart), then those shot in the same direction move with four degrees of speed and those shot in the opposite direction move with two degrees of speed [F196.13–196.26], since (G113111) the effective speed of the shots in the opposite direction is the sum of the cart's speed plus the speed imparted to the arrow by the crossbow, whereas for the same direction the effective speed is the difference between them. Therefore, (G1131) if the earth's rotation implies unequal range for east-west gunshots, then if from a cart moving with one degree of speed arrows are shot with equal strength (equivalent to three degrees of speed from a motionless cart) in opposite directions, then both shots would range 300 yards. Therefore, (G113) if the earth's rotation implies unequal ranges for east-west gunshots, then arrows shot in opposite directions from a moving cart range equally [F196.26–196.31]. But (G112) if so, then on a moving earth east-west gunshots would range equally; and (G111) if the latter, then the earth's rotation would not imply unequal ranges for east-west gunshots; therefore, (G11) if the earth's rotation implies unequal ranges for east-west gunshots, then it does not so imply. Therefore, (G1) the earth's rotation does not imply unequal ranges for east-west gunshots. (I.e., A11 is false.)

Rather, (G214) if the earth rotates, then the eastward projectile is moving with a greater (absolute) speed than the westward one, since (G2141) the (absolute) speed of the eastward projectile is the sum of what it receives from the gunpowder and from the moving earth, whereas the (absolute) speed of the westward projectile is the difference between what it receives from the gunpowder and from the earth; and (G213) if the eastward projectile moves (absolutely) faster, then it travels a greater (absolute) distance than the westward one; and (G212) if so, then the distance (relative to the moving earth) traveled by the eastward projectile is the same as that of the westward projectile, since (G2121) for the eastward projectile the relative distance equals its absolute distance *less* the distance traveled by the gun, and for the westward projectile the relative distance equals its absolute distance *plus* the distance traveled by the gun; and (G211) if the relative distances are the same, so are the ranges, since (G2111) on a rotating earth the observed range of a shot would be the difference between the absolute distance traveled by the projectile and that traveled by the gun. Therefore, (G21) if the earth rotates, then the ranges of east-west gunshots would be equal. Therefore, (G2) east-west gunshots would have equal ranges whether the earth rotates or stands still.

(G31) Proposition A111 follows from A1111 and A1112 if and only if one assumes that, A1113, the distance between the gun and the point hit equals the sum of the distances traveled by the projectile and by the gun for a westward shot, and their difference for an eastward shot; and that, A1114, in this sum and this difference the distance traveled by the projectile on a rotating earth equals the distance traveled on a motionless earth. But (G32) this equation, in A1114, amounts to assuming that the earth stands still. Therefore, (G3) in the step from A1111 and A1112 to A111, the argument begs the question. [F196.34–197.1]

Comments. The G1 criticism amounts to showing that in the object argument the step from A12 to A1 is invalid, i.e., that A1 does not follow from A12. It also means that (the facts about) east-west gunshots do not imply that the earth stands still and do not constitute a valid objection to the earth's motion. G1 as such does *not mean* that the earth's rotation implies equal ranges, but in the argument supporting G1 there are implicit propositions which suggest this implication, which is made explicit in the G2 argument (proposition G21). Because of G21, east-west gunshots do not even *support* the earth standing still, since though those facts could be explained by the earth standing still, they could also be explained by its rotation, so that an explanatory argument from east-west gunshots to the earth standing still would be invalid, because arbitrary. In short, the argument from east-west gunshots is deductively invalid (G1), inductively invalid (G2), and rhetorically invalid (G3).

12. INFERENTIAL VS MATERIAL 'IF', AND THE VERTICAL GUNSHOT ARGUMENT (F197.1–202.16)

The object argument. (A11) In vertically upwards gunshots the projectile returns to the gun. But (A12) if the earth were rotating, the projectile would return to a place west of the gun, since (A12a1) if the earth rotates then during the flight of the projectile the gun has been carried a long distance toward the east, and since (A12b1) if the earth rotated and the projectile returned to the gun then the projectile would have moved transversally, and (A12b2) if the projectile had moved transversally, we would observe its transversal motion (because (A12b21) otherwise we would have a denial of the senses), but (A12b3) we do not observe any transversal motion. Therefore, (A1) the earth does not rotate. [F197.6–197.14, and 200.6–200.13]

The meta-argument. (G1) The vertical gunshot argument begs the question since (G11) in grounding A12 on A12a1 the argument assumes the final conclusion, A1, that the earth does not rotate; for (G111) the rest of this subargument (its latent structure) would have to be the following: "(A12a1) if the earth rotates then during the flight of the projectile the gun has been carried a long distance toward the east, and (A12a2) if the gun is carried a long distance toward the east then the projectile returns to a place west of the gun, because (A12a21) the projectile is not carried the same distance toward the east"; and (G112) the only justification that could be given for A12a21 is that: "(A12a211) the projectile is on a motionless earth, since (A12a2111) the earth stands still"; and (G113) proposition A12a2111 is identical to A1. [F197.1–197.10, and 200.13–200.19]

Moreover, (G2) proposition A12 is false since (G21) if the earth were rotating the projectile would indeed return to the gun; for (G21a111) on a rotating earth both the projectile and the gun would be moving eastward, and hence (G21a11) the projectile would not start its up and down motion from rest but rather would have its up and down motion combined with its eastward rotation; since (G21a12) its eastward rotation is neither eliminated nor impeded by its up and down motion, it follows that (G21a1) the projectile remains always over the gun. Also, (G21b1) observation would show that on a ship an arrow shot upwards returns to the bow whether or not the ship is moving. [F200.19–200.29]

Third, (G3) proposition A12b2 is groundless and false, groundless because (G3a1) there would be no denial of the senses if the projectile moved transversally and we did not observe its transversal motion (i. e., because its supporting reason, A12b21, is false), and false because (G3b1) we could not observe the transversal motion of the projectile if it was so moving. In fact, (G3b11) on a rotating earth when a cannon is fired at AC, by the time its ball reaches its mouth, it will have been carried to DE, the ball will have traveled the transversal CD, but we would have observed it to go along the cannon's barrel, which is not transversal [F200.30–202.16] :

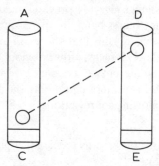

and there is no denial of the senses here because (G3a11) it is the nature of motion to be such that shared motion is imperceptible, and (G3a12) in this situation the horizontal rotatory motion is shared by us and by the observed projectile. [F197.15–199.26]

Comments. If the G1 criticism is to be correct in alleging that the object argument presupposes what it is trying to prove, then the subargument causing this circularity ought to be a valid deduction, and if so A12 would

follow from its final reasons A12a1 and A12a2111, so that if (as G2 argues)
A12 is false then either A12a1 or A12a2111 should be false. Since A12a1
would seem obviously true, it follows that A12a2111 must be false, which
amounts to a proof of the earth's motion. This is puzzling and problematic,
the problem being analogous to the one that arose in Chapter 12 in connection with the circularity argument from actual vertical fall.

Let us therefore ask whether A12a1 is really true. Its apparent truth
derives, I believe, from an ambiguity: 'east' in this context can mean either
"east from the point of view of absolute space" or "east from the point of
view of the earth's surface". The gun would be carried toward the east only in
the former sense, not in the latter. Now, let us see what is the sense of 'east'
and 'west' in A12a2. This would be the sense in which it would follow from
A12a2111, which is the sense of absolute space for 'east', and the sense of
terrestrial surface for 'west'. It follows that A12a1 is indeed true; hence the
above mentioned paradoxical proof of the earth's motion remains.

Of course, this proof would evaporate if one cannot assert the falsehood
of A12, which is asserted in proposition G2, so one might question the
correctness of its supporting argument. The step from G21 to G2 presupposes
that 'if p then q' implies 'not-(if p then not-q)', which as before could be
questioned for being truth-functionally invalid, but which I would not want
to question; the other steps seem correct. G21a12 could of course be
questioned, but Galileo could give his previous justification of it, involving
a claim about two distinct causes being in operation; at any rate there would
be *no* trace of presupposing a claim that the earth moves (which trace might
explain the paradox). G21a111 seems either obvious or dependent merely on
conservation of motion. So our puzzle remains.

Let us look once again, and more closely, at the deduction that generates
the paradox, using the following abbreviations:

R = the earth *r*otates;
S = the projectile returns to the *s*ame place (on the earth's
surface);
Eg = the *g*un is carried to the *e*ast.

The question-begging inference analyzed in G1 reduces to:

(1′)   if A12a1 and A12a2111, then A12.

The crucial conclusion in G2 is:

(2′)   not-A12.

These may be rephrased as:

(1)   if (if R then Eg) and not-R, then (if R then not-S);

(2)   not-(if R then not-S);

which seem to entail

(3)   not-[(if R then Eg) and not-R] ,

by *modus tollens,* and since

(4)   if R then Eg

is true, we get

(5)   R.

I believe that the problem is with the ambiguous meaning of 'if R then not-S'. There are several reasons why in (2) its meaning is inferential, that is, 'not-S can be inferred from R': first, in the context of criticism G2, Galileo wants to deny the validity of the inference from S (=A11) to not-R (=A1), which denial is what (2) asserts; second, (2) is justified by proposition G21 asserting that 'if R then S', which presupposes that 'if R then S' and 'if R then not-S' are not both true, which is only inferentially and not truth-functionally correct (given the self-consistency of R, obvious in this case); third, if (2) had a truth-functional meaning, R would follow immediately in one step by the definition of the material conditional, which is false if and only if the antecedent is true and the consequent false. On the other hand, in (1) the meaning of 'if R then not-S' is not inferential since the context, G1, of this conditional is one where it is being treated as a single proposition, in order to show that its justification presupposes not-R; moreover, in (1) it is obvious that it is the *main* 'if then' connective which is the inferential one, so that the whole meaning of (1) is that "(if R then not-S) can be inferred from (if R then Eg) and not-R". This shows that the above paradox was the result of a fallacy of equivocation.[11]

Finally, the G21b argument presupposes an analogy between the moving earth and a moving ship, which Galileo criticized in the context of the ship experiment argument; however, I do not think there is any real conflict because the disanalogy is destructive only with respect to the alleged observation of different phenomena on a moving and on a motionless ship; given the *same* effects, the disanalogy *reinforces* the inference from the ship to the earth since on a moving ship there would be more possible causes for the arrows to be left behind.

13. EXPLANATION VS INFERENCE, AND THE
HUNTER'S ARGUMENT (F203.34–205.29)

The object argument. (A14) Hunters can hit birds while they are flying;
(A13) the explanation of this fact is not the hunter's anticipation of the bird's
motion when aiming at it, that is, the practice of aiming at the point where
the bird will be when the bullet gets there rather than aiming at the bird,
since (A131) hunters do *not* aim at the target by anticipating it; rather, (A12)
by moving their gun hunters follow the bird in its motion while they aim, and
(A11) in going from the gun to the bird the bullet retains the gun's motion
parallel to that of the bird; therefore, (A1) the explanation is the fact that
hunters aim by moving their gun to follow the bird.

The meta-argument. (G11) The explanation cannot be the practice of
aiming by moving the gun to follow the bird, since (G1111) the gun's motion
is very small compared to the parallel motion of the bird, and hence (G111)
the bullet cannot keep up with the bird; therefore, (G1) the explanation must
be the combination of several factors: (a) aiming by moving the gun, (b)
aiming by anticipation to a small extent, (c) the use for charge of a large
number of small pellets rather than a single large ball, and (d) the great speed
with which the shot moves toward the bird.

Comments. The object argument is *not* an explanatory argument in the
sense that its conclusion states an alleged explanation for some fact stated in
the premises. Rather the conclusion is saying that the fact mentioned in
premise A12 explains the fact mentioned in A14, so that the argument is one
that tries to establish an explanatory connection, taking both the explicandum, A14, and the explicans, A12, as premises. Since it is clear that the
net effect of the whole argument is to account for (i.e., to explain) A14, we
might regard the whole argument itself as constituting an explanation. We
may then define an explanation as an argument whose premises include both
the explicandum and the explicans, and whose conclusion asserts that the
explicans explains, accounts for, or renders comprehensible the explicandum.

The meta-argument refutes the object argument explanation by refuting
some intermediate proposition presupposed by the step from A14, A12, and
A11 to A1, namely that the bullet's motion parallel to the bird and imparted
to it by the hunter's moving of the gun barrel is equivalent to the bird's
parallel motion (and thus sufficient to make the bullet keep up with the bird).
G1111 and G111 are denying these propositions. Thus we may say that the
justification of an explanation would be a justification of the correctness of
the argument constituting the explanation.

The next important point pertains to the structure of the explicandum *per se*. Here it was stated merely as a proposition, but it is clear that not every proposition will generate an explicandum; in fact, in this case the use of the modal qualifier 'can' in A14 constitutes an implicit reference to the possibility that perhaps hunters should be unable, or have difficulty with, hitting moving birds. In short, what constitutes an explicandum is really a factually asserted proposition, plus an argument having its denial as conclusion. In our case this argument would be: "(A221) when a hunter aims and shoots at a flying bird, the bird usually has a component of motion perpendicular to the line of sight at any given moment; therefore, (A22) if a hunter shoots *at* the bird, where he sees it, he will miss; but (A21) hunters do aim at birds when they shoot; therefore, (A2) it should be impossible to hit a flying bird". The explanation given in the object argument really amounts to criticizing the inference from A221 to A22 by pointing out that all that follows from A221 is that if a hunter shoots at the moving bird *without moving his gun* then he should miss; whereas explanation A13 was inferring merely that if A21 is true then A2 is true, but denying A21. The G1 meta-argument introduces additional refinements in the step from A221 to A22.

14. SCIENTIFIC, AD HOMINEM, LOGICAL INVALIDITY, AND THE POINT-BLANK ARGUMENT (F205.30–209.13)

The object argument. (A12) If the earth were rotating then point-blank gunshots toward the east would hit high and toward the west low, since (A121) point-blank gunshots move along the tangent, WE, to the point of firing, O, and (A122) the earth's rotation would cause an eastern target, T_E, to descend, e.g. to T_E', and a western target, T_W, to rise, e.g. to T_W', relative to the tangent, WE; but (A11) there are no up and down deviations in point-blank gunshots; therefore, (A1) the earth does not rotate. [F205.32–206.7]

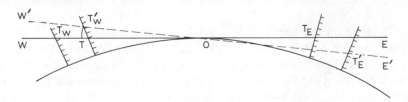

The meta-argument. Since (G111) on a rotating earth the gun as well as the target are moving with respect to the tangent, (G11) the gun and the target are not moving relative to one another, and hence (G1) the projectile will travel from one to the other in the same way as on a motionless earth.

(G211) If A12 is true then the up and down deviations would be imperceptibly small, because (G2111) it can be shown that they would be of the order of an inch; therefore, (G21) we do not know whether or not point-blank gunshots actually exhibit these deviations; therefore, (G2) proposition A11 is groundless.

(G3) The lack of deviation, A11, does not imply that the earth stands still, A1, because (G3112) if the earth rotated then artillerymen would have learned to aim point-blank gunshots as would be required on a moving earth to hit on target, and (G3111) if artillerymen had learned to aim on a moving earth then there would be no deviation, and hence (G311) if the earth rotated then there would be no deviation, and hence (G31) the earth's rotation does not imply a deviation. [F208.20–208.25]

Rather, (G4) one might say that there would be a deviation if the earth stood still. For (G41122) if artillerymen had learned to aim on a moving earth and if the shot follows a tangential path, then artillerymen would be aiming higher to hit lower toward the west, and lower to hit higher toward the east, but (G41121) if artillerymen were aiming higher to hit lower (toward the west) and lower to hit higher (toward the east), then if the earth stood still point-blank gunshots would hit high toward the west and low toward the east, and hence (G4112) if artillerymen had learned to aim on a moving earth and the shot followed a tangential path, then if the earth stood still then point-blank gunshots would hit high toward the west and low toward the east; but (G4111) if the earth were rotating then artillerymen would have learned to aim on a rotating earth; therefore, (G411) if the earth were rotating (and shots followed a tangential path) then the earth standing still would imply a deviation; therefore, (G41) to show that the earth standing still would not imply a deviation you have to show that the earth *is* standing still (or else that the shots do not follow a tangential path). [F208.25–208.28]

Comments. The effect of G1 is to invalidate the argument from A11 to A1 by discrediting A12. But how correct is this criticism? Certainly it would be wrong to say that it presupposes 'circular inertia',[12] for Galileo does not say that the path of the ball also curves downward (for the eastward shot), but rather that the gun is continuously being inclined (F206.9); hence the projectile would have an additional component of velocity 'downwards' with respect to the (motionless) tangent, and it will be conserved (given the principle of conservation of motion, justified earlier). Rather, the situation here is like that alleged for the shooting at flying birds in the hunter's argument (F202–5), with the difference that the parallel speeds of both gun barrel and target are here essentially the same. I say 'essentially' because if the surface distance between the gun and the eastern target is of the order of 90 degrees then the 'downward' speed of the latter is approximately equal to the rotational speed, and hence much greater than the 'downward' speed of the gun, which, being at the point of tangency (by definition), is only a fraction of the rotational speed; however, the equivalence is approximately correct for relatively small distances between the gun and the target. Thus the situation is really like that of the north-south gunshots, concerning which Galileo himself makes the explicit qualification that the answer he gives applies only for ordinary relatively short ranges (F205.10–205.13). This gives additional significance to the quantitative considerations made in G2.

Galileo's criticism G2 concentrates explicitly on what might be called the groundlessness of the crucial premise A11. However, one could pursue the same line of reasoning as follows:

If there is a deviation however small, then point-blank gunshots do not in fact hit correctly; hence, because of G211, if A12 is true then A11 is not; but A12 means that the earth's rotation implies not-A11; it follows that if the earth's rotation implies not-A11, then not-A11; i.e., if A11 then the earth's rotation does not imply not-A11; i.e., if A11 then A11 does not imply that the earth stands still; i.e., A11 does not imply A1. Thus G2 can also be interpreted as a critique of the validity of the point-blank argument, and a more effective one than G1.

The justification of G2111 is given explicitly by Galileo and amounts to the following:

(1)   Assuming that to reach its target a cannon ball takes about as long as it takes a pedestrian to walk two paces, the ball will take about *one second* to reach its target.

(2a)   The western horizon rises 15 seconds of arc in one second of time, since it turns 360 degrees in 24 hours (assuming we are at the equator).[13]

(2b)   Therefore, a western target rises 15 seconds of arc in one second of time.

(2c)   Assuming the distance from the cannon to the target is 500 *braccia*, the 15 seconds of arc are measured along a circle with a radius of 500 *braccia*.

What Galileo seems to have in mind here (2a, b, c) may be explained by referring to the above figure. The cannon is resting at O; E is the eastern and W the western horizon. On a stationary earth the cannon ball would travel in one second from O to the target T_W. On a moving earth the western horizon rises and the eastern one falls so that in one second WE would pivot around O to a new position $W'E'$, where angle WOW' is 15 seconds; the target is now at T_W'; disregarding the eastward component of the target's motion (since this argument is considering only deviations above and below the tangent), OT_W' is also 500 *braccia*.

(3a)   The chord for one minute, on a radius of 100 000 units, is 30 units.

(3b)   Therefore, the chord for one second, on a radius of 100 000 units, is $(1/60) \times 30 = 1/2$ units.

(3c)   That is, the chord for one second, on a radius of 200 000 units, is 1 unit.

(3d)   Therefore, the chord for 15 seconds, on a radius of 200 000, is 15 units.

(3e)   Therefore, the chord for 15 seconds, on a radius of 500 *braccia* = $(500/200\,000) \times 15$ *braccia* = $15/400$ *braccia* = approximately $4/100$ *braccia*.

(3f)   Therefore, the target rises approximately $4/100$ *braccia*, which is approximately one inch.

What Galileo seems to be doing here (3a–f) is to find the distance from T_W' to the line OW. He is approximating this distance by the arc TT_W', whose center is O. The approximation is correct since the angle TOT_W' is very small (15 seconds). Galileo's answer checks with the following calculation:

$$\text{arc } TT_W' = \left(\frac{15}{360 \times 60 \times 60}\right) \times (2\,\pi\,500) = \frac{15\pi}{36 \times 36} = \frac{5\pi}{12 \times 36}$$

$$= \text{approx. } \frac{\pi}{12 \times 7} = \frac{\pi}{84} = \text{approx.} \frac{1}{25} \, .$$

(4)   Therefore, the deviation is about $1/25$ *braccia*, or about one inch.

What Galileo does in (4) is to reason that, since the target 'rises' one inch, the shot will be one inch 'low'.

An interesting question to ask about this part of the critique is whether it is *ad hominem*, in the seventeenth century meaning of this term, that is whether the argument draws consequences of ideas accepted by one's opponents, but not necessarily acceptable or accepted by oneself. The idea in question here would be that involved in thinking of the horizon as rising and falling, on a moving earth. It seems un-Galilean to think that the motion of the earth would involve the kind of shift of the horizon suggested by our figure. And yet this helps rendering the passage comprehensible. All this is evidence that the argument is indeed *ad hominem*.

Galileo's criticism G3 constitutes another reason for the invalidity of the step from A11 to A1.

I believe that G4 amounts to an argument to the effect that if the point-blank argument is valid, then it is incomplete insofar as there is no justification that the earth's motionlessness would *not* imply a deviation; one is entitled to this proof in view of G411, A1, and A11.

It is interesting to contrast the three reasons G1, G2, and G3, for the invalidity of the object argument. Both G1 and G2 involve scientific substantive considerations, whereas G3 is a rather rhetorical kind of criticism; this rhetoric pertains to the implicit consideration that what would count as deviation and as proper point-blank aim depends partly on whether or not the earth rotates. The reasoning in G1 is rather external insofar as there is an appeal to both composition and conservation of motion, which the propounders of the object argument do not accept; whereas criticism G2 is wholly internal to the object argument, and this easily explains the greater effectiveness of G2.

15. THE LOGIC OF PUNCTUATION, AND THE BIRDS ARGUMENT
(F209.13–212.30)

The object argument. (A121) When birds fly they are moving at will; therefore, (A12) if the earth were rotating then flying birds would have to keep up with the terrestrial rotation through their own efforts; but (A11) it is incredible that they would have enough energy to move so fast; therefore, (A1) the earth cannot move. [F209.13–209.25]

The meta-argument. (G11) If the earth were rotating then the lower regions of the air would be rotating with it; (G12) if so, then flying birds would keep up with the terrestrial rotation through the agency of the air; therefore, (G1) proposition A12 is not correct. [F209.26–210.20]

(G2) The birds argument seems stronger than the projectile arguments because (G211) birds can move at will against the innate tendencies of terrestrial bodies (for example, (G2111) a live bird can fly upwards whereas a dead one will fall downwards like any heavy body), and hence (G21) the reasoning that applies to projectiles does not apply to birds. [F212.3–212.10]

(G311) The fact that the reasoning that applies to projectiles does not apply to birds means that on a rotating earth we would not see birds do the same thing as projectiles, first because (G311a1) if a dead bird were dropped from a tower it would like a stone continue to rotate and also start falling, but (G311a2) if the bird were alive then it would continue to rotate and also use its wings to go where it wanted, and (G311a3) if the latter happened then we would only see the motion due to its wings (since (G311a31) this would be the only motion not shared by us), and second because (G311b1) on a rotating earth the bird's ability to move at will would amount to an ability to add or subtract degrees of speed with respect to the fixed rotational speed; hence (G31) there would be no observable difference between the flight of birds on a rotating earth and on a motionless one; hence (G3) the greater strength of the birds argument over the projectile arguments does not make it correct. [F212.10–212.26]

Comments. The reconstruction of G2 and G3 was made very difficult by the punctuation found in Favaro's standard critical edition, which, as usual, is different from that of the 1632 edition; though normally an improvement over the original, in this case Favaro's punctuation was not. The relevant Favaro text can be seen in the following literal translation, which is also a line-by-line translation, so that the line numbers noted here correspond to those on p. 212 of the Favaro edition:

3 Your having more difficulty for this than for the other objections,
4 seems to me to depend on birds being animate, and being able
5 therefore to use force at will against the primary motion innate
6 in earthly things, in such a way precisely that we see them, while they are
7 alive, flying even upwards, a motion impossible to them as heavy bodies, whereas
8 dead they cannot but fall downwards; and therefore you judge
9 that the reasons that take place for all sorts of projectiles mentioned
10 above, cannot take place for birds; and this is very true, and
11 because it is true, therefore we do not see, Mr. Sagredo, those projectiles doing
12 what birds do: for if from the top of a tower you let
13 fall a dead bird and a live one, the dead one will do the same that
14 a stone does, that is it will follow first the general diurnal motion, and then
15 the downward motion, as a heavy body; but if the released bird is alive, who

16 will forbid it, the diurnal motion always staying with it, from going,
17 by a beating of wings, toward whatever part of the horizon it likes most?
   and
18 this new motion, being its own in particular and not shared by us,
19 must become perceptible to us. And if it had by its flight moved toward
20 the west, who is to forbid it from returning to the tower
21 by the same beating of wings? Because, finally, to fly up
22 toward the west was nothing other than to subtract from the diurnal
   motion, which has,
23 for example, ten degrees of speed, a single degree, so that it was left
   with
24 nine, while it was flying; and if it had alighted to earth, it would have
   regained
25 the ten common degrees, to which by flying toward the east it could
   have
26 added one, and with the eleven return to the tower: and in sum,
27 if we well consider and more intimately contemplate the effects
28 of the flying of birds, they do not differ in anything from projectiles
   thrown in
29 all directions, except that these are moved by an
30 external projector, and those by an internal principle.

The punctuation in this passage suggests that the main break comes on line 19, because of the period on that line. This would suggest that lines 19—30 constitute either the second of two main arguments in the passage (the first being presumably on lines 3—19), or else a self-contained subargument of the main argument. Either alternative runs into the problem that lines 26—30 do not connect with anything else in this passage. If we mentally delete these lines, and if we follow our logical intuition telling us that lines 19—26 do not constitute one of two main arguments, then we are left with the following subargument interpretation: (lines 19—21) on a rotating earth birds would be able to fly since (lines 21—26) flying on a rotating earth would involve being able to add and subtract degrees of speed with respect to the fixed rotational speed. (This two-proposition argument was collapsed into a single proposition, G311b1, above.) Now, how would this subargument connect with the rest of the passage? Looking at lines 3—19, the colon strikes one as the most important punctuation mark; it indicates either a break intermediate in force between a period and a semicolon, or else that an explanation or justification follows. Whichever is the case, since lines 12—19 are obviously reasons for something, the next question is what is the claim which they support. The suggested choice is the proposition (lines 11–12) that we do not see projectiles doing what birds do. Unfortunately, logical intuition vetoes this interpretation, since such a conclusion (lines 11–12) is about what we see, and the propositions on lines 12–19 are about what would happen on a rotating earth, and hence lines 11–12 do not connect with lines 12–19. The problem is apparently alleviated by the fact that the reasoning indicators on lines 10–11 seem to provide the reason for the claim on lines 11–12. This would lead one to think that the main break in the entire passage comes with the colon on line 12, for after all lines 12–19 connect pretty well with lines 19–26, better than they connect with lines 3–12. The second part of the passage (lines 12–26) would then involve the minor problem of finding a proposition supported by the claims made therein. The first part (lines 3–12) would involve the major problems that lines 11–12 do not connect with lines 9–11, since on the one hand the 'this' of line 10 should refer to lines 9–10, but on the other hand one cannot conclude that we do not see birds and projectiles behave similarly (lines 11–12) from the fact that "the reasons which take place for all sorts of projectiles mentioned above, cannot take place for birds" (lines 9–10). The situation would now be that lines 11–12 connect neither with what follows nor with what precedes; because of this, one might try an appropriate reformulation of its literal meaning, namely that we *would* not see *on a rotating earth* those projectiles doing what birds *would* do. This proposition does connect very well with lines 12–26; hence lines 11–26 would now make sense as a unit; hence a main break should have been indicated on line 10 or 11, by a period perhaps; this is especially true since lines 3–10 can be easily interpreted as a distinct argument, as it is done in G2 above. The only residual problem that this division of the passage at line 10 would generate stems from lines 10–11, which strongly suggest an inferential connection; the problem is solved by interpreting lines 10–11 ("and this is very true, and because it is true, therefore . . . ") as a way of indicating that the immediately preceding proposition (lines 9–10) can be rephrased to mean what lines 11–12 state (according to the nonliteral interpretation worked out above).

The preceding account describes the manner in which I arrived at my reconstructions G2 and G3. But it would probably have been impossible for me to arrive at them if I had not had the benefit of the more amorphous punctuation in the 1632 edition of the *Dialogue*. Among other differences, the latter has a semicolon on line 12 instead of a colon, a semicolon on line 19 instead of a period, and a period on line 26 instead of a colon.[14]

16. IGNORATIO ELENCHI, QUANTITATIVE INVALIDITY,
INCOMPLETENESS, AND THE CENTRIFUGAL FORCE ARGUMENT
(F214–244)

The object argument. (A12) If the earth were rotating, objects on its surface would be scattered away from it, toward the heavens, because (A121) rotation has the power of extruding objects lying on the rotating body; but (A11) objects on the earth's surface are not observed to be scattered toward the heavens; therefore, (A1) the earth does not rotate.

The meta-argument. (G122) Proposition A12 does not follow from A121 because (G1221) what follows is either that (a) if the earth had always been rotating then there would (now) be no objects on its surface, or that (b) if the earth were to begin rotating then objects on its surface would be scattered toward the heavens; but (G121) proposition A11 connects only with (b) above; therefore, (G12) what follows from A121 and A11 is that the earth has not (recently) begun to rotate. But (G11) propositions A121 and A11 are the final reasons of the object argument. Therefore, (G1) what the object argument really proves is not its conclusion, A1, but rather the conclusion that the earth has not (recently) begun to rotate. [F214.30–216.11]

Second, (G2) the argument is quantitatively invalid because (G21) in order for A12 to follow from A121 the centrifugal tendency due to the extruding power has to be sufficiently great; but (G22) it is not because (G221) the downward tendency due to weight could overcome the centrifugal tendency due to rotation; for (G2211) the centrifugal tendency is measured by the tangential displacement (that would result if the body were extruded), and (G2212) the downward tendency is measured by the displacement along the secant from the body to the earth's surface, and (G2213) the geometry of the situation is such that a body can remain on the earth's surface by undergoing a very small secant displacement while it undergoes a very large tangential displacement, and (G2214) the secant speed, however small, is always sufficient to compensate for the tangential speed, however large, since (G22141) from the center of a circle, O, one can draw a secant, OI, intersecting a tangent, TT', so close to the point of tangency, T, as to yield an arbitrarily small ratio between the portion of the secant outside the circle, IC, and the portion of the tangent to the point of tangency, IT. [F217.4–224.28]

Third, (G31111) though it is true that the cause of extrusion increases as the speed when the radius is constant [F238.2–238.6], (G31112) when the speeds are equal the cause of extrusion decreases as the radius increases

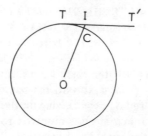

[F243.25–243.26], so that (G3111) the cause of extrusion increases directly with the speed but inversely with the radius [F238.25–238.27]; thus (G311) it may be that the cause of extrusion remains constant when the speed is increased as much as the radius, namely when equal number of rotations are made in equal times [F244.6–244.11]; hence, (G31) the earth's rotation would cause as much extrusion as a wheel which rotated once in twenty four hours [F244.11–244.15]; therefore, (G3) a rotating earth would not scatter its surface bodies toward the heavens. In support of G31112 note that (G311121) the cause of extrusion is equal to the force needed to prevent the body from escaping along a tangent, that (G311122) this force is that required to compensate for the tangential displacement if the body had been extruded, and that (G311123) this compensation is smaller for greater circles

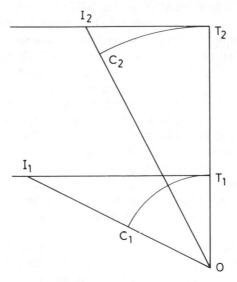

since (G3111231) when the linear speeds are the same (e.g., along T_1C_1 and T_2C_2) the deviation from the tangent is measured by the distance from the tangent to the circumference (I_1C_1 and I_2C_2), and (G3111232) this distance decreases as the circle becomes larger ($I_2C_2 < I_1C_1$) [F242.11–243.26].

Comments. (1) In accordance with criticism G1, the argument from whirling may be regarded as a classic instance of *ignoratio elenchi* (or 'irrelevant conclusion'), in Aristotle's own sense of an argument alleged to prove one thing, but which at best proves something else.[15]

(2) I have used the term 'quantitative invalidity' in my reconstruction of the G2 critique. The suggestiveness of this term is obvious; its importance derives from the fact that it is a very important special case of failure of implication. It is the case when *the reasons* why one proposition does not follow from another involve quantitative considerations. In other words, it is true that (A12) on a rotating earth objects would be scattered toward the heavens, but the amount of such scattering is very small and also smaller than the amount by which they move toward the earth's center due to the weight. Similar considerations would apply for G3.

(3) As historians of science have not tired of pointing out, in G2 the step from G22141 to G2214 is incorrect. However, it is one thing to say *that* it is invalid, and it is another to *understand* what is wrong, to explain *why* it is invalid. The theoretically astonishing thing is that in the same context it is Galileo himself who provides us with the means for such an explanation. It is as if he had deliberately perpetrated the error in order to be able to make a point in the theory of reasoning, which is to say not that he is using rhetorically improper procedures, but rather that this passage has an essentially logical component.

Why, then, is the step erroneous? It is not clear how, for example, one would use the method of counter-example; one would have to find the form of both propositions, and then produce a pair such that the one having the form of G22141 is obviously true, and the one having the form of G2214 obviously false. This would involve reformulating the argument, and the faithfulness of the reformulation would present a very difficult, if not insurmountable problem.

Another approach would be to say that G22141 as it stands is indisputably true, whereas G2214 is false, this falsehood being uncontroversial nowadays, but also such that it could or should have been admitted by Galileo himself. After all, in stating the argument he asserts as a truth which he accepts, that "heavy bodies, whirled with speed around a fixed center, can acquire impetus to move away from that center, even when they are such as to have a propension to go there naturally" (F216.19–216.21). Such an explanation of this argument's invalidity would be using the essential idea of the counter-example method. However, in bypassing the troublesome notion of form, it would fail to provide understanding of what error is being committed.

If we turn to Galileo's own account, we find that the argument is followed by a discussion of the applicability of mathematics to physical reality (F229–237), in which Salviati makes many subtle and crucial methodological points.[16] One of these is that the real problem is to use a correct mathematical model for the physical situation, i.e., to approximate the physical processes by the right type of mathematical entity (F234). Applying this discussion to the present problematic inference, we may begin by saying that proposition G22141 is a mathematical truth, but that G2214 is a physical proposition; to say this is perhaps not to make a claim about the 'logical form' of these propositions, but it is to say something about their nature or character. It would follow that only to the extent that the physical entities and processes mentioned in proposition G2214 are correctly approximated by the abstract entities of the other proposition, is the inference correct. Hence it is not, since the various geometrical segments mentioned in the premise correspond to physical displacements, not to the speeds associated with these displacements, which speeds are the physical entities mentioned in the conclusion.

(4) The manner in which Galileo begins to state his G3 criticism reminds one of the following equivocation charge: (FG414) The step from A121 to A12 presupposes two propositions: A122, that the cause of extrusion increases as the rotational speed does, and, A123, that the speed of terrestrial rotation would be very great. Now, (FG4132) it is true that the cause of extrusion increases with the rotational speed when the radius is constant, namely that it increases as the number of rotations per unit time; but (FG4131) it is false that the cause of extrusion increases with the rotational speed when the number of rotations in equal times is constant, namely that it increases as the radius; thus, (FG413) 'rotational speed' may refer either to angular speed or to linear speed. Now, (FG412) if angular speed is meant then A122 would be true but A123 false, since (FG4121) the earth makes only one rotation in twenty four hours. Whereas, (FG411) if linear speed is meant then A123 would be true and A122 false. Therefore, (FG41) not all premises of the subargument to A12 can be simultaneously true. Therefore, (FG4) A12 is groundless.

One problem with this argument is that though FG414, FG4132, and FG4131 are in the text (F237.29–238.15), the other propositions are not. Moreover, the step to FG413 is questionable since it interprets the denial of an increase with the radius mentioned in FG4131 as a denial of an increase with the linear speed, whereas it really amounts to a denial of an increase *purely* with linear speed. Third, proposition FG4131 is not in fact true. Hence it would be uncharitable to attribute this argument to the text. Nevertheless it does make (in the abstract) an interesting equivocation.

(5) Criticism G3 amounts partly to a critique of quantitative invalidity for the object argument, but it also suggests a charge of incompleteness, insofar as proposition G3111 shows that it is wrong to consider speed as the only variable. Galileo charges incompleteness almost in so many words (F237.37–238.2); moreover, he has a long discussion (F238.13–242.11) designed to show that rotation is not the only phenomenon where speed is not the only effective variable: for example, the resistance to being moved and to being stopped increases as the speed *and* the weight (F241.1–241.11).

(6) I believe that the subargument to G311123 avoids the problem of misapplying a mathematical model besetting the subargument to G2214. Whereas in the earlier argument the model falsely represents a static situation, in the present case the model is properly dynamic.

NOTES

[1] As before, this is an abbreviated reference to G. Galilei, edited by Favaro, *Opere* 7, 43–57. All unqualified references in this chapter refer to this book.

[2] As in earlier chapters, numerals after the period refer to lines.

[3] In such contexts, 'argument x' means 'the argument supporting x'.

[4] The intermediate steps in this inference are:

$$(13.1) \ [\text{if not-}H_C \text{ and } X, \text{ then not-}H_A] \text{ and } [\text{if } X \text{ then } H_A];$$
$$(13.2) \ \text{if } [(\text{not-}H_C \text{ and } X) \text{ and } X] \text{ then } (\text{not-}H_A \text{ and } H_A).$$

[5] Cf. the discussions at the end of Chapter 12 and in Sections 8 and 12 of this chapter. See also the remarks in Note 8 of Chapter 12.

[6] B. Mates, *Elementary Logic*, p. 81; I. M. Copi, *Introduction to Logic*, p. 252; W. C. Salmon, *Logic*, pp. 49–50; and D. Kalish and R. Montague, *Logic*, p. 92.

[7] My claim here is one of comparative accuracy, for there are some contexts where the conventional translations would be acceptable, and many others where neither would be adequate.

[8] See M. Scriven, *Reasoning*, pp. 71–77.

[9] Cf. the discussion in Chapter 14, Section on 'Interdependence of Reasons'

[10] If this is equated with formal implication (in the conventional model-theoretic sense), then we would need, as an underlying contextual assumption, to exclude from our object language those nonmolecular sentences that are inconsistent, otherwise we would have a counterexample to (F3.1) when R is inconsistent. Another way out of the problem might be to give a probabilistic analysis of these conditionals in accordance with E. W. Adams's *Logic of Conditionals*. A further possibility would be to explore the relevance of the various 'relevance-entailment' systems found in A. R. Anderson and

N. D. Belnap, Jr.'s *Entailment.* Cf. our discussions above at the end of Chapter 12, and in Sections 4 and 12 of this chapter. See also our remarks in Note 8 of Chapter 12.

[11] I realized this while comparing the two deductions below, the first one of which is in the object language and correct, and the second one of which is in the metalanguage and incorrect. The notation used is from B. Mates, *Elementary Logic.*

I

{1}	(1) $P \longrightarrow Q$	P
Λ	(2) $(-P \& P) \longrightarrow -Q$	TH
{3}	(3) $-[(P \longrightarrow Q) \& (P \longrightarrow -Q)]$	P
{3}	(4) $(P \longrightarrow Q) \longrightarrow -(P \longrightarrow -Q)$	3, R
{1, 3}	(5) $- (P \longrightarrow -Q)$	1, 4, MP
Λ	(6) $-P \longrightarrow (P \longrightarrow -Q)$	2, TH
{1, 3}	(7) P	5, 6, TH
{3}	(8) $(P \longrightarrow Q) \longrightarrow P$	7, C

II

(1) Assume: $\sigma \vDash \phi$
(2) Not-σ and $\sigma \vDash$ not-ϕ
(3) Assume: not both $\sigma \vDash \phi$ and $\sigma \vDash$ not-ϕ
(4) Not-($\sigma \vDash$ not-ϕ); from (1) and (3)
(5) Not-$\sigma \vDash (\sigma \vDash$ not-ϕ); (?) from (2) (?)
(6) σ ; from (4) and (5).

In II, step (5), corresponding to step (6) of I, is wrong since (5) means: "not-$\sigma \vDash$ '$\sigma \longrightarrow$ not-ϕ' is valid", which is not true, a counterexample being $\sigma = P$, $\phi = Q$.

[12] Cf. G. Galilei, *Dialogue,* ed. Salusbury-Santillana, p. 193, note 64.

[13] Galileo claims to give maximum advantage to his opponents by assuming the location to be at the equator. Even Drake (*Dialogue,* p. 477, footnote to p. 181) claims that there is really no such advantage, but it seems to me that there is; for, whereas at the equator the horizon can be imagined to rise and fall 360 degrees in 24 hours, at the poles there would be merely an horizontal turning with no rising and falling, and at intermediate latitudes we would have a mixture of horizontal turning and rising and falling, the latter being less than at the equator. Drake also does not seem to be entirely right in claiming that "the demonstration is quantitatively worthless" (*ibid.*), for I show below that Galileo's figures seem to check.

[14] See the section on Favaro in Chapter 10 above.

[15] Aristotle, *On Sophistical Refutations,* 167a21; cf. Hamblin, *Fallacies,* pp. 31–32.

[16] See Chapter 5 above.

CHAPTER 17

CRITICISM, COMPLEXITY, AND INVALIDITIES:
THEORETICAL CONSIDERATIONS

SOME DATA ANALYSIS

The results of the previous chapter may be regarded as a data basis in need of further analysis. Let us begin by making an inventory of the logical topics found in Galileo's book, as seen from the section titles of the last chapter. The following table will serve our purpose.

PAGES	ARGUMENT	LOGICAL TOPIC
33–38.24	Three-dimensionality	Conclusion vs reasons vs causes
38.25–57.4	Natural motion	Logical evaluation
57.5–62.27	Universe center	Equivocation, composition, and circularity
62.28–71.33	Contrariety	Contradiction, ad hominem, and equivocation
71.34–83.1i	A posteriori	Counter-example method
83.12–87.14	Teleological	Charity
159.29–162.14	Violent motion	Equivocation
162.15–164.25	Two motions	Implication vs support
166.30–169.21	Mixed motion	Groundlessness
169.22–180.18	Ship experiment	Analogy
194.12–197.1	East-west gunshots	Deduction, induction, and rhetoric
197.1–202.16	Vertical gunshot	Inferential vs material 'if'
203.34–205.29	Hunter's	Explanation vs inference
205.30–209.13	Point-blank	Scientific, *ad hominem,* and logical invalidity
209.13–212.30	Birds	Logic of punctuation
214–244	Centrifugal force	Ignoratio elenchi, quantitative invalidity, and incompleteness

The first impression is that the logical topics are as varied as the physical ones.

Next, it is interesting to count the number of simple subarguments of the main arguments. The first number we arrive at is 32, there being 16 sections each containing one object argument and one meta-argument. Denoting by n-A the object argument of the nth section, and by n-G the meta-argument of the same section, these 32 arguments would consist of the following number of simple subarguments:

413

1-A: 3	9-A: 5
1-G: 3	9-G: 5+1+2=8
2-A: 3	10-A: 3
2-G: 6+3+1+3+4+6=23	10-G: 16
3-A: 4	11-A: 3
3-G: 5+2+1+10=18	11-G: 7+5+1=13
4-A: 3	12-A: 4
4-G: 1+2+2+4+5=14	12-G: 2+5+4=11
5-A: 1	13-A: 2
5-G: 3+1+1+5+2=12	13-G: 3
6-A: 6	14-A: 2
6-G: 3	14-G: 2+3+3+4=12
7-A: 4	15-A: 2
7-G: 5	15-G: 1+3+5=9
8-A: 3	16-A: 2
8-G: 7+6+3=16	16-G: 3+4+6=13.

Thus we have at least 229 self-contained arguments whose features may be made the topic of further investigations.

Next, by examining the content of the various sections of our data basis, we find that many more individual logical topics are discussed or mentioned than is apparent from the section titles. The following index shows this, where the numbers without hyphen or preceding an hyphen refer to the section numbers of the last chapter, the 'G' numbers refer to the various parts of the meta-arguments, and the 'C' numbers in parenthesis refer to my comments:

INDEX OF INDIVIDUAL LOGICAL TOPICS

Active evaluation of reasoning, 2, 3-G1, 3-G2, 4-G1, 5-G2(C2)

Ad hominem argument, 4-G4(C6)

Ad hominem invalidity, 14

Analogical reasoning, 10

Analogy, 10

Argument, *ad hominem,* 4-G4(C6)

Argument, explanatory, 5-G1 (C1), 8-G2, 9-(C), 11-G2

Argument, self-contradictory, 4-G3, 8-G2

Begging the question, 3-G4, 8-G2, 11-G3, 12-G1

Causes vs reasons vs conclusions, 1

Charity, principle of, 6

Circularity, 3-G4, 8-G2, 11-G3, 12-G1

Composition, fallacy of, 3-G3

Conclusions vs reasons vs causes, 1

Converse of a given proposition, 10

Counter-example method, 3-G4, 5-G2(C2), 5-G3(C3)

Deduction, 11

Deduction, induction, and rhetoric, 11

Deductive invalidity, 11-G1

Equivocation, 3-G1, 4-G5, 4-G5(C7), 7, 11-(C), 16-G3(C4)

INDEX OF INDIVIDUAL LOGICAL TOPICS

Errors in reasoning, explanation vs
  description of, 16-(C3)
Evaluative nature of reasoning, 5-G4(C4)
Explanation vs inference, 13
Explanatory argument, 5-G1(C1), 8-G2,
  9-(C), 11-G2
Fallacy of composition, 3-G3
Groundlessness, 7, 8-G2, 9, 10, 11-G3,
  12-G2
'If-then', 8-G2, 12
'If-therefore', 4-(C3)
Ignoratio elenchi, 16-G1
Implication vs support, 8
Incompleteness, 16-G3(C5)
Induction, 11
Induction, deduction, and rhetoric, 11
Inductive invalidity, 11-G2
Inference vs explanation, 13
Infinitely progressing argument, 4-G3(C5)
Invalidity, 1, 2, 3-G2, 4-G1, 5-G1, 5-G2,
  6-G1, 7, 8-G2, 11-G1
Invalidity, *ad hominem,* 14
Invalidity, deductive, 11-G1
Invalidity, inductive, 11-G2
Invalidity, logical, 14
Invalidity, quantitative, 16-G2
Invalidity, rhetorical, 4-C7, 11-G3

Invalidity, scientific, 14
Irrelevant conclusion, 16-G1
Liar's paradox, 4-G3(C4)
Logical evaluation, 2
Logical invalidity, 14
Method of alternative conclusion, 2, 3-G1,
  3-G2, 4-G1, 5-G2(C2)
Method of counterexample, 3-G4,
  5-G2(C2), 5-G3(C3)
Nonterminating argument, 4-G3(C5)
Paradox, liar's, 4-G3(C4)
Punctuation, logic of, 15
Quantitative invalidity, 16-G2
Reasoning and evaluation, 5-G4(C4)
Reasons vs causes vs conclusions, 1
Rhetoric, 11
Rhetorical invalidity, 4-(C7), 11-G3
Rhetorical strengthening of arguments,
  5-G3(C3)
Rhetoric, induction, and deduction, 11
Scientific invalidity, 14
Self-contradictory argument, 4-G3, 8-G2
Support, 9-G1, 9-G2, 9-G3
Support vs implication, 8
'Unless', 4-(C8)
Uselessness, 4-G2(C2)
Truth-functions, 12

The next step is to formulate a number of main topics by which to group
the individual ones. The following index is a possible elaboration:

INDEX OF MAIN LOGICAL TOPICS

Active evaluation, 2, 3-G1, 3-G2, 4-G1, 5-G2(C2)
Ad hominem: *ad hominem* argument, 4-G4(C6); *ad hominem* invalidity, 14
Begging the question and circularity, 3-G4, 8-G2, 11-G3, 12-G1
Counter-example, method of, 3-G4, 5-G2(C2), 5-G3(C3)
Equivocation, 3-G1, 4-G5, 4-G5(C7), 7, 11-(C), 16-G3(C4)
Evaluation, methodology of: logical evaluation, 2; evaluative nature of reasoning,
  5-G4(C4); principle of charity, 6; explanation of errors in reasoning, 16-(C3)
Evaluation, principles of: fallacy of composition, 3-G3; uselessness, 4-G2(C2); infinitely
  progressing argument, 4-G3(C5); irrelevant conclusion, 16-G1; incompleteness,
  16-G3(C5)

INDEX OF MAIN LOGICAL TOPICS

The above topics can, in turn, be reduced to three, as follows:

Evaluation categories:
circularity
equivocation
fallacy of composition
groundlessness
incompleteness
infinite progression
invalidity
irrelevant conclusion
question-begging
self-contradiction
uselessness

Evaluation methods:
active evaluation
ad hominem argument
explanation of error in reasoning
method of alternative conclusion
method of counter-example
principle of charity
Elements of reasoning:
deduction
evaluation
explanation
induction
linguistic expression
persuasion.

The difference between categories and methods of evaluation is that the former are various types of attributes with which one may evaluatively characterize reasoning, whereas the latter are techniques that the logical theorist may use for arriving at and justifying the attribution of the various categories. Hence categories and methods are also related insofar as the use of a given method yields an evaluation in terms of some corresponding category, and conversely, the evaluation in terms of some given category requires the use of some correspondingly appropriate method. It also follows that besides examining the internal relationships among the various concepts in each list, we need to examine the external relationships among concepts from different lists.

For the moment it suffices to note that the list of evaluation categories is merely an expansion of the notion of principles of evaluation used at the previous stage of theorizing, in the Index of Main Logical Topics, for the purpose of including therein a number of miscellaneous terms. Similarly, the list of evaluation methods is an expansion of the previous topic of methodology of evaluation.

Note also, that it now seemed better to list question-begging and circularity as separate topics. Furthermore, the notion of an *ad hominem* argument has been treated as a method of evaluation rather than a fault of reasoning since, as clarified earlier, what is meant here is not the *ad hominem* fallacy of contemporary logic textbooks, but rather the technique of criticizing an opponent by deriving from his own assumptions conclusions which are unacceptable to him. Similarly, though the method of explanation of errors in reasoning may remind one of the so-called genetic fallacy, an examination of the context in which this term was conceived [16-(C3)] indicates that what we have here is a method for criticizing an argument by giving an analysis of it which allows us to understand what error is being committed.

It is immediately apparent that all evaluation categories listed above are negative or unfavorable ones. In part this is the result of our previous choice of a data basis, which was constructed from Galileo's critiques of Aristotelian arguments. This represents no limitation, however, since all terms in the list of evaluation categories are merely one side of a coin the other side of which is a positive term; in other words, there is no conceptual difference between the question of whether or not an argument is, for example, invalid and the question of whether or not it is valid. By negating each term in the list we get the corresponding positive term; in reality, it is such a pair of contradictory terms that defines each distinct category.

The elements of reasoning are constitutive aspects of it in the sense that reasoning is normally a mixture of the various activities listed. How exactly they interrelate will be examined below, but it should be mentioned now that there may be some overlap (e.g. between induction and explanation), and there may be related activities that have not been but should be put in the list (e.g., justification, probably related to persuasion). The term *linguistic expression* refers to the use of such words as *if, therefore, because,* which use may be regarded as a minimal (perhaps 'operational') definition of reasoning. Evaluation was put in the list, not only because our evidence is clear that the evaluation of reasoning is itself a form of reasoning, but also to explore the more interesting possibility that reasoning is a form of evaluation. This would be so if and to the extent that all other listed elements could be reduced to it.

Finally, this list of elements has the following *prima facie* relationships to the categories and to the methods. There may be a correspondence between evaluation methods and elements of reasoning; for example, deduction and method of counter-example seem to go together. On the other hand, some evaluation categories may subdivide into as many special cases as there are elements of reasoning; for example, we have already seen that invalidity may be of a deductive, inductive or rhetorical kind. Other evaluation categories may instead correspond to a definite element of reasoning; for example, it seems that the uselessness of an argument is essentially a category relating to persuasion.

CRITICISM, COMPLEX STRUCTURE, AND OPENNESS

Before proceeding with any further analysis, a number of obvious but far-reaching conclusions supported by our data ought to be discussed. The most pervasive features for the classification of arguments pertain to the distinction of simple vs complex, and of critical vs constructive. The former is meant in the rather precise sense defined in Chapter 14, namely that a simple argument is one whose propositional structure contains no intermediate propositions, whereas a complex argument contains at least one intermediate proposition. The critical/constructive distinction partly corresponds to that of meta-argument vs object argument of last chapter, and partly to concepts and terminology used in Henry Johnstone's theory of argumentation.[1] The pervasiveness of these two distinctions contrasts with traditional logical theories according to which the basic distinctions are those between deductive and inductive, truth-functional and quantificational, and predicate and relational arguments.

A second overwhelming fact about our data is that arguments are normally complex. Though it is true that complex arguments are in a sense a series of simple ones, it is also true that simple arguments are elements of complex ones. So one may expect great differences between an approach that takes complex arguments as fundamental, and one that takes simple ones as fundamental. The usual accounts follow the latter, and hence they are probably not merely a simplification, but an over-simplification. These considerations accord very well with Hamblin's theory of argument.[2]

Our data also indicate that most arguments are critical, rather than constructive. It is not clear whether this is merely the result of our selection of data, for after all they were compiled as an attempt to follow a practical approach to the evaluation of reasoning in general, and to its criticism in particular.[3] It must be remembered, however, that our data are not biased with respect to Galileo's *Dialogue,* for in our earlier explanation of our principles of selection, it was shown that the whole book consists of critical arguments. The question remains, of course, how typical Galileo's book is. But this, too, has been decided, when it was argued[4] that Galileo's book probably constitutes the best and richest collection of arguments available anywhere; though this makes it untypical in the sense of uniquely valuable, it does not make it untypical with respect to the generalization presently being considered. If a qualification is needed for this generalization, it is rather the remark that, to use Johnstone's expression, "critical arguments are more fundamental than constructive ones".[5] Again, the great stress on constructive arguments found in standard logic books probably introduces a second distorting oversimplification in the theory of reasoning.

Besides such empirical support for the primacy of complex and of critical arguments, there are also the following fundamental reasons. The basic purpose of an argument is to make a proposition more acceptable than it would be in the absence of the argument.[6] This means that without the argument its conclusion would be less acceptable. But how could the acceptability of a proposition be less, when there is no supporting argument? Only if there are present certain objections to the proposition; the argument is then an attempt to remove such objections. We may then say that an argument is a defense of its conclusion from actual or potential objections. No argument would be needed if there were no need to make the conclusion more acceptable, i.e. if there were no previous objections. Thus the objections are prior, and objections are nothing but critical arguments.

Next we might note, that complexity fits very well with critical character, for our data show that the meta-arguments were much more complex than the object arguments, and in the context of Galileo's book the former are critical and the latter constructive. It is probably the case that complexity is a *consequence* of critical character, though it is not clear how this is to be shown. But this brings us to the primacy of complexity vis-à-vis simplicity.

A simple argument may be viewed as a piece of reasoning where the acceptability gap and the inferential gap between final premises and final conclusion are minimal. But as this gap decreases, so does the need for argument. Hence we may expect arguments to be given mostly when the gap is considerable. But if the gap is considerable, several intermediate steps are probably required to establish a connection between final premises and final conclusions; and this is to say that the argument will be complex.

Now the following objection comes to mind. How can the logical theorist

be missing anything of importance by taking short, simple gaps as fundamental, in view of the fact that complex arguments owe their success to the
reduction of a bigger gap to a series of short and simple ones? If the intermediate steps are removed from a complex argument, so that only the final
reasons and final conclusions are left, then what remains is likely to be no
argument, because no connection will be visible, and hence the acceptability
of the conclusion will not be increased by the presence of the premises.

In order to answer this objection, we need to call attention to another
fundamental feature of reasoning, pervasive throughout our data, which I
shall call its openness. This refers to the fact that, under normal conditions,
much of the propositional structure of reasoning is latent (in the sense of
Chapter 14); in other words, there are always implicit propositions whose
formulation and statement would make its structure more explicit. One
might say that arguments are normally incomplete, but because of the
negative connotations of the notion of incompleteness, it is preferable to say
that arguments normally have an open structure. Or one might say that
almost all arguments are inductive, in one sense of this term, according to
which the conclusion states more than is contained in the premises; but
because of the confusions surrounding the concept of inductive argument,
and because I wish to reserve the term induction to one of the elements or
aspects of (normal) reasoning, I will here avoid this term.

Now, the understanding of reasoning requires that its structure be made
(more) explicit. But to make the structure more explicit complicates it (in
the precise sense relevant here) since, as has been noted by Gilbert Ryle and
by Johnstone in another context,[7] when one argues from explicit premises,
the debate instantly moves a step back, namely to the acceptability of the
premises themselves. It follows that to understand some piece of actual
reasoning, the chances are that it needs to be reconstructed as a complex
argument; therefore, the simple steps making up the latter are theoretical
abstractions; hence if one considers them in isolation, one is missing the point
of the reasoning; but in standard logical theories, the preoccupation with
simple arguments constitutes such consideration in isolation.

Structural complexity is then partly a consequence of structural openness.
But it is easy to see that the latter derives from the critical nature of reasoning. For if reasoning primarily involves the removal of actual or potential
objections, then an argument will contain explicitly only those assertions
which are needed for such a specific purpose; the rest will be left implicit, in
the sense that it can be brought out as a result of new counter-arguments, of
theoretical analysis, or of theoretical evaluation.

To summarize, three of the most important facts about reasoning are: (1) critical arguments are prior to constructive ones, (2) arguments with a complex propositional structure are prior to arguments with simple structure, and (3) arguments normally have an open structure, namely are always susceptible to further enlargement and complication of their structure. Moreover, the critical dimension of reasoning seems even more fundamental than its complexity and its openness.

THE PRIMACY OF NEGATIVE EVALUATION

Since the concept of a critical argument seems so important in the theory of reasoning, let us pause for some general reflection on this and related matters. So far, a critical argument has meant an argument which finds fault with, or passes unfavorable judgment upon, some proposition or some other argument; that is, an argument whose conclusion is a negative evaluation. We suggested in an earlier chapter,[8] and our empirical data basis established, that the negative evaluation of reasoning promised to be methodologically very fruitful since it involves both good reasoning (constituting the critical argument itself) and bad reasoning (constituting the subject matter of the evaluation). Furthermore, it emerged in the last section that negative evaluation has something like a constitutive function for reasoning; this is implied by the priority of critical over constructive arguments; the implicit and intuitively-made negative evaluations in the process of reasoning tell us where not to go, which conclusions not to draw. It is as if negative evaluations and prescriptions, and prohibitive rules were more accessible and effective than positive ones both at the level of the practice of reasoning and at the level of logical theory.

This status for negative evaluations corresponds to trends and results discernible in other approaches to logic and other fields of scholarship and of culture in general.

For example, in the context of art history and criticism, E. H. Gombrich claims that whereas "I do not think that there are any wrong reasons for liking a statue or a picture ... there *are* wrong reasons for disliking a work of art".[9] This means that there are objective standarrds for negative evaluation, but not for positive evaluation. Hence, critical arguments presumably obey certain principles of evaluation, whereas constructive arguments do not.

A second example comes from psychotherapy, where one of the most interesting approaches is the so-called rational-emotive psychotherapy, originated by Albert Ellis.[10] Its central doctrine is that neurotic disturbances are the result of errors in reasoning, of which the victim is unaware and which he is unable to overcome. The therapy consists of discussing with the patients the problems of the latter, in order to determine the specific pattern in accordance with which they reason about personal and emotional matters. One of its doctrines is that though the pursuit of happiness is not systematizable, the avoidance of unhappiness is: "If we can't tell you how to be happy, can we tell you how not to be unhappy? Paradoxically, yes. Because while human beings differ enormously in what brings them positive contentment, they are remarkably alike in what makes them miserable."[11] What this means is that, the criticism of arguments whose conclusions trigger negative emotions is feasible, but the construction of arguments triggering positive emotions is not feasible.

A third manifestation of the same trend is the philosophy of science of Popper and his followers. In the Popperian approaches, the essential feature of a scientific theory is its falsifiability or testability rather than its provability or confirmability;[12] the essential feature of rationality lies in the critical attitude, i.e. in being open to criticism, rather than in being right or in being in possession of the truth;[13] the most significant feature of the growth of scientific knowledge is the occurrence of errors and the struggle for their elimination;[14] and the most important thing about a scientific theory or any belief in general is what one does with it, rather than how one acquired it. We need not share their reasons for such emphasis on criticism and refutation, namely that universal generalizations possess the property of falsifiability in a sense in which they do not possess that of provability or probable confirmability; that the same holds for hypothetical explanatory theories; that the critical attitude is self-referentially consistent in a way in which the justificationist attitude is not; and that the stress on error redeems from obscurity and neglect significant portions and episodes of the development of actual science. Instead it suffices to reinterpret their main doctrine from our present point of view. This can be done best by using one of the most recent and clearest formulations, given by Noretta Koertge: "A principle P is held rationally relative to the knowledge available at time t if and only if (a) P is held open to criticism [and] (b) there are no known cogent criticisms of P at time t."[15] If we speak of arguments, rather than principles, statements, or propositions, then we can automatically take care of the fact that this theory relativizes the rationality of P to the available knowledge, since an argument's conclusion is always (by definition) relative to its premises. Second, to speak of arguments is a way of objectifying the psychologistic reference to the 'holding' of a principle open to criticism. Thus the Popperian nonjustificationist theory of rationality can be reformulated in argument-theoretical terminology as the claim that a rational argument is one which has no known cogent objections; in turn, a cogent objection is a (meta) argument about some aspect of an (object) argument such that the meta-argument has no known cogent objections; and so on.

I think it is obvious that such a definition makes sense of the considerable evidence in our data basis, for the Aristotelian arguments are invalid ('irrational') because and insofar as the Galilean critical meta-arguments are unanswerable ('cogent'), while the latter are correct for the same reason.

Such a definition of validity may appear circular since it defines the validity of one argument in terms of the validity of another. But this objection would not be exactly right, since what happens is that the validity of one argument is defined in terms of the validity of a higher level argument; one might say that validity$_1$ is defined in terms of validity$_2$, which is defined in terms of validity$_3$, etc., where validity$_1$ is validity at the object level, validity$_2$ is validity at the next higher meta-level, and so on. Formal circularity is thus trivially avoided.

Moreover, and more importantly, in practice there is seldom any circularity, for the fact is that when the discussion is transferred to the metalevel, it becomes quite effective, and although the first meta-argument may not be decisive, the next step would be to criticize the meta-argument. In other words, the above negative definition of validity provides a very effective framework for the methodology of reasoning and for the methodology of the evaluation of reasoning. This is the sense in which it is claimed to be correct. In such a framework, critical arguments are crucial and more basic than constructive ones, since the search for the former, and the failure of the search, give worth to constructive, justificatory arguments.

Next, the best-articulated and most-widely accepted concept of validity is the model-theoretic one, according to which an argument is valid if and only if there is no model which satisfies the premises but does not satisfy the conclusion.[16] This corresponds to the less precise ideas that a valid argument is one for which it is impossible that its premises be true and its conclusion false, or one such that there is no argument of the same logical form having true premises and false conclusion. If we call *counter-examples* such models and arguments mentioned in the definiens of these definitions, then we may say that a valid argument is one for which no counter-example exists. There is no question that this concept of validity has great explanatory power, especially in mathematical reasoning and in mathematical logic. Moreover, the search for counter-examples, together with proofs that none exist, is not merely a hit and miss affair, and there is great power in the method of assuming the existence of a counter-example and either deriving a contradiction or constructing an appropriate satisfying model. However, it is easy to overestimate both the precision and the power of the definition, for ultimately the concept of a model is relatively open and subject to transformations, as the development of model theory shows. Nevertheless, from the present point of view, the model-theoretic definition of validity is very interesting since it can be interpreted as a special case of criticism or negative evaluation of an argument; a disproof by counter-example is an extreme, clear-cut type of objection to the original argument, while proof of nonexistence of counter-example is an extreme clear-cut type of failure to find any objections.

Finally, one of the most widespread approaches to the teaching of introductory logic is what may be called the fallacy-avoidance or critical thinking approach. This is reflected in the titles and subtitles of a large segment of contemporary textbooks.[17] The underlying idea is to teach how to reason validly, or how to improve one's reasoning, by teaching how to recognize instance of invalid or fallacious reasoning. This stress on criticism is often abused and frequently superficial, as we saw in an earlier chapter, and its practical justification lies in the preponderance of fallacious reasoning in the contemporary American scene. Its theoretical justification, however, must be in the critical-evaluative nature of reasoning itself, in the fact that reasoning is correct when it lacks specifiable faults, and hence reasoning is implicitly the evaluation of evidence and of connections among propositions, together with the avoidance of fallacious conclusions.

EVALUATION AND INVALIDITIES

I have been arguing in support of the conclusion that, on the one hand, both the understanding and the evaluation of reasoning are forms of reasoning, and that, on the other hand, both understanding and reasoning are themselves forms of evaluation. The ratiocinative character of understanding is supported by the fact that the understanding of reasoning is essentially the reconstruction of an inferential structure of propositions out of the actual linguistic material expressing the reasoning under consideration, and the process of reconstruction is conceived as higher level reasoning whose premises are empirical propositions about this material and whose conclusion is the reconstructed argument. The ratiocinative character of evaluation is supported both by the fact the evaluation of arguments is normally expressed as a meta-argument about the argument being evaluated, and by the fact that such evaluation has often an active character, namely it involves reasoning at the same level as the argument being evaluated and arriving at different conclusions from the ones drawn in it. The evaluative character of understanding is supported by the central role played by the principle of charity, for to attempt seriously to understand a piece of reasoning is implicitly to display a favorable attitude toward it, and without a certain amount of such appreciation one would be unable even to detect the alleged connections whose presence is necessary for reasoning. Finally, the evaluative character of reasoning has been supported by the fact that so much of the reasoning in our data basis consists of evaluations, by a relatively general argument exhibiting the fundamental importance of the notion of a critical argument, and by showing that this stress on criticism (negative evaluation) corresponds to ideas accepted in a number of more or less related areas. We have not yet shown the mutual relationship between reasoning and evaluation by a detailed analysis of our data basis. To this we now turn.

Let us begin our more detailed analysis with evaluation categories. The most striking fact about them is their number and variety, by contrast with the mere (formal) invalidity and falsehood of premises of standard logical textbooks. It may be, of course, that the multitude of evaluations that we have found to inhere in actual reasoning and in logical practice can be reduced to these two, though this is prima facie doubtful, for after all our list already includes invalidity. Of course, what we have called invalidity is only the intuitive analogue of formal invalidity and not equivalent to it. On the other hand, it did turn out that invalidity was by far the most common evaluation category, so it seems promising to examine it first.

Invalidity is basically the failure of one proposition to follow from others: a simple argument or inferential step is invalid if and only if the conclusion does not follow from the premise(s) or reason(s). There are several different reasons for such failure. It may be that the conclusion does not follow because a counter-example exists (which would yield the special case of formal invalidity). Clearly, such criticism is relevant only when the propositional structure of this inferential step is viewed as closed; since we have seen that this structure is normally open, formal invalidity seems relatively unimportant.

A conclusion may not follow because it does not follow any more than some other different propositions. This occurs primarily with explanatory arguments, where the conclusion is an explanation of the premises, and there is no reason to prefer it to some specifiable alternative explanation. There is active evaluation here insofar as the alternative conclusion needs to be formulated and shown capable of accounting for the premises. We may call this type inductive invalidity, though of course one would need to show that other types of inductive arguments (such as analogy and generalizations) can be reduced to explanatory arguments.

It may be that the conclusion does not follow because of the falsity of some presupposed premise (i.e., some latent proposition required to make the structure more explicit and less open). Since it is rare, or rather unrealistic, for such falsity to be clear, obvious, or uncontroversial, it is normally the conclusion of some argument constructible in the context. Thus the falsity of such a presupposed premise corresponds to the correctness of some counter-argument adducible in the context, and the correctness of the latter is nothing but its contextual unobjectionableness. The construction of such counter-arguments means that this type of evaluation is active, in the sense defined earlier.

It may be that the conclusion does not follow because some presupposed premise is groundless, i.e., because there is no reason in the context to assert that presupposed proposition, i.e., because in the context no argument supporting this proposition is correct. This corresponds to complicating the structure of the argument under consideration by making the premise under consideration an intermediate proposition supported by some subargument(s) in a bigger argument; the incorrectness of these subarguments is equivalent to the groundlessness of the premise in question. This type of criticism is much more significant when the critic himself takes up the burden of proof, in the sense that, instead of merely asserting that no reason is explicitly provided within the original argument for the presupposed premise, he shows that some reason(s) could be given in the context, but that this reason is subject to some insuperable objection. In turn, the construction and criticism of these subarguments involves reasoning at the same level as the argument being evaluated; thus we have here a type of active evaluation.

The last two cases may be labeled *structural invalidity* because the invalidity derives essentially from the openness and complexity of the propositional structure of reasoning.

Another reason for invalidity may be that what does follow from the premise(s) is some specifiable proposition different from or inconsistent with the conclusion. Two special cases fall under this heading. One occurs when the counter-conclusion can be seen to follow rather immediately and directly from the premise(s) explicitly stated in the original argument. The more important case occurs when the counter-conclusion can be arrived at on the basis of these explicit premises together with other propositions which are independently justifiable or contextually acceptable; here we have more active involvement in the sense of more elaborate reasoning at the object level. We may label these cases of *constructive invalidity* since the invalidity is the result of the contextual construction of counter-arguments indicating which counter-conclusion does follow.

How does invalidity relate to the other evaluation categories? The groundlessness of an argument reduces to groundlessness of its final reasons, and the latter corresponds to the invalidity of contextual arguments supporting them. An incomplete argument is essentially one which is constructively invalid. The fallacy of irrelevant conclusion is also a case of constructive invalidity. The fallacy of composition is a special case of formal invalidity. Uselessness refers to the fact that the given argument is in the context part of a bigger one formed by using the final conclusion of the given argument as an intermediate proposition to arrive at some other conclusion; the original argument is useless when the subargument added to it has some type of invalidity; that is, the uselessness of an argument corresponds to the invalidity of the constructed additional subargument grafted onto it to define a fuller contextual argument.

Equivocation is really a type of structural invalidity, but deserves a special name because it is especially important. Its importance can be seen from the frequency of this evaluation category in our data basis. Its structural basis can be seen as follows. The simplest case of a complex argument will have one final conclusion C, one intermediate proposition I, and one final reason R:

$$
\begin{array}{c}
C \\
| \\
I \\
| \\
R
\end{array}
$$

Equivocation occurs when it cannot be both that I follows from R and C follows from I, since if I follows it acquires a meaning such that C is seen rather obviously not to follow, whereas if C follows from I then I must have a meaning such that it rather obviously does not follow from R. We may say also that equivocation derives from the semantical ambiguity of some term(s) t in I, where the ambiguity is such that if t has one meaning then one of the inferential connections holds but not the other, whereas if t has the second meaning then the other inferential connection holds but not the first.

Thus we might regard equivocation as *semantical invalidity*. This term has the advantage of stressing the connection between equivocation and invalidity, and the special significance of the structural problem generating the invalidity. However, we must not forget that the semantical ambiguity of I does not derive from the abstract possession of a double meaning by t, but rather from the contextual double meaning of t, one that connects it to C and the other with R.

The primacy of complex structure is beautifully illustrated by equivocation that occurs with simple arguments. Suppose we have two premises P1 and P2, and one conclusion C:

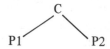

Here equivocation can occur when some term(s) t of one of the premises, say P1, has two meanings such that, if given one meaning, C does follow from P1 and P2 but P1 is rather obviously not true, whereas given the second meaning, P1 is true but C rather obviously does not follow. The problem then arises, how can anyone ever fall into such equivocation which involves giving an argument which is obviously either valid but unsound, or sound but invalid. I believe that if this were all that was happening in equivocation, it would seldom occur. However, the fallacy becomes more comprehensible when we realize that the above simple argument would normally be an injudicious abstraction from a complex one, where P1 would be an intermediate proposition supported by some reason, say P11. Then it is easy to see how in the context of the subargument 'P11, ∴ P1' one will attach to P1 an appropriate meaning so that it does follow from P11, whereas in the context of combining P1 with P2, if there is a crucial term with a double meaning, that the combination (which yields C) will be possible only if one uses the second meaning of the term, different from the previous one.

The other evaluation categories in our list were *not* cases of invalidity. An infinitely progressing argument (e.g., some versions of the liar's paradox) is one which has no nonarbitrary final conclusion; the explicit conclusion in the actual argument does indeed follow, but from it there follows its denial as well, and so on without end. One might say that we have a degenerate case of invalidity where the conclusion does not follow because there is no *final* conclusion.

A self-contradictory argument is one with inconsistent premises. Such arguments automatically excape formal invalidity, though they may not escape the other types. However, since not all premises can be true simultaneously, it will never happen that they must (all) be accepted, and hence it will never happen that the conclusion must be accepted because of the premises. So we might say that the conclusion does not follow *from the premises* because there are *no* (practically realizable) premises. This would yield another degenerate case of invalidity, the source of the degeneracy ('no premises') being opposite to that which generates infinite progression ('no conclusion').

Question-begging is a case of circularity in the sense that a question-begging argument has some final premise(s) whose justification requires in the context that one assume what is being proved (i.e., the final conclusion of the question-begging argument), and this means that the bigger argument consisting of the original one plus the justification of the crucial premise(s) is such that the same proposition appears both as final conclusion and as a final reason. The latter is precisely a circular argument.

Circularity, too, may be regarded as a degenerate case of invalidity, for we may say that the conclusion does not *follow* from the premise because it *is* one of them. This would yield a type of degeneracy intermediate between the previous two.

We thus have six types of invalidity, each deriving from the reasons why the conclusion may not follow from the premise(s): formal, inductive, structural, constructive, semantical, and degenerative.

Let us now examine the elements of reasoning listed above as one of the main topics extracted from our data basis. Deduction may be viewed as the process of avoiding formal invalidity, and induction as the process of avoiding inductive invalidity.

Explanation may be equated with induction in two senses. First many inductive arguments may be viewed as explanatory ones; for example, generalizations may be conceived as inferences to the best explanation,[18] whereas arguments from analogy may be conceived as generalizations about the properties shared by two similar entities, the range of the generalization being restricted to a special class of relevant properties.[19] Moreover, inductive arguments are often, and perhaps normally, reports (from the context of discovery, rather than justification) of how one arrived at the conclusion in the first place.[20] This means that they are (causal) explanations of one's own thoughts.

The linguistic expression of reasoning, by means of reasoning indicator words, or by the more elaborate reconstructions and structure diagrams found in our data basis, probably corresponds to the avoidance of structural invalidity. This is to be taken in the sense of reducing the amount of structural invalidity and disciplining the attempts to avoid it, for instances of structural invalidity are always potentially present, so that its elimination is an indefinitely long process and can be achieved only temporarily and contextually.

Persuasion corresponds to constructive invalidity in the sense that the explosure of constructive invalidity is the attempt to persuade (oneself or others) that even if we accept the premises in question, we need not accept the conclusion. The articulation of constructive invalidity is a form of persuasion in the sense that the criticism begins by accepting the same premises and attempts to show that something different follows.

Persuasion is also partly the attempt to avoid degenerative invalidity since the problem with the latter is that the reasoning being criticized can never lead from a state of nonacceptance to a state of acceptance of the conclusion.

Regarding equivocation or semantical invalidity, it has been pointed out by Hamblin[21] that equivocation typically occurs when there is a failure of persuasion in spite of any discernible invalidity. That is, the premises seem acceptable, the inferential connections seem also acceptable, but the conclusion is not. The way to solve the problem is to detect an ambiguous term, and a consequent equivocation.

Finally, evaluation corresponds to the avoidance of no one particular type of invalidity, but to all of them collectively. For the process of avoiding these invalidities is implicitly a form of evaluation, and so in that sense deduction, induction, persuasion, and expression are forms of evaluation.

The evaluation methods listed earlier may be analyzed as follows. The method of counter-example is obviously the method for showing formal invalidity; the method of alternative conclusion is the one that shows inductive invalidity. Active evaluation and *ad hominem* argument are essentially the same, and they are the method for proving constructive invalidity.

The principle of charity is not exactly a method, but it does relate to structural invalidity, for the exposure of the latter is impossible without a good deal of structural reconstruction work, which requires application of the principle. And conversely, if the argument under criticism has first been reconstructed in a charitable way, then the argued falsity or groundlessness of one of its propositions acquires force.

The methods for showing semantical and degenerative invalidity (i.e., equivocation and circularity) do not appear in our list above because they are equivalent to the method of analyzing reasoning into its propositional structure. Nothing special is involved here other than the principles for understanding propositional structure, and an understanding of the concepts of equivocation and degenerative invalidity, especially its most important case of circularity.

In summary, our main result seems to be that reasoning is a form of evaluation in which propositions (ideas, thoughts) are arranged in such a way as to avoid invalidities of the kinds discussed above.

NOTES

[1] *Philosophy and Argument*, pp. 57–92.
[2] *Fallacies*, p. 229.
[3] See Chapter 15.
[4] In Chapters 11 to 13.
[5] *Philosophy and Argument*, p. 82.
[6] Cf. Hamblin's *Fallacies*, p. 245.
[7] H. W. Johnstone, Jr., *Philosophy and Argument*, p. 58; G. Ryle, 'Proofs in Philosophy', p. 321.
[8] See above, chapter on 'The Evaluation of Reasoning'.
[9] *The Story of Art*, p. 5; italics in the original.
[10] See his *Reason and Emotion in Psychotherapy, A Guide to Rational Living*, and *Growth Through Reason*.
[11] A. Ellis, *Guide to Rational Living*, pp. 69–70.
[12] Cf. K. Popper, *The Logic of Scientific Discovery*, and *Objective Knowledge*.
[13] K. Popper, *The Open Society and Its Enemies*, Vol. 2, pp. 224–58; W. W. Bartley, III, *The Retreat to Commitment*, and 'Rationality versus the Theory of Rationality'.
[14] J. Agassi, *Towards an Historiography of Science*.
[15] 'Bartley's Theory of Rationality', p. 79.
[16] See, for example, A. Taski, *Introduction to Logic*, and B. Mates, *Elementary Logic*.
[17] See, for example, M. Black, *Critical Thinking*; D. B. Annis, *Techniques of Critical Reasoning*; W. E. Moore, *Creative and Critical Thinking*; A. Michalos, *Improving Your Reasoning*; M. C. Beardsley, *Thinking Straight*; S. M. Engel, *With Good Reason: An Introduction to Informal Fallacies*; W. W. Fearnside and W. B. Holther, *Fallacy, The Counterfeit of Argument*; and N. Capaldi, *The Art of Deception*.
[18] R. H. Ennis, 'Enumerative Induction and Best Explanation'.
[19] See M. Scriven, *Reasoning*, pp. 210–15.
[20] The essence of this suggestion may be found in Hamblin, *Fallacies*, pp.. 224–52.
[21] *Fallacies*, pp. 283–303.

TOWARD A GALILEAN THEORY OF RATIONALITY

The preceding investigations have an internal structure and unifying topic. That is, they constitute a concrete theory of rationality, concrete because the case study of Galileo is considered; theoretical because this Galilean material is studied not for its own sake, or with a tabula rasa, but with an eye toward the theoretical lessons one can learn; and rationality-oriented because these theoretical conclusions and suggestions deal with the nature of human rationality. I speak here of rationality *simpliciter*, and not of *scientific* rationality, because the latter has been treated not as something sui generis, but merely as a special case of rationality in general. In this regard, Chapters 7–17 show in what sense and exactly how there is nothing above and beyond ordinary reasoning in Galileo's *Dialogue*.

From this point of view, the Introduction argues that Galileo's *Dialogue* has a special place for anyone interested in being rational and in reflecting and learning about rationality. Chapters 8 and 9 may be regarded as a formulation of the problem of rationality by reference to Galileo's book, as well as a discussion of the problem in two selected but especially important fields, the philosophy and the history of science. Chapters 1–7 deal concretely with those features of rationality that might be labeled *macroscopic*, and Chapters 10–17 with those features that might be labeled *microscopic*. The macroscopic features are primarily three, namely rhetoric, method, and reasoning. Rhetoric refers separately to communication with others (Chapter 1), to the structuring of one's throughts (Chapter 2), and to the discursive handling of emotions (Chapter 3). Method pertains collectively to effective means for the establishment of aims and achievement of results (Chapter 4); to self-reflection on one's practice (Chapter 5); and to the judicious balancing of potentially conflicting activities, aims, and requirements (Chapter 6). Reasoning refers to a special activity of crucial importance (Chapter 7).

The microscopic features of rationality may be referred to as its microstructure. Thus the structure of reasoning studied in Chapters 10–17 is also the microstructure of rationality, since reasoning is one of the elements of rationality. Insofar as reasoning is the central element to which all others can be reduced (Chapter 7), then its structure is all there is to the microstructure of rationality. Elements of the (macro) structure of reasoning are deduction,

432

explanation, persuasion, expression, and evaluation (Chapter 17). Of course, it would be possible to continue and refine the analysis, and deal with the finer features of reasoning, which would be simultaneously the microstructure of reasoning and the micro-microstructure of rationality. This has not been done in my investigations, except implicitly and incidentally in the course of the analysis in Chapter 16. The reason was that I thought one needed to understand the bigger features of reasoning before going on to the finer ones.

What is the general relationship between these two levels, macro and micro, of the structure of rationality? I believe that the facts of the *micro*-structure of rationality (i.e., the *macro*structure of reasoning) are less immune, if not completely so, to historical change and development and to social conditions and situations, than the facts about rhetoric and method. That is, rhetorical techniques and methods may change without any change in the ratiocinative techniques underlying them. Relativism would thus be false at the level of reasoning, but possible at the level of rhetoric and method. It might be useful to make this definitionally true by defining ratiocinative techniques to be those that remain and have remained constant in the midst of changes in rhetoric and methodology.

Similarly with progress. My study provides no evidence that at this level (reasoning) Galileo's rationality was in any way inferior to present-day rationality. However, the fact that he had to use the rhetorical techniques that he did, and to emphasize the methods he did, is primarily a reflection of his historical and social conditions. Our rhetorical and methodological procedures can be taken to have grown since Galileo's time and perhaps to excel his, at least in sophistication and complexity. But no similar growth is discernible in reasoning.

Though theoretically reducible to reasoning, rhetoric and methodology are not dispensible in practice, however. Hence a growth of rationality is definable, deriving from the progress in methodology and rhetoric. I would also be willing to admit, as suggested above, that there is a deeper level of rationality than that of reasoning considered here. For example, deduction, besides being one of the macroscopic elements of reasoning, might be the basis of all others. Hence this distinction between the macro- and the micro-structure of rationality should not be taken as absolute, but rather as a matter of depth of analysis, with the understanding that such depth is a matter of degree. However, just as the rhetorical and methodological levels of rationality are not dispensable for human beings, in favor of mere reasoning, neither is the logical level here explored, i.e. the one relating to reasoning, dispensable in favor of mere deduction. And this is perhaps one of the most important

conclusions supported by these investigations, namely that there is, both in
theory and in practice, a logic which is intermediate between mere rhetoric
and mere methodology on the one hand, and pure deduction on the other.
This is the logic most operative both in science and in daily life, and hence
the truly central feature of rationality.

Such a double-tiered theory of rationality could also properly be labeled
"Galilean" in the double sense that Galileo's *Dialogue* has provided us both
with the data, study materials, and subject matter of rationality, and with
the inspiration, model, and approach to follow. That is, our investigations
follow a demonstrably Galilean approach because we have applied to our
study of rationality the "dialectical" methodology that Galileo used in his
study of natural phenomena (Chapter 6). To this demonstration we now turn.

We began with an examination of various aspects of Galileo's *Dialogue*
which are interesting and important both for their own sake and for their
relevance to the problem of rationality.

In Chapter 1 were analyzed the book's rhetoric, in the sense of the term
'rhetoric' which refers to the appearances and pretensions found in verbal
discourse, and the various *impressions* conveyed by it, as opposed to its
substance. We examined the evidence giving the impression that Galileo is
engaged in religious apologetics relating to the Church's anti-Copernican
decree of 1616, the evidence hinted at in the book's title giving the impression
that it is a two-sided presentation of all the available arguments *for and
against* Copernicanism so that the issue could be decided by the proper
authorities, and the evidence that the book is an attempted demonstrative
proof of the earth's motion. We concluded that it is in reality a justification
of Copernicanism in the sense of an attempt to produce adherence or to
increase assent to it. The justification proceeds on three levels, rhetorical,
logical-scientific, and philosophical-methodological.

In Chapter 2 we presented the evidence that in the *Dialogue* Galileo gives
arguments and evidence designed to *support* the theory that the earth moves;
this is the logical aspect of Galileo's justification, in the sense that logic is the
theory and practice of reasoning; it is also its 'scientific' aspect in the sense of
the term that refers to the objective presentation and analysis of arguments
and evidence; consequently we have here the logical-scientific part of the
book's 'rhetoric', in the sense of the latter term that refers to the *general*
theory of argumentation. This chapter also included a discussion of the intrinsically interesting and important problem of the unity and structure of the
Dialogue, and its logical structure was presented as a solution to the problem.

In Chapter 3 we discussed the rhetorical aspect of Galileo's justification of

Copernicanism, in the sense of the term 'rhetoric' that refers to verbal techniques of persuasion which operate on human feelings and emotions. These are alogical but not illogical techniques, from the point of view of a strict sense of logic which would restrict it to the purely intellectual aspects of reasoning. However, these techniques are logical in the broad sense in which logic refers to the theory and practice of any type of argumentation, designed to induce assent, as long as a distinction is made between what's proper and what's improper. The distinction is made by reference to how the arguments operate on human feelings and emotions. For this reason the discussion here had aesthetic overtones and implications, to the extent that aesthetics is the study of the proper linguistic expression of feelings and emotions. The discussion was thus also a study of the literary aspects of Galileo's book since it proceeded by analyzing all passages significant from a literary point of view. This chapter also relates to one aspect of the problem of scientific rationality discussed in Chapter 8, since it suggests that rhetoric (in the sense of Chapter 3) and literary art have a role to play in science, specifically to make possible radical scientific changes involving transitions from one very fundamental theory to another.

In Chapter 4 we analyzed the book's scientific content, scientific in the sense of pertaining to the concerns of present-day scientists. We argued that the book has a richer scientific content than most historians of science are inclined to believe, that their judgments contrast somewhat with scientists', that this contrast is partly the result of a tension that exists between scientific relevance and historical accuracy, and that there are general reasons for thinking that this problem is insuperable. We then gave a concrete discussion of this tension by analyzing various interpretations of Galileo's account of the tides in the Fourth Day of the book, and we concluded that the acuteness of the problem is thereby confirmed. In an attempt to combine scientific relevance and historical accuracy, we gave a reconstruction of the First Day; then we argued that, in spite of the novelty of such a reconstructed content, its accuracy compares favorably with typical interpretations, using as a concrete example the various attempts to view the beginning passages of the First Day from the point of view of the law of inertia. We concluded that the most important scientific content that the book possesses is its methodological content, that is to say the lessons, ideas, and suggestions that scientists can get from it about the nature and proper methods for the acquisition of knowledge.

Chapter 5 provided a systematic and complete reconstruction of Galileo's book from the point of veiw of methodology, exhibiting the interplay between scientific practice and philosophical reflection. It is also a reconstruction of Galileo's clarification of the epistemological and methodological concepts needed, above and beyond the objective presentation of the evidence, for a full justification of Copernicanism.

After these analyses of the rhetorical, logical, literary, scientific, and methodological aspects of Galileo's *Dialogue,* an attempt was made at a systematization of Galileo's methodology, with an eye both toward gaining some theoretical understanding and toward making some practical use of it.

In Chapter 6 we argued that the essential feature of Galileo's methodology is the judicious synthesis of such opposites as experience and reflection, observation and speculation, quantitative analysis and qualitative considerations, causal explanation and phenomenological description, and anti-verbalism and logical analysis. This is the sense in which Galileo's methodology may be termed 'dialectical'. Chapter 7 argued that at a deeper level Galileo's dialectical methodology is logical, in the sense that the just-mentioned judicious synthesis of opposites ultimately reduces to reasoning. Thus reasoning may be regarded as a microstructural level of rationality, while at a macrostructural level rationality takes the form of various contrasting procedures, such as quantitative and qualitative analysis, all of which need to be kept in mind but none of which must be allowed to gain absolute predominance.

These theoretical lessons in methodology were then put to use in examining a number of leading Galilean interpreters and a series of concrete methodological problems, and in suggesting critiques and solutions from the point of view of the theory and practice of reasoning. Chapter 8 examined the problem of the role of reason in science and Paul Feyerabend's view that Galileo's work shows that 'anything goes' in scientific inquiry; we argued that Feyerabend sees a conflict between rhetoric and reason which does not exist, and that he fails to distinguish between proper and improper rhetoric; we also argued that the sound part of his account shows that rhetoric has a non-negligible role to play in science, while at the same time logic has an essential role to play in rhetoric.

Chapter 9 suggested the application of the principle that "a science which fails to forget its founders is lost" to the study of the history of science and to Alexandre Koyré, whose pioneering *Galileo Studies* has exerted a formative influence on the discipline. Without denying the past value of Koyré's work, we argued that now is no longer the time for continued reverence, but rather for a display of Galilean independent-mindedness. And just as Galileo was able to argue that he was being more Aristotelian than the Peripatetics, by refuting some of Aristotle's substantive conclusions while following the essence of his approach, so in this chapter we refuted Koyré's claim that Galileo's work shows that good physics is made a priori, by following more rigorously Koyré's own method of logical analysis.

Chapter 10 showed that the logical neglect of Galileo's work has deep scholarly and historical roots by criticizing certain aspects of the erudition of four leading Galilean scholars, Emil Strauss, Antonio Favaro, Stillman Drake, and Maurice Clavelin. It was suggested that my critique points in the direction of a way to overcome the dichotomy between logic and history.

These critiques (Chapters 8, 9, and 10) were conducted primarily from the point of view of the most basic feature of Galileo's methodology, namely its logical character or emphasis on reasoning. In fact, these philosophers and historians were criticized primarily for their inattention to matters of reasoning. Such neglect, though perhaps understandable in the past, is no longer justified. Next we undertook a critique of several groups of scholars who certainly cannot be blamed for their neglect of reasoning; they are logicians and other theorists of reasoning, who are vitally concerned with the topic. However, it turned out that their work could be criticized from the point of view of the other essential feature of Galileo's methodology, namely its dialectical component (discussed in Chapter 6).

From this point of view, that methodology suggested the following approach to the scientific study of reasoning: identify a number of relevant opposites and then attempt a judicious arbitration, and identify existing excesses in current practice and then attempt to correct them. Since the greatest contemporary excess in the scientific study of reasoning is the abstractness of formal logic, we generally advocated a more realistic, concrete, historical, and pragmatic approach (Chapter 13), analogous to what is found in the psychology of reasoning (Chapter 11), in the 'new rhetoric' (Chapter 12), in the linguistics of reasoning (Chapter 13), and in works by Stephen Toulmin and by Michael Scriven (Chapter 13). We argued, however, that all of these particular approaches need certain emendations in order to be acceptable. In fact, the relevant dichotomies in this field are such opposites as logic and psychology, logic and rhetoric, understanding and evaluation, and theory and practice, but they have not been properly handled.

Chapter 11 made a methodological criticism of recent work in the psychology of reasoning by arguing that experimentation and formal logic cannot be combined in the way attempted by psychologists of reasoning, whose combination turns out to be both insufficiently formal and improperly empirical; it was argued that, as a way out of their own difficulties, the correct synthesis of a logical and a psychological approach would be a historical orientation. Chapter 12 made a methodological criticism of the exponents of the "new rhetoric" by arguing that they fail to achieve a proper synthesis of logical and of rhetorical analysis. Chapter 13 explored the possibility that the science of reasoning may lie in the *practice* of reasoning, at the metalevel of reasoning about reasoning.

Finally, in Chapters 14–17 we attempted a constructive sketch of such a concrete, Galilean theory of reasoning. Chapter 14 formulated a number of basic principles for the logical analysis of reasoning by using as illustrative examples various arguments and passages from Galileo's *Dialogue*. The main result was that for the understanding of reasoning, which is a pre-condition for its evaluation, one needs to identify its propositional structure, which refers to the inferential interrelationships among its propositions. The concept of proposition here developed was one derivative from that of reasoning, so that the identity of individual propositions is dependent on the role they play in the various steps of an argument.

Chapter 15 contains a criticism of textbook accounts of fallacies, since it is perhaps the most widely-practiced approach to the criticism of arguments. The inadequacy of the fallacy approach led us to suggest a more practical-oriented one, in accordance with the program of Chapters 11–13. The suggestion was to examine, in the light of the principles of Chapter 14, the critical arguments contained in Galileo's book.

In Chapter 16 we carried out this suggestion for 16 main arguments which constitute about the first half of the book, thus providing a data basis or reasoning materials for further theorizing later (Chapter 17). The principles of selection of these arguments were stated and defended, and it was also shown how the arguments in the rest of the book do not differ significantly from these sixteen. The chapter consists of sixteen sections, each containing a discussion of an argument together with relevant logical concepts and theoretical ideas. Each discussion was systematically divided into three parts, the first being a reconstruction of an object-argument criticized by Galileo, the second being a reconstruction of Galileo's own critical meta-argument, and the third being a series of comments by myself designed to clarify concepts, formulate principles, and illustrate ideas, relevant for the theory of reasoning.

Chapter 17 conducted an analysis of the data of the previous chapter and led to such conclusions as that the most important distinction for the classification of arguments is the distinction between critical and constructive arguments, that critical arguments are prior, that reasoning normally consists of complex arguments, that criticism (negative evaluation) has an essential and constitutive role to play in reasoning, and that several related but distinct types of invalidities are definable.

To summarize, we may say that Galileo's *Dialogue* suggests new ways in which science is logical and logic can be scientific. The evidence from Galileo indicates that science is primarily a method rather that a set of abstract truths; that method involves not the fixed adherence to some formal universal rules, but rather the judicious balancing of such opposites as speculation and observation, quantitative analysis and qualitative considerations, causal explanation and phenomenological description, anti-authoritarianism and traditionalism, and practical involvement and philosophical reflection; that in particular such judgment does not automatically exclude rhetorical persuasion or aesthetically expressed emotion; but that such exercise of judgment reduces ultimately to a matter of reasoning, namely the drawing of conclusions from premises and the formulation of reasons for claims.

Moreover, Galileo's book constitutes a unique source material and data basis for a reform of logic, in the sense of the study of reasoning. I have argued that in order to become more scientific, the study of reasoning needs a greater empirical orientation toward reasoning as it actually occurs in the world; that such empirical emphasis cannot be satisfied by the experimental approach of psychology, but rather by a broadly conceived historical approach; that such a historical approach must be able to use but ultimately to transcend high standards of scholarly erudition; that it must recognize the distinct importance of rhetorical analysis (i.e., the theory and practice of persuasion), while retaining the supremacy of logical analysis; and that it must not shy away from attempts at concrete theoretical systematizations, however provisional or limited their value may be.

PAGE CONCORDANCE OF *DIALOGUE* EDITIONS

The purpose of the table below is to facilitate cross-references among the following editions of Galileo's *Dialogue*; the original edition of 1632, the one published as volume VII of the 'National Edition' of his works edited by Favaro, the Pagnini edition, Drake's English translation, and Santillana's revision of Salusbury's 17th century English translation. Though Favaro's edition may be regarded as definitive, the 1632 edition is still useful since it is now easily available in a fascimilie reprint and since Favaro's modernization of the original punctuation (though usually an improvement) occasionally obscures the logical structure of the original text,[1] so that consultation of the 1632 edition is often valuable for a better understanding of the more intricate passages. The Pagnini edition, though its text is taken from Favaro, can still be read with profit because of its excellent notes. Drake's translation is the most easily available one in any language and not easily excelled; nevertheless its imperfections[2] can often be remedied for the English-speaking reader by judiciously consulting Santillana's edition, which also contains excellent notes. In paginating the latter I discovered that it is somewhat incomplete and that many passages (ranging from a few lines to a few dozen lines) are simply not translated; occasionally Santillana notes the fact, but usually he does not.[3]

The table has been constructed as follows. The first column contains page numbers from the Favaro edition, and the other columns the corresponding page numbers for the other editions; thus anyone in possession of any of these other editions can write on their pages the Favaro page numbers, which may be taken as standard. What I have done is to indicate the line of the page of those other editions that corresponds to the *first line* on the corresponding Favaro page. In the case of the two Italian editions this line correspondence is exact, down to the particular word or part of a word with which the page begins; these beginning words are found in the column after the 1632 edition page and line numbers, and though they were taken from that edition they of course refer also to the Pagnini edition. In the case of the two English translations the line numbers are often approximate to plus or minus one line, on account of the difference in sentence structure; however, on lines where a new interlocutor intervenes in the dialogue the correspondence is exact. The line numbers are denoted as decimals, so that the number preceding the decimal point is a page number, and the number following it is a line number; the line number is arrived at by counting from the top of the page, except for those cases in the Santillana column where a 'b' follows the decimal point, in which case the counting begins with the last line of text. (This extra notation was unnecessary for the other editions since, unlike Santillana's, their lines are evenly spaced and so can be counted 'mechanically' from the top, e.g. by means of an appropriately graduated scale.) In the partial words after the 1632 edition numbers, left hand hyphens indicate that the first part of the word appears on the last line of the preceding Favaro page, and right hand hyphens indicate that the word either continues on the following line of the 1632 edition or is too long to be printed in this table in its entirety. The word 'missing' in the Santillana column indicates that the first line of the corresponding Favaro page is part of a passage left untranslated. Other clarifications are made, as the need arises, by means of explanatory footnotes.

Favaro	Edition of 1632		Pagnini	Drake	Salusbury-Santillana
33	1.1	GIORNATA	93.1	9.1	12.1
34	2.8	di lunghezza	94.25	9.37	12.23
35	3.9	SALV.	96.13	11.1	14.17
36	4.9	prodotte	98.2	12.3	15.19
37	5.12	dal	99.19	13.5	missing
38	6.11	retti	100.31	14.3	missing
39	7.15	chiama	102.11	15.5	17.10
40	8.14	-rarmi	104.1	16.5	19.1
41	9.15	-che	105.17	17.8	20.20
42	10.14	-colare	106.30	18.8	21.b5
43	11.15	potessimo	108.6	19.5	22.28
44	12.15	de	109.25	20.6	24.16
45	13.16	-nevolmente	111.8	21.6	25.30
46	14.3	per	112.26	21.39	26.19
47	15.3	-desima	114.8	22.36	28.7
48	16.8	SALV.	116.3	23.36	30.4
49	17.14	SIMP.	117.17	24.35	31.6
50	18.15	difettosa	118.28	25.31	32.2
51	19.21	dimonstrate	120.9	26.35	33.1
52	20.27	tirando	121.23	27.35	34.4
53	21.28	qual	122.36	28.35	34.30
54	22.30	velocità	124.25	29.36	36.2
55	23.2	là	127.29[7]	31.21[11]	36.13
56	23.6	mobile	127.35	31.24	36.16
57	24.8	globo	129.19	32.25	38.3
58	25.9	del	131.11	33.25	40.17

Favaro	Edition of 1632		Pagnini	Drake	Salusbury-Santillana
59	26.8	non	132.33	34.24	41.b2
60	27.8	-bricare	134.11	35.23	43.10
61	28.8	dire	135.24	36.22	44.5
62	29.8	per	137.16	37.22	45.11
63	30.9	natura	138.33	38.20	46.b13
64	31.9	-trari	140.11	39.19	47.b12
65	32.11	-guire	141.26	40.20	48.23
66	33.10	-cedervi	143.2	41.18	49.b5
67	34.7	li	144.16	42.18	50.b6
68	35.7	del	145.31	43.16	52.9
69	36.7	SIMP.	147.13	44.16	53.12
70	37.8	di-	148.26	45.15	54.8
71	38.10	della	150.7	46.14	55.2
72	39.10	è	151.20	47.13	56.13
73	40.9	*cum*	152.34	48.14	57.15
74	41.11	altro	154.11	49.15	58.14
75	42.11	il	155.23	50.14	59.14
76	43.11	-tagora	157.2	51.14	60.10
77	44.13	perchè	158.18	52.15	61.b3
78	45.13	macchia	160.26	53.16	62.b1
79	46.14	vi	162.3	54.13	63.b7
80	47.15	vuole	164.3	55.13	65.12
81	48.12	mercè	166.15	56.11	66.15
82	49.14	silenzio	167.30	57.12	67.14
83	50.14	prossime	169.6	58.15	68.13
84	51.16	-che	170.22	59.15	69.8
85	52.16	non	172.4	60.16	70.b4
86	53.17	piante	173.20	61.14	71.b3
87	54.18	enco-	175.1	62.12	72.b3
88	55.17	e-	176.16	63.10	73.b10
89	56.18	Luna	177.33	64.9	74.21
90	57.19	della	179.8	65.10	75.16
91	58.20	-l'incidenza	180.25	66.11	76.17
92	59.21	-camento	183.11	67.11	77.b2
93	60.21	acquisto	185.5	68.10	78.b1
94	61.22	suo	186.20	69.9	79.b5
95	62.22	opache	187.37	70.8	80.b2
96	63.23	-tuosa	189.15	71.10	82.18
97	64.24	persuaso	190.29	72.9	83.14
98	65.25	specchio	192.7	73.10	84.20
99	66.26	-cessitata	193.21	74.10	85.14
100	67.25	la riflessione	194.35	75.7	86.6
101	68.25	in	196.8	76.4	87.3
102	69.26	a	197.22	77.3	87.b3

Favaro	Edition of 1632		Pagnini	Drake	Salusbury-Santillana
103	70.23	SIMP.	199.5	77.36	88.27
104	71.23	-streras-	200.21	78.36	89.24
105	72.25	neri	202.1	79.33	90.19
106	73.26	SAGR.	203.18	80.30	91.b14
107	74.28	e	205.12	81.28	92.23
108	75.28	larga	206.26	82.25	93.24
109	76.30	SAGR.	208.3	83.26	94.12
110	77.28	-rata	209.31	84.23	95.9
111	78.28	può	211.11	85.23	96.10
112	79.28	e meno	212.26	86.24	97.9
113	80.30	-minazione	214.5	87.24	99.2
114	81.31	ch'io	215.18	88.23	99.35
115	82.31	arriva	216.33	89.24	100.b5
116	83.33	SIMP.	218.12	90.24	101.b1
117	84.33	esser	219.20	91.21	102.b1
118	85.32	della	220.31	92.21	104.4
119	86.31	perchè	222.9	93.22	105.1
120	87.32	nostro	223.25	94.23	105.b3
121	88.33	miglia	225.2	95.22	107.1
122	89.33	facendo	226.16	96.22	107.b2
123	90.33	più	227.27	97.22	108.b5
124	91.34	volerne	229.18	98.23	109.b8
125	92.23	le	230.23	99.15	110.b18
126	93.25	-samento	232.6	100.16	111.19
127	94.26	la	233.31	101.17	112.15
128	95.25	questa	235.10	102.14	113.13
129	96.26	matematiche	236.24	103.14	114.b7
130	97.27	-tate	238.12	104.14	115.b5
131	98.28	una	240.7	105.16	117.4
132	99.1	GIORNATA	255.1[8]	106.1	118.1
133	99.31[4]	a	256.9	107.2	119.b5
134	100.36	-mento	257.21	108.1	120.b4
135	101.36	e	259.11	109.3	121.b5
136	102.36	SALV.	261.3	110.3	123.4
137	103.36	a	262.17	111.2	124.7
138	104.38	perchè	263.30	112.4	125.5
139	105.38	-tre	265.13	113.5	126.7
140	106.40	-sario	266.29	114.5	127.5
141	108.1	SALV.	269.12	115.3	127.b2
142	109.2	ha	271.10	116.3	128.b2
143	110.3	si	272.26	117.2	129.b1
144	111.4	SALV.	274.8	118.4	130.b5
145	112.4	-terato	275.21	119.3	131.b2

Favaro	Edition of 1632		Pagnini	Drake	Salusbury-Santillana
146	113.6	di	277.9	120.2	132.b4
147	114.7	-sano	278.26	121.3	134.5
148	115.6	-sime	280.6	122.4	135.4
149	116.7	tosto	281.21	123.4	136.b16
150	117.8	-due	282.36	124.10	137.b16
151	118.7	Con-	284.12	125.10	139.2
152	119.6	tempo	285.27	126.8	140.11
153	120.7	-glieria	287.5	127.7	141.18
154	121.6	-clusioni	288.20	128.4	142.4
155	122.7	Copernico	290.4	129.4	143.1
156	123.8	SIMP.	291.18	130.3	144.4
157	124.6	non	292.33	131.1	145.b11
158	125.2	-sentazione	294.4	131.36	146.b3
159	126.3	foglie	295.20	132.35	147.b9
160	127.3	modi	296.34	133.36	148.b13
161	128.3	SIMP.	298.11	134.37	149.b4
162	129.4	bisogna	299.26	135.38	150.b8
163	130.2	*eadem*	300.37	136.35	151.b9
164	131.5	apparisco-	302.15	138.4	152.b5
165	132.7	Terra	303.30	139.6	153.b4
166	133.9	SALV.	305.6	140.5	154.b7
167	134.9	ò	306.19	141.3	155.b13
168	135.10	così	307.35	142.4	156.b15
169	136.11	-sasse	309.11	143.5	157.23
170	137.11	stà	311.2	144.5	158.21
171	138.13	SALV.	312.18	145.7	159.b10
172	139.14	-posizioni	313.33	146.7	160.b13
173	140.13	piano	315.8	147.4	161.b19
174	141.14	SIMP.	316.22	148.7	162.11
175	142.15	di	318.1	149.3	163.8
176	143.15	oltre	319.14	150.6	164.b1
177	144.13	-der	320.33	151.8	165.b1
178	145.13	leggieri	322.12	152.4	166.b11
179	146.14	SAGR.	323.25	153.4	167.b6
180	147.14	nave	325.2	154.5	168.b12
181	148.15	l'havrà	326.15	155.5	169.b12
182	149.16	nave	328.2	156.5	170.21
183	150.16	-giu-	329.13	157.3	171.17
184	151.16	de	330.26	158.3	172.b13
185	152.14	l'inferiori	331.37	159.3	173.b11
186	153.13	SAGR.	333.10	160.3	174.b14
187	154.13	SAGR.	334.21	161.3	175.b17
188	155.13	lasci	335.33	162.1	176.10
189	156.15	-vimento	337.24	163.2	177.6

Favaro	Edition of 1632		Pagnini	Drake	Salusbury-Santillana
190	157.14	altro	338.36	164.2	178.3
191	158.15	Stante	340.17	165.3	179.12
192	159.20	-nità	341.30	166.4	180.16
193	160.22	è	343.8	167.4	181.b8
194	161.22	della	346.9	168.7	183.8
195	162.24	SIMP.	347.25	169.7	184.7
196	163.24	per	348.36	170.6	184.b10
197	164.23	-sofare	350.11	171.6	185.b19
198	165.23	fusse	351.25	172.6	186.b16
199	166.23	parte	352.37	173.4	187.12
200	167.25	SAGR.	354.14	174.3	188.1
201	168.26	SIMP.	355.35	175.3	188.b3
202	169.33	terra	357.15	176.2	190.4
203	170.33	è eguale;	358.26	177.8	190.b8
204	171.34	-ginato	360.2	178.7	191.b11
205	172.36	SAGR.	361.18	179.10	192.18
206	173.37	-lela	362.33	180.9	193.17
207	174.38	SALV.	364.10	181.6	194.14
208	175.40	trenta	365.25	182.2	195.8
209	177.1	perchè	366.34	182.41	195.b3
210	178.1	con-	368.17	183.38	196.b4
211	179.2	perchè	369.30	184.38	198.6
212	180.2	-tamente	371.6	185.39	199.2
213	181.3	-gusta	372.22	186.37	199.b9
214	182.4	di	373.34	187.35	200.8
215	183.4	-tire	375.10	188.34	201.2
216	184.5	per	376.25	189.31	201.20
217	185.5	-chè	378.1	190.27	202.15
218	186.4	SALV.	379.14	191.27	203.b9
219	187.4	co'l	380.26	192.28	204.b8
220	188.1	nuova	382.1	193.27	205.b4
221	188.40	SALV.	383.13	194.27	207.2
222	189.40	-zione	385.3	195.28	208.1
223	191.3	SALV.	386.18	196.28	209.3
224	192.4	SALV.	387.27	197.28	210.1
225	193.14	-gente	389.8	198.26	211.8
226	194.18	-stato	390.22	199.28	212.7
227	195.19	fatta	392.3	200.31	213.b8
228	196.21	per	393.16	201.31	214.b8
229	197.22	pure	394.28	202.29	215.b10
230	198.23	e	396.4	203.27	216.b7
231	199.25	SALV.	397.20	204.27	217.b8
232	200.25	SALV.	398.36	205.30	219.3
233	201.29	SIMP.	400.12	206.30	220.11

Favaro	Edition of 1632		Pagnini	Drake	Salusbury-Santillana
234	202.29	-gioni	401.27	207.29	222.13
235	203.28	sopra	403.2	208.28	223.8
236	204.28	è	404.17	209.23	224.11
237	205.26	for-	405.30	210.20	225.10
238	206.27	noi	407.7	211.21	226.b19
239	207.27	-peto	408.21	212.18	227.12
240	208.29	SALV.	409.37	213.16	228.7
241	209.30	SAGR.	411.15	214.21	229.14
242	210.28	romano	412.26	215.22	230.17
243	211.33	SAGR.	414.17	216.27	232.4
244	212.33	ritenere	416.11	217.26	233.4
245	213.32	che	417.23	218.23	233.b1
246	214.32	à	419.4	219.20	235.5
247	215.32	per	420.16	220.22	236.4
248	216.31	si	421.28	221.22	237.2
249	217.33	SAGR.	423.13	222.23	238.3
250	218.31	-tete	424.26	223.19	239.8
251	219.31	SALV.	426.12	224.17	240.9
252	220.37	Luna	427.28	225.21	241.b23
253	221.39	cosa	429.7	226.21	242.15
254	222.40	-menti	430.23	227.20	243.17
255	224.2	anco	431.36	228.20	244.b15
256	225.7	al	433.14	229.22	245.b12
257	226.6	-rebbe	434.30	230.23	246.b3
258	227.11	sia	436.14	231.29	247.b2
259	228.12	72	437.34	232.37	249.1
260	229.4	medesima	438.36	233.28	249.b3
261	230.5	eccettuatone	440.13	234.31	250.b5
262	231.5	SIMP.	441.28	235.35	252.8
263	232.5	sopra	443.5	236.37	253.b6
264	233.6	SIMP.	444.17	238.1	254.b2
265	234.8	viventi	445.30	239.10	256.7
266	235.8	*hoc*	447.6	240.15	257.6
267	236.10	al	448.19	241.23	258.9
268	237.13	cioè	449.32	242.39	259.13
269	238.14	non	451.8	244.4	260.11
270	239.16	-torno	452.21	245.4	261.10
271	240.17	nè	453.36	246.15	262.12
272	241.9	uno	455.4	247.8	263.12
273	242.10	da	456.23	248.14	264.14
274	243.12	-tissima	457.37	249.14	265.19
275	244.6	perchè	459.12	250.11	266.5
276	245.7	del	460.28	251.12	266.b2
277	246.8	SAGR.	462.5	252.12	missing

Favaro	Edition of 1632		Pagnini	Drake	Salusbury-Santillana
278	247.7	la	463.15	253.10	267.b12
279	248.9	-conferenza	464.28	254.14	268.b15
280	249.10	se	466.16	255.13	269.b21
281	250.11	tal	467.32	256.19	270.15
282	251.12	-che	469.11	257.18	272.3
283	252.13	a	471.2	258.21	273.7
284	253.15	SIMP.	472.21	259.22	274.12
285	254.15	SIMP.	474.1	260.24	275.13
286	255.16	-nente	475.16	261.24	276.7
287	256.14	somma	476.29	262.24	277.6
288	257.15	a	478.7	263.25	278.8
289	258.16	SIMP.	480.32	264.27	279.b17
290	259.16	-vengo-	482.11	265.25	280.b19
291	260.17	convengono	483.27	266.25	281.19
292	261.18	SIMP.	485.6	267.26	282.16
293	262.18	-stare	487.2	268.32	283.b16
294	263.19	-plero	488.16	269.36	284.21
295	264.21	-sario	489.35	271.1	285.b14
296	265.22	avvenga	491.15	272.1	286.b15
297	266.23	contro	493.21	273.1	287.b12
298	267.25	-nen-	495.3	274.4	288.b11
299	269.1	GIORNATA	5.1	276.1	290.1
300	269.31	loro	6.11	277.1	291.17
301	270.37	il tempo	7.23	278.2	292.14
302	271.36	traportate	9.7	278.39	293.b5
303	272.30	necessario	10.16	279.31	295.3
304	273.26	che	11.24	280.26	296.11
305	274.24	SIMP.	13.11	281.25	297.b4
306	275.19	nel	14.16	282.17	298.15
307	276.17	3	15.36	283.17	299.19
308	277.3	12	17.3	284.7	299.b2
309	278.1	SAGR.	18.13	285.6	300.b5
310	278.38	farne	19.21	286.6	301.b8
311	280.5	ben	20.31	287.5	302.b6
312	281.3	che	22.6	287.41	303.b10
313	282.2	svanisca	23.20	288.38	304.b15
314	283.1	nota	25.5	289.34	305.16
315	284.1	SALV.	26.17	290.33	306.b20
316	284.29	dal	27.14	291.20	307.7
317	285.29	egli	28.26	292.20	308.2
318	286.27	accadere	29.34	293.17	308.b3
319	287.27	Ticone	31.11	294.17	310.1
320	288.19	Agecio	32.11	295.1	310.b18

Favaro	Edition of 1632		Pagnini	Drake	Salusbury-Santillana
321	289.17	SIMP.	33.30	295.37	311.17
322	290.11	Dove	35.11	296.35	312.10
323	291.10	della	36.18	297.36	312.b1
324	292.1	Ma	37.15	298.20	313.20
325	293.1	Regola	38.22	299.10	314.12
326	294.1	Quest-	39.32	300.1	315.1
327	295.1	Regola	41.5	300.33	315.b9
328	296.1	SAGR.	42.15	301.30	316.23
329	296.30	94910 17200	43.25	302.27	317.11
330	298.19[5]	bisogna	45.18	303.22	318.6
331	299.13	Nella	46.1	304.17	318.b5
332	300.23	Nell-	47.1	305.6	319.b17
333	301.26	cioè	48.5	306.1	320.20
334	302.37	La	49.14	306.37	321.15
335	304.1	Di	50.19	307.28	322.1
336	304.38	Ticone	51.35	308.25	322.b18
337	305.39	Landgr.	53.20	309.22	323.16
338	306.33	qual	54.30	310.14	323.b1
339	307.32	-siopea	56.5	311.14	324.b12
340	308.30	l'eccesso	57.17	312.19	326.1
341	309.39	e	58.28	313.19	327.3
342	311.10	Angoli	60.4	314.19	327.b16
343	312.13	sopra	61.12	315.17	328.b12
344	313.11	appunto	62.22	316.13	329.22
345	314.11	-golo	63.33	317.15	330.b17
346	315.6	-stuma	65.2	318.9	331.16
347	316.2	sin	67.11	319.8	332.14
348	317.2	niente	70.8	320.8	333.b9
349	317.40	-cipio	71.24	321.7	334.b10
350	318.40	e	73.10	322.5	335.b12
351	320.1	SIMP.	75.3	323.6	336.b13
352	321.25	-chè	76.23	324.22	338.6
353	322.25	-gior	78.12	325.19	339.1
354	323.25	maniera	79.29	326.20	339.b4
355	324.24	-star	81.8	327.22	340.b3
356	325.24	ragione	82.23	328.22	342.2
357	325.31	Copernicano	90.3[9]	334.2[12]	342.8
358	325.34	ci	90.8	334.5	342.11
359	325.37	-nissimo	90.12	334.7	342.13
360	325.40	congiunzio-	90.16	334.10	342.17
361	326.4	&	90.21	334.13	342.20
362	326.7	che	90.25	334.17	342.23
363	326.35	strut-	91.28	335.6	343.19
364	327.36	si	93.7	336.6	344.14

Favaro	Edition of 1632		Pagnini	Drake	Salusbury-Santillana
365	328.34	-tro	94.27	337.6	345.14
366	329.34	libero	96.10	338.2	346.9
367	330.34	oltre	97.28	339.4	347.7
368	331.35	solamente	99.7	340.2	348.10
369	332.37	SALV.	100.21	341.4	349.b6
370	333.37	-trarii	102.1	342.6	351.5
371	334.37	terra	103.27	343.30	352.b15
372	336.26	la	104.33	344.24	353.b5
373	337.27	erano	106.32	345.26	355.3
374	338.27	alcun	108.8	346.26	356.5
375	339.29	-feriore	109.24	347.27	357.b11
376	340.30	e	111.2	348.26	358.b13
377	341.39	i	112.16	349.30	359.23
378	342.40	prima	113.28	350.30	missing
379	344.5	andar	115.4	351.35	361.11
380	345.5	abitatori	116.20	352.34	362.8
381	346.7	a	118.4	353.35	363.7
382	347.7	-renze	119.17	354.35	364.5
383	348.10	ci	120.31	355.38	365.8
384	349.12	-nico	122.13	356.39	366.11
385	350.13	-sime	123.27	358.2	367.b17
386	351.13	esser	126.5	359.3	368.b17
387	352.9	una	127.18	359.39	369.16
388	353.5	quali	128.32	360.32	370.b5
389	354.6	verso	130.33	361.34	371.b5
390	355.8	SAGR.	132.11	362.35	372.b1
391	356.9	carta	134.7	363.32	373.b3
392	357.10	SIMP.	135.21	364.32	375.3
393	358.10	appunto	137.11	365.31	376.6
394	359.10	-tribuirli	138.26	366.31	377.7
395	360.10	ci	140.10	367.34	378.7
396	361.13	per	141.26	368.34	379.b14
397	362.15	stella	143.8	369.35	380.b15
398	363.16	che	144.25	370.36	381.b17
399	364.16	adoperiamo	146.5	371.37	382.b13
400	365.16	-ture	147.22	372.38	383.b13
401	366.18	SALV.	148.36	374.1	385.3
402	367.20	SAGR.	150.14	375.6	385.b1
403	368.20	che	151.25	376.3	386.b6
404	369.18	la	153.3	377.1	387.b8
405	370.18	-ticolare	154.17	378.2	388.b8
406	371.19	sì	156.15	379.2	390.2
407	372.20	obliquo	157.32	380.4	391.2
408	373.35[6]	punto	159.12	381.4	392.4

Favaro	Edition of 1632		Pagnini	Drake	Salusbury-Santillana
409	374.38	SALV.	160.27	382.8	393.3
410	375.38	di	162.9	383.8	394.2
411	377.6	remoto	163.27	384.5	395.5
412	378.25	raggio	165.22	385.11	396.1
413	379.24	che	167.5	386.12	396.b5
414	380.22	-cate	168.19	387.9	397.b10
415	381.23	necessario	170.1	388.11	398.b7
416	382.25	-zione	171.15	389.13	400.2
417	383.25	-tano	173.4	390.13	401.8
418	385.7	Hora	174.32	391.34	402.3
419	386.1	Cap.	175.37	392.29	403.10
420	387.1	archi	177.13	393.30	missing
421	388.2	niente	178.27	394.33	404.b12
422	389.1	e	180.2	395.31	missing
423	390.2	le velocità	181.16	396.31	406.2
424	391.4	SIMP.	182.31	397.33	407.5
425	392.4	acquista	184.10	398.36	408.2
426	393.5	ogni	185.29	399.37	409.3
427	394.5	conceduto	187.18	400.38	410.7
428	395.3	-riori	189.6	401.38	411.13
429	396.2	essere	190.16	402.32	412.10
430	397.3	della	192.2	403.31	413.10
431	398.5	-verta	193.21	404.32	414.7
432	399.5	ne	195.10	405.31	415.4
433	400.7	esercitar	196.26	406.34	416.9
434	401.8	-teria	198.6	407.35	417.6
435	402.9	del	199.21	408.37	418.6
436	403.9	-stenu-	201.11	409.38	419.1
437	404.10	-pernico	202.27	410.38	419.b1
438	405.10	-trina	204.3	411.37	420.b8
439	406.9	-trasse	205.22	412.39	421.b8
440	407.7	-tiquattr-	206.32	413.38	422.b13
441	408.7	vasi	208.18	414.37	423.b10
442	409.1	GIORNATA	223.1[10]	416.1	424.1
443	409.30	globo	224.13	417.1	425.18
444	410.36	-vamento	226.9	417.40	426.18
445	411.36	il	227.24	418.40	427.15
446	412.37	Luna	229.6	419.39	428.14
447	413.39	le	231.12	421.1	429.b16
448	415.1	SALV.	232.29	421.40	430.b9
449	416.1	-vilii	234.6	422.38	431.b5
450	417.3	SALV.	235.19	423.40	432.b5
451	418.4	-riamoci	236.36	425.2	433.21

Favaro	Edition of 1632		Pagnini	Drake	Salusbury-Santillana
452	419.5	SIMP.	238.14	426.7	434.16
453	420.23	destra	239.33	427.7	435.b7
454	421.25	Hor	241.13	428.8	436.b19
455	422.26	in	242.31	429.6	437.b11
456	423.27	e	244.31	430.3	437.b3
457	425.7	SAGR.	246.32	431.5	439.1
458	426.7	-menti	248.9	432.3	439.b11
459	427.8	osserva	249.23	433.5	440.18
460	428.9	Sicilia	251.3	434.5	441.19
461	429.9	altre	252.16	435.6	442.14
462	430.11	è	253.30	436.9	443.15
463	431.12	ò	255.8	437.10	444.14
464	432.12	monti	256.24	438.14	445.10
465	433.11	d'	258.2	439.13	446.4
466	434.12	effetti	260.7	440.14	446.b2
467	435.13	differenza	261.23	441.13	448.5
468	436.13	il	262.37	442.14	449.7
469	437.14	sia	264.17	443.15	450.b18
470	438.14	di	265.31	444.17	451.21
471	439.16	co'l	267.9	445.21	452.b10
472	440.18	ritener	268.27	446.23	453.30
473	441.19	de	270.3	447.21	454.b7
474	442.21	dall'	271.26	448.27	455.b6
475	443.20	e	273.4	449.28	456.b2
476	444.22	si	274.28	450.30	457.b1
477	445.24	e	276.5	451.33	459.b2
478	446.25	senza	278.3	452.33	460.4
479	447.26	cagione	279.16	453.35	461.1
480	448.27	desidero	281.6	454.35	461.b2
481	449.29	al	282.22	455.36	463.2
482	450.30	-lissima	284.2	456.39	464.23
483	451.31	intenda	285.28	457.37	465.b5
484	453.7	incammi-	286.35	459.12	466.b10
485	454.9	Ma	288.10	460.14	467.b10
486	455.10	haveste	289.23	461.17	468.b13
487	456.12	-dere	291.7	462.21	469.b13
488	457.13	-metter	292.27	463.24	470.b10
489	458.14	SALV.	294.12	463.23	471.b2

NOTES

[1] See above Chapters 10 (Favaro section) and 16 (Section 15).

[2] See above Chapter 10. Moreover, in the course of compiling this appendix I found two more, which, though perhaps not as logically consequential as the imperfections previously noted, nevertheless are relatively clearcut: (1) *'scienze matematiche pure'* (F128.37–129.1) does not mean 'mathematical sciences alone' (D103), but rather 'pure mathematical sciences'; and (2) 'in un altro vaso di angusta bocca' (F212.37–213.1) does not mean 'into a wide vessel' (D186), but rather 'into a narrow vessel'.

[3] Santillana indicates (p. 15) his omission of the second half of the discussion of three-dimensionality with which the *Dialogue* begins. (The passage corresponds to F36.10–38.15.) The following omissions, however, are not noted (where 'SS' denotes page references to the Salusbury-Santillana translation, and 'b' indicates that lines are counted from the bottom):

F156.32–157.3	cf.	SS145.b11	F276.13–277.29	cf.	SS267.12
F164.7–165.5	cf.	SS153.b4	F310.33–311.1	cf.	SS302.b6
F172.35–173.8	cf.	SS161.b19	F342.24–343.1	cf.	SS328.b12
F174.3–174.4	cf.	SS162.13–162.14	F377.32–378.3	cf.	SS360.b20
F195.5–195.13	cf.	SS184.11–184.12	F418.36–419.1	cf.	SS403.10
F195.21–195.30	cf.	SS184.17	F419.34–420.2	cf.	SS403.b1
F199.13–199.24	cf.	SS187.b15	F421.34–422.4	cf.	SS405.b17
F202.29–202.35	cf.	SS190.b11	F422.10–422.24	cf.	SS405.b11
F204.6–204.7	cf.	SS192.17–192.18	F422.28–422.30	cf.	SS405.b7
F208.2–209.9	cf.	SS195.9	F450.25–450.34	cf.	SS433.18
F212.20–212.26	cf.	SS199.20	F457.17–457.21	cf.	SS439.18
F213.15–213.34	cf.	SS200.4	F457.26–457.30	cf.	SS439.22
F214.33–215.1	cf.	SS201.1	F458.8–458.16	cf.	SS439.b5
F215.18–215.33	cf.	SS201.15–201.16	F475.35–475.37	cf.	SS457.b1–457.b2

[4] Here (and in other similar circumstances) I am counting in such a way that the title is taken to occupy two lines, and the dialogue is taken to begin on the third line.

[5] The computation on the top half of this page (above line 19) is also on p. 330 in Favaro.

[6] Here (and in other similar circumstances) I am counting in such a way that the diagram is taken as occupying the same number of lines of text that would be printed in its place if it were not there.

[7] The text in the Pagnini edition from p. 125.9 to p. 126.23 appears in Favaro in a footnote on p. 54, and the text in Pagnini from p. 126.23 to 127.29 appears in Favaro in a footnote on p. 55.

[8] Pp. 241–254 in Pagnini contain a note on Aristotelian physics.

[9] Pp. 83.1–90.2 in Pagnini correspond to the footnote text in Favaro pp. 356–362, which in turn correspond to Galileo's addendum handwritten in his own copy of the *Dialogue*. Pagnini's p. 83.1–83.36 appears on p. 356 in Favaro; Pagnini's pp. 83.37–85.1 on Favaro's p. 357; Pagnini's pp. 85.2–86.8 on Favaro's p. 358; and so on for Pagnini's pp. 86.8–87.15, 87.15–88.21, 88.21–89.27, and 89.27–90.2.

[10] Pp. 209–222 in Pagnini contain a note on Copernicus' third motion.

[11] Pp. 30.8–31.19 in Drake correspond to the footnote text on pp. 54–55 in Favaro.

[12] Pp. 328.29–333.40 in Drake correspond to Galileo's addendum handwritten in his copy of the *Dialogue*, which is printed as a footnote in Favaro on pp. 356–362. The correspondence is as follows:

Page in Favaro	Pages in Drake
356	328.29–329.15
357	329.16–330.9
358	330.9–331.3
359	331.3–331.39
360	331.40–332.34
361	332.34–333.30
362	333.30–333.40

SELECTED BIBLIOGRAPHY

Adams, Ernest W., *The Logic of Conditionals*. Dordrecht: Reidel, 1975.

Agassi, Joseph, 'The Role of Corroboration in Popper's Methodology', *Australasian Journal of Philosophy* 39 (1961), 82–91.

Agassi, Joseph, *Towards an Historiography of Science. History and Theory*, Beiheft 2. The Hague: Mouton, 1963.

Agazzi, Evandro, 'Fisica galileiana e fisica contemporanea', in *Nel quarto centenario della nascita di Galileo Galilei*, pp. 1–51. Milan: Società Editrice Vita e Pensiero, 1966.

Aiton, E. J., 'Galileo and the Theory of the Tides', *Isis* 56 (1965), 56–61.

Aiton, E. J., 'Galileo's Theory of the Tides', *Annals of Science* 10 (1954), 44–57.

Aiton, E. J. 'On Galileo and the Earth-Moon System', *Isis* 54 (1963), 265–6.

Anderson, Alan R., Nuel D. Belnap, Jr., *et al.*, *Entailment: The Logic of Relevance and Necessity*, Vol. 1. Princeton: Princeton University Press, 1975.

Angell, Richard B., *Reasoning and Logic*. New York: Appleton, 1964.

Annis, David B., *Techniques of Critical Reasoning*. Columbus, Ohio: Charles E. Merrill, 1974.

Antoni, Carlo, *Commento a Croce*. Venice: Neri Pozza, 1965.

Barenghi, Giovanni, *Considerazioni sopra il Dialogo*. Pisa, 1638.

Bar-Hillel, Yehoshua, 'Argumentation in Natural Languages', in *Akten des XIV Internationalen Kongresses fur Philosophie*, II, 3–6. Vienna: Verlag Herder, 1969, Rpt. *Aspects of Language*, pp. 202–5.

Bar-Hillel, Yehoshua, 'Argumentation in Pragmatic Languages', in *Aspects of Language*, pp. 206–21.

Bar-Hillel, Yehoshua, *Aspects of Language. Essays and Lectures on Philosophy of Language, Linguistic Philosophy, and Methodology of Linguistics*. Jerusalem: The Magnes Press, The Hebrew University; Amsterdam: North-Holland, 1970.

Bar-Hillel, Yehoshua, 'Comments', in 'Formal Logic and Natural Language (A Symposium)', edited by J. F. Staal.

Bar-Hillel, Yehoshua, 'A Neglected Recent Trend in Logic', *Logique et Analyse* 39 (1967), 235–8.

Bar-Hillel, Yehoshua, (ed.), *Pragmatics of Natural Languages*. New York: Humanities Press, 1971.

Bartley, William, W., III, 'Rationality versus the Theory of Rationality', in *The Critical Approach to Science and Philosophy*, edited by M. Bunge, pp. 3–31. New York: Free Press, 1964.

Bartley, William W., III, *The Retreat to Commitment*. New York: Knopf, 1962.

Beardsley, Monroe C., *Thinking Straight*. 3rd edition. Englewood Cliffs, New Jersey: Prentice-Hall, 1966.

Black, Max, *Critical Thinking*. 2nd edition. Englewood Cliffs, New Jersey: Prentice-Hall, 1952.

Bréhier, E., *The History of Philosophy: The Seventeenth Century*. Translated by W. Baskin. Chicago: University of Chicago Press, 1966.

Brown, Harold I., 'Galileo, the Elements, and the Tides', *Studies in History and Philosophy of Science* 7 (1976), 337–51.

Burstyn, Harold L., 'Galileo and the Earth-Moon System: Reply to Dr. Aiton', *Isis* 54 (1963), 400–1.

Burstyn, Harold L., 'Galileo and the Theory of the Tides', *Isis* 56 (1965), 61–3.

Burstyn, Harold L., 'Galileo's Attempt to Prove that the Earth Moves', *Isis* 53 (1962), 161–85.

Burtt, E. A., *Metaphysical Foundations of Modern Physical Science*. Garden City, New York: Doubleday, 1954.

Butts, Robert E., and Joseph C. Pitt (eds.), *New Perspectives on Galileo*. Dordrecht: Reidel, 1978.

Chiaramonti, Scipione, *Difesa . . . al suo Antiticone e Libro delle tre nuove Stelle*. Florence: Landini, 1633.

Clavelin, Maurice, *The Natural Philosophy of Galileo*. Translated by A. J. Pomerans. Cambridge, Massachusetts and London: M.I.T. Press, 1974.

Clavelin, Maurice, *La philosophie naturelle de Galilée*. Paris: Colin, 1968.

Coffa, José A., 'Galileo's Concept of Inertia', *Physis* 10 (1968), 261–81.

Cohen, I. Bernard, 'Galileo, Newton, and the Divine Order of the Solar System', in *Galileo, Man of Science*, edited by E. McMullin, pp. 207–31.

Cohen, I Bernard, 'Newton's Attribution of the First Two Laws of Motion to Galileo', in *Symposium Internazionale di Storia, Metodologia, Logica e Filosofia della Scienza*, pp. XXV–XLIV.

Cohen, L. Jonathan, *The Probable and the Provable*. Oxford: Oxford University Press, 1977.

Cohen, Morris R. and Ernest Nagel, *An Introduction to Logic and Scientific Method*. New York: Harcourt, 1934.

Copi, Irving M., *Introduction to Logic*. 4th edition. New York: Macmillan, 1972.

Copleston, F., *History of Philosophy*, Vol. III, *Late Medieval and Renaissance Philosophy*. Westminster, Maryland: Newman Press, 1963.

Croce, Benedetto, *Historical Materialism and the Economics of Karl Marx*. Translated by C. M. Meredith. New York: Russell and Russell, 1966. First English edition, 1914. First Italian edition, 1900.

Croce, Benedetto, *History: Its Theory and Practice* [1920]. Translated by D. Ainslie, New York: Russell and Russell, 1960.

Croce, Benedetto, *Philosophy of the Practical*. Translated by D. Ainslie. 1913; rpt. New York: Biblio and Tannen.

Croce, Benedetto, *La poesia*. Bari: Laterza, 1936.

Croce, Benedetto, *Storia dell'età barocca in Italia*. Bari: Laterza, 1929.

Croce, Benedetto, *Teoria e storia della storiografia*. Bari: Laterza, 1917.

Croce, Benedetto, *Theory and History of Historiography*. Translated by D. Ainslie. London: Harrap, 1921.

Crombie, A. C., 'Galileo Galilei: A Philosophical Symbol', in *Actes du VIIIᵉ Congrès International d'Histoire des Sciences*, 1956, III, 1089–95.

Drake, Stillman, 'The Evolution of *De Motu*', *Isis* 67 (1976), 239–50.

Drake, Stillman, 'Galileo and the First Mechanical Computing Device,' *Scientific American* 234 (1976), 104–13.

Drake, Stillman, 'Galileo and the Law of Inertia', *American Journal of Physics* 32 (1964), 601–8.

Drake, Stillman, *Galileo at Work: His Scientific Biography*. Chicago: University of Chicago Press, 1978.

Drake, Stillman, 'Galileo in English Literature of the Seventeenth Century', in *Galileo, Man of Science*, edited by E. McMullin, pp. 415–31.

Drake, Stillman, 'Galileo's Discovery of the Law of Free Fall', *Scientific American* 228 (1973), 84–92.

Drake, Stillman, 'Galileo's Experimental Confirmation of Horizontal Inertia: Unpublished Manuscripts', *Isis* 64 (1973), 291–305.

Drake, Stillman, 'Galileo's New Science of Motion', in *Reason, Experiment, and Mysticism in the Scientific Revolution*, edited by M. L. Righini Bonelli and W. R. Shea, pp. 131–56.

Drake, Stillman, 'Galileo's "Platonic" Cosmogony and Kepler's *Prodromus*', *Journal for the History of Astronomy* 4 (1973), 174–91.

Drake, Stillman, *Galileo Studies*. Ann Arbor: University of Michigan Press, 1970.

Drake, Stillman, 'Impetus Theory and Quanta of Speed Before and After Galileo', *Physis* 16 (1974), 47–65.

Drake, Stillman, 'Review of Clavelin's *Philosophie naturelle de Galilée*', *Isis* 61 (1970), 275–7.

Drake, Stillman, 'Semicircular Fall in the *Dialogue*', *Physis* 10 (1968), 89–100.

Drake, Stillman, 'The Uniform Motion Equivalent to a Uniformly Accelerated Motion from Rest', *Isis* 63 (1972), 28–38.

Drake, Stillman, 'Velocity and Eudoxian Proportion Theory', *Physis* 15 (1973), 49–64.

Drake, Stillman (ed. and trans.), *Discoveries and Opinions of Galileo*. Garden City, New York: Doubleday, 1957.

Drake, Stillman (ed. and trans.), *Galileo Against the Philosophers*. Los Angeles: Zeitlin and Ver Brugge, 1976.

Drake, Stillman and C. D. O'Malley (eds. and trans.), *The Controversy on the Comets of 1618*. Philadelphia: University of Pennsylvania Press, 1960.

Drake, Stillman and I. E. Drabkin (eds. and trans.), *Mechanics in Sixteenth Century Italy*. Madison: University of Wisconsin Press, 1969.

Drake, Stillman, and J. MacLachlan, 'Galileo's Discovery of the Parabolic Trajectory', *Scientific American* 232 (1975), 102–10.

Duhem, Pierre, *The Aim and Structure of Physical Theory*. Translated by P. P. Wiener. Princeton, New Jersey: Princeton University Press, 1954.

Duhem, Pierre, *Essai sur la théorie physique de Platon a Galilée*. Paris: Hermann, 1908.

Duhem, Pierre, *Études sur Léonard de Vinci*. 3 Vols. Paris: Hermann, 1905–1913.

Duhem, Pierre, *Les origines de la statique*, 2 Vols. Paris: Hermann, 1905–1906.

Duhem, Pierre, *To Save the Phenomena*. Translated by E. Doland and C. Maschler. Chicago: University of Chicago Press, 1969.

Duhem, Pierre, *Le systéme du monde*. 10 Vols. Paris: Hermann, 1913ff., and 1954ff.

Einstein, Albert, Foreword to *Dialogue Concerning the Two Chief World Systems*, by G. Galilei. Edited by S. Drake. Berkeley: University of California Press, 1953.

Ellis, Albert, *Growth Through Reason: Verbatim Cases in Rational-Emotive Psychotherapy*. Palo Alto, California: Science and Behavior Books, 1971.

Ellis, Albert, *Reason and Emotion in Psychotherapy*. New York: Lyle Stuart, 1962.

Ellis, Albert, and R. A. Harper, *A Guide to Rational Living*. Hollywood, California: Wilshire Book Company, 1968.

Engel, S. Morris, *With Good Reason: An Introduction to Informal Fallacies*. New York: St. Martin's Press, 1976.

Ennis, Robert H., 'Enumerative Induction and Best Explanation', *Journal of Philosophy* 65 (1968), 523–29.

Fearnside, W. Ward and W. B. Holther, *Fallacy: The Counterfeit of Argument*. Englewood Cliffs, New Jersey: Prentice-Hall, 1959.

Feyerabend, Paul K., 'Against Method', in *Minnesota Studies in the Philosophy of Science*, Vol. 4, edited by M. Radner and S. Winokur, pp. 17–130. Minneapolis: University of Minnesota Press, 1970.

Feyerabend, Paul K., *Against Method*. Atlantic Highlands, New Jersey: Humanities Press, Inc., 1975.

Feyerabend, Paul K., *Ausgewählte Aufsätze*. Brunswick: Vieweg, 1974.

Feyerabend, Paul K., 'Consolations for the Specialist', in *Criticism and the Growth of Knowledge*, edited by I. Lakatos and A. Musgrave, pp. 197–230.

Feyerabend, Paul K., *Contro il metodo*. Milan: Lampugnani Nigri, 1973.

Feyerabend, Paul K., *Einführung in die Naturphilosophie*. Brunswick: Vieweg, 1974.

Feyerabend, Paul K., 'Explanation, Reduction, and Empiricism', in *Minnesota Studies in the Philosophy of Science*, Vol. 3, edited by H. Feigl and G. Maxwell, pp. 28–97. Minneapolis, Minnesota: University of Minnesota Press, 1962.

Feyerabend, Paul K., 'How to be a Good Empiricist', in *Philosophy of Science: The Delaware Seminar*, Vol. 2, edited by B. Baumrin, pp. 3–39. New York: Interscience, 1963.

Feyerabend, Paul K., 'Machamer on Galileo', *Studies in History and Philosophy of Science* 5 (1975), 297–304.

Feyerabend, Paul K., *I problemi dell'empirismo*. Milan: Lampugnani Nigri, 1971.

Feyerabend, Paul K., 'Problems of Empiricism', in *Beyond the Edge of Certainty*, edited by R. Colodny, pp. 145–260. Englewood Cliffs, New Jersey: Prentice-Hall, 1965.

Feyerabend, Paul K., 'Problems of Empiricism, Part II', in *The Nature and Function of Scientific Theories*, edited by Robert G. Colodny, pp. 275–353. Pittsburgh, Pennsylvania: University of Pittsburgh Press, 1970.

Feyerabend, Paul K., *Wider den Methodenzwangtheorie*. Frankfurt: Suhrkamp, 1976.

Feynman, Richard P., R. B. Leighton, and M. Sands, *The Feynman Lectures on Physics*. 3 Vols. Reading, Massachusetts: Addison-Wesley, 1963.

Finocchiaro, Maurice A., 'The Concept of *Ad Hominem* Argument in Galileo and Locke', *The Philosophical Forum* 5 (1974), 394–404.

Finocchiaro, Maurice A., 'Dialectical Aspects of the Copernican Revolution: Conceptual Elucidations and Historiographical Problems', in *The Copernican Achievement*, edited by Robert S. Westman, pp. 204–12. Berkeley: University of California Press, 1975.

Finocchiaro, Maurice A., 'Essay-review of *Criticism and the Growth of Knowledge*', in *Studies in History and Philosophy of Science* 3 (1973), 357–72.

Finocchiaro, Maurice A., 'Galileo and the Philosophy of Science', in *PSA 1976*

458   SELECTED BIBLIOGRAPHY

(Proceedings of the 1976 Biennial Meeting of the Philosophy of Science Association),
   Vol. 1, edited by F. Suppe and P. D. Asquith, pp. 130–9. East Lansing, Michigan:
   Philosophy of Science Association, 1976.
Finocchiaro, Maurice A., 'Galileo as a Logician', *Physis* **16** (1974), 129–48.
Finocchiaro, Maurice A., 'Galileo's Space-Proportionality Argument: A Role for Logic in
   Historiography', *Physis* **15** (1973), 65–72.
Finocchiaro, Maurice A., *History of Science as Explanation.* Detroit, Michigan: Wayne
   State University Press, 1973.
Finocchiaro, Maurice A., 'Logic and Rhetoric in Lavoisier's Sealed Note: Toward a
   Rhetoric of Science', *Philosophy and Rhetoric* **10** (1977), 111–22.
Finocchiaro, Maurice A., 'Philosophizing About Galileo', *British Journal for the
   Philosophy of Science* **26** (1975), 255–64.
Finocchiaro, Maurice A., 'Review of Butts and Pitt's *New Perspectives on Galileo*',
   Philosophy of the Social Sciences, forthcoming.
Finocchiaro, Maurice A., 'Review of Clavelin's *Natural Philosophy of Galileo*', *Review of
   Metaphysics* **29** (1976), 544.
Finocchiaro, Maurice A., 'Review of G. Gentile's *La filosofia di Marx*', *The Thomist* **39**
   (1975), 423–26.
Finocchiaro, Maurice A., '*Vires Acquirit Eundo*: The Passage Where Galileo Renounces
   Space-Acceleration and Casual Investigation', *Physis* **14** (1972), 125–45.
Fogelin, Robert J., *Understanding Arguments: An Introduction to Informal Logic.* New
   York: Harcourt Brace Javanovich, 1978.
Funkenstein, Amos, 'The Dialectical Preparation for Scientific Revolutions', in *The
   Copernican Achievement*, edited by Robert S. Westman, pp. 165–203. Berkeley:
   University of California Press, 1975.
Galilei, Galileo, *Dialogo di Galileo Galilei Linceo ...* Florence, 1632; rpt. Brussels:
   Editions Culture et Civilisation, 1966.
Galilei, Galileo, *Dialogo sopra i due massimi sistemi del mondo, tolemaico e
   copernicano.* Edited by Libero Sosio. Turin: Einaudi, 1971.
Galilei, Galileo, *Dialog über die beiden hauptsächlichsten Weltsysteme.* Translated by
   Emil Strauss. Leipsig: Teubner, 1891.
Galilei, Galileo, *Dialogue Concerning the Two Chief World Systems.* Translated by S.
   Drake. Berkeley: University of California Press, 1953, 1962, and 1967.
Galilei, Galileo, *Dialogue on the Great World Systems.* Salusbury's translation revised by
   G. de Santillana. Chicago: University of Chicago Press, 1953.
Galilei, Galileo, *On Motion and on Mechanics.* Translated by I. E. Drabkin and S. Drake.
   Madison, Wisconsin: University of Wisconsin Press, 1960.
Galilei, Galileo, *Opere.* 5 Vols. Edited by P. Pagnini. Florence: Salani, 1964.
Galilei, Galileo, *Opere.* Volumes II and III, *Dialogo sopra i due massimi sistemi.* Edited
   by P. Pagnini. Florence: Salani, 1964.
Galilei, Galileo, *Le opere di Galileo Galilei.* 20 Vols. Edizione Nazionale by A. Favaro
   et al. Florence: Barbera, 1890–1909, 1929–1939, and 1968.
Galilei, Galileo, *Le opere di Galileo Galilei.* Vol. VII, *Dialogo sopra i due massimi sistemi.*
   Edizione Nazionale by A. Favaro *et al.* Florence: Barbera, 1890–1909, 1929–1939,
   and 1968.
Galilci, Galileo, *Two New Sciences.* Translated with Introduction and Notes by S. Drake.
   Madison, Wisconsin: University of Wisconsin Press, 1974.

Garin, Eugenio, 'Chi legga di A. Koyré . . . ', *Giornale critico della filosofia italiana* **11** (1957), 406–408.

Gellner, Ernest, 'Beyond Truth and Falsehood', *British Journal for the Philosophy of Science* **26** (1975) 331–42.

Gentile, Giovanni, *La filosofia di Marx*. 5th edition by V. Bellezza. Florence: Sansoni, 1974. First edition, 1899.

Geymonat, L. *Galileo Galilei: A Biography and Inquiry into his Philosophy of Science*. Translated by S. Drake. New York: McGraw-Hill, 1965. First Italian edition, 1957.

Gillispie, Charles C., *The Edge of Objectivity: An Essay in the History of Scientific Ideas*. Princeton, New Jersey: Princeton University Press, 1960.

Gombrich, Ernst H., *The Story of Art*. 11th edition. New York: Phaidon Publishers, 1966.

Hall, A. R., 'Essay Review of Clavelin's *La philosophie naturelle de Galilée*', *British Journal for the History of Science* **6** (1972), 80–4.

Hamblin, C. L., *Fallacies*. London: Methuen, 1970.

Hartner, Willy, 'Galileo's Contribution to Astronomy', in *Galileo, Man of Science*, edited by E. McMullin, pp. 178–94.

Heilbron, John L., 'Review of Clavelin's *Philosophie naturelle de Galilée*', *Journal of the History of Philosophy* **8** (1970), 341–43.

Hempel, Carl G., *Philosophy of Natural Science*. Englewood Cliffs, New Jersey: Prentice-Hall, 1966.

Hendel, C. W., *Studies in the Philosophy of David Hume*. Indianapolis, Indiana: Bobbs-Merrill, 1963.

Hesse, Mary B., *The Structure of Scientific Inference*. Berkeley: University of California Press, 1974.

Hintikka, Jaakko, *The Semantics of Questions and the Questions of Semantics*. Acta Philosophica Fennica, Vol. 28, No. 4. Amsterdam: North-Holland, 1976.

Hume, David, *Dialogues Concerning Natural Religion*. Edited by N. Kemp-Smith. Indianapolis, Indiana: Bobbs-Merrill, n.d. First Edition, 1779.

Husserl, Edmund, *The Crisis of European Sciences and Transcendental Phenomenology*. Translated by D. Carr. Evanston, Illinois: Northwestern University Press, 1970.

Hurlbutt, R. H., III. *Hume, Newton, and the Design Argument*. Lincoln, Nebraska: University of Nebraska Press, 1965.

Jaspers, Karl, *The Great Philosophers*. 2 volumes. Translated by R. Mannheim. New York: Harcourt, Brace and World, Inc., 1962. First German edition, 1957.

Jones, W. T., *A History of Western Philosophy*. New York: Harcourt, Brace and World, 1952.

Johnstone, Henry W., Jr., *Philosophy and Argument*. University Park, Pennsylvania: Pennsylvania State University Press, 1959.

Kahane, Howard, *Logic and Contemporary Rhetoric: The Use of Reason in Everyday Life*. Belmont, California: Wadsworth, 1971.

Kalish, D. and R. Montague, *Logic: Techniques of Formal Reasoning*. New York: Harcourt, Brace and World, 1964.

Kant, Immanuel, *Critique of Pure Reason*. Translated by N. Kemp Smith. New York: St. Martin's Press, 1965. First German edition, 1781.

Kedrov, B. M. and B. H. Kouznetsov, 'La Logique de Galilée et la logique de la physique actuelle', in *Symposium Internazionale di Storia, Metodologia, Logica e Filosofia della Scienza*, pp. IXC–XCVII.

Koertge, Noretta, 'Bartley's Theory of Rationality', *Philosophy of the Social Sciences* **4** (1974), 75–81.

Koestler, Arthur, *The Sleepwalkers*. New York: Grosset and Dunlap, 1959.

Kouznetsov, Boris, 'Galilée et Einstein. Prologue et Épilogue de la Science Classique', in *Actes du XIIe Congrès International d'Histoire des Sciences (1968)*, Volume 5, pp. 59–63. Published 1971.

Kouznetsov, Boris, 'L'idée d'homogénéité de l'espace dans le *Dialogo* de Galilée', in *Actes du Symposium International des sciences physiques et mathematiques dans la première moitié du XVIIe siecle, Pisa-Vinci, 1958*, pp. 133–41. Paris, 1960.

Kouznetsov, Boris, 'Le soleil comme centre du monde, et l'homogénéité de l'espace chez Galilée', in *Le Soleil à la Renaissance*, pp. 73–88. Brussels: Presses Universitaires de Bruxelles, 1965.

Kouznetsov, Boris, 'Style et Contenu de la Science', *Diogenes*, No. 89, Spring 1975, pp. 55–75.

Koyré, Alexandre, *Études Galiléennes*. 3 Vols. 1939; rpt. Paris: Hermann, 1966.

Koyré, Alexandre, *Galileo Studies*. Translated by J. Mepham. Hassocks, England: Harvester Press, 1978.

Kreyche, Robert J., *Logic for Undergraduates*. 3rd edition. New York: Holt, Reinhart and Winston, 1970.

Kuhn, Thomas S., *The Structure of Scientific Revolutions*. 2nd edition. Chicago: University of Chicago Press, 1970. 1st Edition, 1962.

Kyburg, Henry E., Jr., *Probability and Inductive Logic*. New York: Macmillan, 1970.

Lakatos, Imre and Alan Musgrave (eds.), *Criticism and the Growth of Knowledge*. Cambridge: Cambridge University Press, 1970.

Lyons, John, *Structural Semantics: An Analysis of Part of the Vocabulary of Plato*. Oxford: Basil Blackwell, 1963.

Mach, Ernst, *The Science of Mechanics*. Translated by T. J. McCormack. La Salle, Illinois: Open Court, 1960. 1st German edition, 1883.

Machamer, Peter K., 'Feyerabend and Galileo: The Interaction of Theories and the Reinterpretation of Experience', *Studies in History and Philosophy of Science* **4** (1973), 1–46.

MacLachlan, James, 'The Test of an "Imaginary" Experiment of Galileo's', *Isis* **64** (1973), 374–9.

Mahoney, M. S., 'Galileo's Thought', *Science* **187** (1975), 944–5.

Mates, Benson. *Elementary Logic*. 2nd edition. New York: Oxford University Press, 1972.

McMullin, Ernan, 'The Conception of Science in Galileo's work', in *New Perspectives on Galileo*, edited by Robert E. Butts and J. C. Pitt, pp. 209–57.

McMullin, Ernan, 'Introduction: Galileo, Man of Science', in *Galileo: Man of Science*, edited by E. McMullin, pp. 3–51.

McMullin, Ernan (ed.), *Galileo: Man of Science*. New York: Basic Books, 1968.

Merton, Robert K., *On Theoretical Sociology*. New York: Free Press, 1967.

Michalos, Alex C., *Improving Your Reasoning*. Englewood Cliffs, New Jersey: Prentice-Hall, 1970.

Montague, Richard, *Formal Philosophy*. New Haven, Connecticut: Yale University Press 1974.

Moore, W. Edgar., *Creative and Critical Thinking*. Boston: Houghton, 1967.

Nakayama, Shigeru, 'Galileo and Newton's Problem of World Formation', *Japanese Studies in the History of Science* 1 (1962), 76–82.

Natanson, Maurice and H. W. Johnstone, Jr. (eds.), *Philosophy, Rhetoric, and Argumentation*. University Park, Pennsylvania: Pennsylvania State University Press, 1965.

Naylor, Ronald H., 'Galileo and the Problem of Free Fall', *British Journal for the History of Science* 7 (1974), 105–34.

Naylor, Ronald H., 'Galileo: Real Experiment and Didactic Demonstration', *Isis* 67 (1976), 398–419.

Naylor, Ronald H., 'Galileo's Simple Pendulum', *Physis* 16 (1974), 23–46.

Newton, Isaac, *Mathematical Principles of Natural Philosophy*. Motte's translation revised by F. Cajori. Berkeley: University of California Press, 1934. 1st Latin edition, 1687.

Nobile, Vittorio, 'Sull'argomento galileiano della quarta giornata dei *Dialoghi* e sue attinenze col problema fondamentale della geodesia', *Atti dell'Accademia Nazionale dei Lincei, Classe di Scienze Fisiche* 16 (1954), 426–33.

Oregius, Augustinus, *De Deo uno*. Rome, 1629.

Ortega y Gasset, José, *En torno a Galileo*. Madrid: Revista de Occidente, 1956.

Ortega y Gasset, José, *Man and Crisis*. Translated by M. Adams. New York: Norton, 1958. Spanish title: *En torno a Galileo*.

Osherson, D., 'Models of Logical Thinking', in *Reasoning: Representation and Process in Children and Adults*, edited by Rachel J. Falmagne, pp. 81–91. Hillsdale, New Jersey: Lawrence Erlbaum Associates, 1975.

Pap, A., *An Introduction to the Philosophy of Science*. New York: The Free Press of Glencoe, 1962.

Perelman, Chaïm, *The Idea of Justice and the Problem of Argument*. Translated by J. Petrie. New York: Humanities Press, 1963.

Perelman, Chaïm, 'The New Rhetoric', in *Pragmatics of Natural Languages*, edited by Y. Bar-Hillel, pp. 145–49.

Perelman, Chaïm, 'A Reply to Henry W. Johnstone, Jr.', in *Philosophy, Rhetoric, and Argumentation*, edited by Maurice Natanson and H. W. Johnstone, Jr., pp. 135–7. University Park, Pennsylvania: Pennsylvania State University Press, 1965.

Perelman, Chaïm and L. Olbrechts-Tyteca, *The New Rhetoric: A Treatise on Argumentation*. Translated by J. Wilkinson and P. Weaver. Notre Dame: University of Notre Dame Press, 1969.

Popper, Karl, *The Logic of Scientific Discovery*. New York: Harper and Row, 1959. 1st German edition, 1935.

Popper, Karl, *Logik der Forschung*. Vienna: Julius Springer, 1935.

Popper, Karl, *Objective Knowledge: An Evolutionary Approach*. New York: Oxford University Press, 1972.

Popper, Karl, *The Open Society and Its Enemies*. Vol. 2. Princeton, New Jersey: Princeton University Press, 1971. First published 1945.

Quine, Willard V., *Methods of Logic*. 3rd edition. New York: Holt, Rinehart and Winston, 1972.

Remusat, C. de, *Bacon, sa Vie, son Temps, et son Influence jusqu'à nos Jours*. 2nd edition. Paris, 1858.

Righini Bonelli, M. L. and W. R. Shea (eds.), *Reason, Experiment, and Mysticism in the Scientific Revolution*. New York: Science History Publications, 1975.

Ryle, Gilbert, 'Proofs in Philosophy', *Revue internationale de philosophie* 8 (1954), 152ff. Reprinted in G. Ryle, *Collected Papers*, Vol. 2. New York: Barnes and Noble, 1971.

Saarinen, Esa (ed.), *Game-Theoretical Semantics*. Dordrecht: Reidel, 1978.

Salmon, Wesley C., *Logic*. 2nd edition. Englewood Cliffs, New Jersey: Prentice-Hall, 1973.

Santillana, Giorgio de, *The Crime of Galileo*. Chicago: University of Chicago Press, 1955.

Schmitt, Charles B., 'Review of Shapere's *Galileo*', *Renaissance Quarterly* 28 (1975), 362–3.

Scriven, Michael, 'Definitions, Explanation, and Theories', in *Minnesota Studies in the Philosophy of Science*, Vol. 2, edited by H. Feigl, M. Scriven, and G. Maxwell, pp. 99–195. Minneapolis: Minnesota: University of Minnesota Press, 1958.

Scriven, Michael, 'The Frontiers of Psychology: Psychoanalysis and Parapsychology', in *Frontiers of Science and Philosophy*, edited by Robert G. Colodny, pp. 79–129. Pittsburgh, Pennsylvania: University of Pennsylvania Press, 1962.

Scriven, Michael, 'A Possible Distinction Between Traditional Scientific Disciplines and the Study of Human Behavior', in *Minnesota Studies in the Philosophy of Science*, Vol. 1, edited by H. Feigl and M. Scriven, pp. 330–9. Minneapolis, Minnesota: University of Minnesota Press, 1956.

Scriven, Michael, 'Psychology Without a Paradigm', in *Clinical Cognitive Psychology*, edited by Louis Breger.

Scriven, Michael, *Reasoning*. New York: McGraw-Hill, 1976.

Scriven, Michael, 'Science: The Philosophy of Science', *International Encylopedia of the Social Sciences*, 1968, Vol. 14, pp. 83–92.

Scriven, Michael, 'A Study of Radical Behaviorism', in *Minnesota Studies in the Philosophy of Science*, Vol. 1, edited by H. Feigl and M. Scriven, pp. 88–130. Minneapolis, Minnesota: University of Minnesota Press, 1956.

Scriven, Michael, 'Verstehen Again', *Theory and Decision* 1 (1971), 382–6.

Scriven, Michael, 'Views of Human Nature', in *Behaviorism and Phenomenology*, edited by T. W. Wann, pp. 163–90. Chicago: University of Chicago Press, 1964.

Seeger, Raymond J., *Men of Physics: Galileo Galilei, His Life and Works*. New York: Pergamon Press, 1966.

Seeger, Raymond J., 'On the Role of Galileo in Physics', *Physis* 5 (1963), 5–38.

Settle, Thomas B., 'An Experiment in the History of Science', *Science* 133 (1961), 19–23.

Shapere, Dudley, *Galileo: A Philosophical Study*. Chicago: University of Chicago Press, 1974.

Shea, William R., *Galileo's Intellectual Revolution*. New York: Science History Publications, 1972.

Shea, William R., 'Review of Clavelin's *Philosophie naturelle de Galilée*', *British Journal for the Philosophy of Science* 21 (1970), 124–5.

Skyrms, Brian, *Choice and Chance: An Introduction to Inductive Logic*. Belmont, California: Dickenson Publishing Company, 1966.

Sosio, Libero, 'Galileo e la cosmologia', in G. Galilei, *Dialogo sopra i due massimi sistemi*, edited by L. Sosio.

Staal, J. F. (ed.), 'Formal Logic and Natural Languages (A Symposium)', *Foundations of Language* 5 (1969), 256–84.

Stein, Howard, 'Maurice Clavelin on Galileo's Natural Philosophy', *British Journal for the Philosophy of Science* 25 (1974), 375–97.

Strawson, P. F., *Introduction to Logical Theory*. London: Methuen, 1952.

Strφmholm, P., 'Galileo and the Scientific Revolution', *Inquiry* 18 (1975), 345–53.

Symposium Internazionale di Storia, Metodologia, Logica e Filosofia della Scienza. Atti. 'Galileo nella storia e nella filosofia della scienza', Manifestazioni celebrative del IV centenario della nascita di Galileo (Florence-Pisa, 1964). Collection des Travaux de l'Academie Internationale d'Histoire des Sciences, 16. Florence: Gruppo Italiano di Storia delle Scienze, 1967.

Tarski, Alfred, *Introduction to Logic and to the Methodology of Deductive Sciences*. New York: Oxford University Press, 1965.

Thomason, R. H. (ed.), *Formal Philosophy: Selected Papers of Richard Montague*. New Haven, Connecticut: Yale University Press, 1974.

Toulmin, Stephen, *Human Understanding*. Vol. 1, *The Collective Use and Evolution of Concepts*. Princeton, New Jersey: Princeton University Press, 1972.

Toulmin, Stephen, *The Philosophy of Science: An Introduction*. New York: Harper and Row, 1960. 1st edition 1953.

Toulmin, Stephen, *The Uses of Argument*. Cambridge: Cambridge University Press, 1964. First published, 1958.

Ueberweg, F., *History of Philosophy*. Vol. II, *Modern Philosophy*. Translated by G. S. Morris from the 4th German edition. New York, 1876.

Uzdilek, S. M., 'Galileo Galilei, The Founder of Experimental Philosophy', in *Symposium Internazionale di Storia, Metodologia, Logica e Filosofia della Scienza*.

Vlastos, Gregory (ed.), *The Philosophy of Socrates*. Garden City, New York: Doubleday, 1971.

Wallace, William A., 'Galileo and Reasoning *Ex Suppositione*: The Methodology of the *Two New Sciences*', in *PSA 1974*, edited by R. S. Cohen *et al.*, pp. 79–104. Boston Studies in the Philosophy of Science, Vol. 32. Dordrecht, Reidel, 1976.

Wallace, William A., 'Galileo Galilei and the *Doctores Parisienses*', in *New Perspectives on Galileo*, edited by R. E. Butts and J. C. Pitt, pp. 87–138.

Wallace, William A., 'Mechanics from Bradwardine to Galileo', *Journal of the History of Ideas* 32 (1971), 16–28.

Wason, P. C., 'The Drafting of Rules', *New Law Journal* 118 (1968), 548–9.

Wason, P. C. and P. N. Johnson-Laird, *Psychology of Reasoning: Structure and Content*. Cambridge, Massachusetts: Harvard University Press, 1972.

Weinreich, U., 'Explorations in Semantic Theory', in *Current Trends in Linguistics*, Vol. III, *Theoretical Foundations*, edited by T. A. Sebeok, pp. 395–477. The Hague, 1966.

Westman, Robert S. (ed.), *The Copernican Achievement*. Berkeley, California: University of California Press, 1975.

Wisan, Winifred L., 'Galileo's Scientific Method: A Reexamination', in *New Perspectives on Galileo*, edited by R. E. Butts and J. C. Pitt, pp. 1–57.

Wisan, Winifred L., 'The New Science of Motion: A study of Galileo's *De motu locali*', *Archive for History of Exact Sciences* 13 (1974), 103–306.

INDEX

BOSTON STUDIES IN THE PHILOSOPHY OF SCIENCE

Editors:
ROBERT S. COHEN and MARX W. WARTOFSKY
(Boston University)

1. Marx W. Wartofsky (ed.), *Proceedings of the Boston Colloquium for the Philosophy of Science 1961-1962.* 1963.
2. Robert S. Cohen and Marx W. Wartofsky (eds.), *In Honor of Philipp Frank.* 1965.
3. Robert S. Cohen and Marx W. Wartofsky (eds.), *Proceedings of the Boston Colloquium for the Philosophy of Science 1964-1966. In Memory of Norwood Russell Hanson.* 1967.
4. Robert S. Cohen and Marx W. Wartofsky (eds.), *Proceedings of the Boston Colloquium for the Philosophy of Science 1966-1968.* 1969.
5. Robert S. Cohen and Marx W. Wartofsky (eds.), *Proceedings of the Boston Colloquium for the Philosophy of Science 1966-1968.* 1969.
6. Robert S. Cohen and Raymond J. Seeger (eds.), *Ernst Mach: Physicist and Philosopher.* 1970.
7. Milic Capek, *Bergson and Modern Physics.* 1971.
8. Roger C. Buck and Robert S. Cohen (eds.), *PSA 1970. In Memory of Rudolf Carnap.* 1971.
9. A. A. Zinov'ev, *Foundations of the Logical Theory of Scientific Knowledge (Complex Logic).* (Revised and enlarged English edition with an appendix by G. A. Smirnov, E. A. Sidorenka, A. M. Fedina, and L. A. Bobrova.) 1973.
10. Ladislav Tondl, *Scientific Procedures.* 1973.
11. R. J. Seeger and Robert S. Cohen (eds.), *Philosophical Foundations of Science.* 1974.
12. Adolf Grünbaum, *Philosophical Problems of Space and Time.* (Second, enlarged edition.) 1973.
13. Robert S. Cohen and Marx W. Wartofsky (eds.), *Logical and Epistemological Studies in Contemporary Physics.* 1973.
14. Robert S. Cohen and Marx W. Wartofsky (eds.), *Methodological and Historical Essays in the Natural and Social Sciences. Proceedings of the Boston Colloquium for the Philosophy of Science 1969-1972.* 1974.
15. Robert S. Cohen, J. J. Stachel and Marx W. Wartofsky (eds.), *For Dirk Struik. Scientific, Historical and Political Essays in Honor of Dirk Struik.* 1974.
16. Norman Geschwind, *Selected Papers on Language and the Brain.* 1974.
18. Peter Mittelstaedt, *Philosophical Problems of Modern Physics.* 1976.
19. Henry Mehlberg, *Time, Causality, and the Quantum Theory* (2 vols.). 1980.
20. Kenneth F. Schaffner and Robert S. Cohen (eds.), *Proceedings of the 1972 Biennial Meeting, Philosophy of Science Association.* 1974.
21. R. S. Cohen and J. J. Stachel (eds.), *Selected Papers of Léon Rosenfeld.* 1978.
22. Milic Capek (ed.), *The Concepts of Space and Time. Their Structure and Their Development.* 1976.
23. Marjorie Grene, *The Understanding of Nature. Essays in the Philosophy of Biology.* 1974.

24. Don Ihde, *Technics and Praxis. A Philosophy of Technology.* 1978.
25. Jaakko Hintikka and Unto Remes, *The Method of Analysis. Its Geometrical Origin and Its General Significance.* 1974.
26. John Emery Murdoch and Edith Dudley Sylla, *The Cultural Context of Medieval Learning.* 1975.
27. Marjorie Grene and Everett Mendelsohn (eds.), *Topics in the Philosophy of Biology.* 1976.
28. Joseph Agassi, *Science in Flux.* 1975.
29. Jerzy J. Wiatr (ed.), *Polish Essays in the Methodology of the Social Sciences.* 1979.
32. R. S. Cohen, C. A. Hooker, A. C. Michalos, and J. W. van Evra (eds.), *PSA 1974: Proceedings of the 1974 Biennial Meeting of the Philosophy of Science Association.* 1976.
33. Gerald Holton and William Blanpied (eds.), *Science and Its Public: The Changing Relationship.* 1976.
34. Mirko D. Grmek (ed.), *On Scientific Discovery.* 1980.
35. Stefan Amsterdamski, *Between Experience and Metaphysics. Philosophical Problems of the Evolution of Science.* 1975.
36. Mihailo Marković and Gajo Petrović (eds.), *Praxis. Yugoslav Essays in the Philosophy and Methodology of the Social Sciences.* 1979.
37. Hermann von Helmholtz: *Epistemological Writings. The Paul Hertz/Moritz Schlick Centenary Edition of 1921 with Notes and Commentary by the Editors.* (Newly translated by Malcolm F. Lowe. Edited, with an Introduction and Bibliography, by Robert S. Cohen and Yehuda Elkana.) 1977.
38. R. M. Martin, *Pragmatics, Truth, and Language.* 1979.
39. R. S. Cohen, P. K. Feyerabend, and M. W. Wartofsky (eds.), *Essays in Memory of Imre Lakatos.* 1976.
42. Humberto R. Maturana and Francisco J. Varela, *Autopoiesis and Cognition. The Realization of the Living.* 1980.
43. A. Kasher (ed.), *Language in Focus: Foundations, Methods and Systems. Essays Dedicated to Yehoshua Bar-Hillel.* 1976.
46. Peter L. Kapitza, *Experiment, Theory, Practice.* 1980.
47. Maria L. Dalla Chiara (ed.), *Italian Studies in the Philosophy of Science.* 1980.
48. Marx W. Wartofsky, *Models: Representation and the Scientific Understanding.* 1979.
50. Yehuda Fried and Joseph Agassi, *Paranoia: A Study in Diagnosis.* 1976.
51. Kurt H. Wolff, *Surrender and Catch: Experience and Inquiry Today.* 1976.
52. Karel Kosík, *Dialectics of the Concrete.* 1976.
53. Nelson Goodman, *The Structure of Appearance.* (Third edition.) 1977.
54. Herbert A. Simon, *Models of Discovery and Other Topics in the Methods of Science.* 1977.
55. Morris Lazerowitz, *The Language of Philosophy. Freud and Wittgenstein.* 1977.
56. Thomas Nickles (ed.), *Scientific Discovery, Logic, and Rationality.* 1980.
57. Joseph Margolis, *Persons and Minds. The Prospects of Nonreductive Materialism.* 1977.
58. Gerard Radnitzky and Gunnar Andersson (eds.), *Progress and Rationality in Science.* 1978.

59. Gerard Radnitzky and Gunnar Andersson (eds.), *The Structure and Development of Science*. 1979.
60. Thomas Nickles (ed.), *Scientific Discovery: Case Studies*. 1980.
61. Maurice A. Finocchiaro, *Galileo and the Art of Reasoning*. 1980.